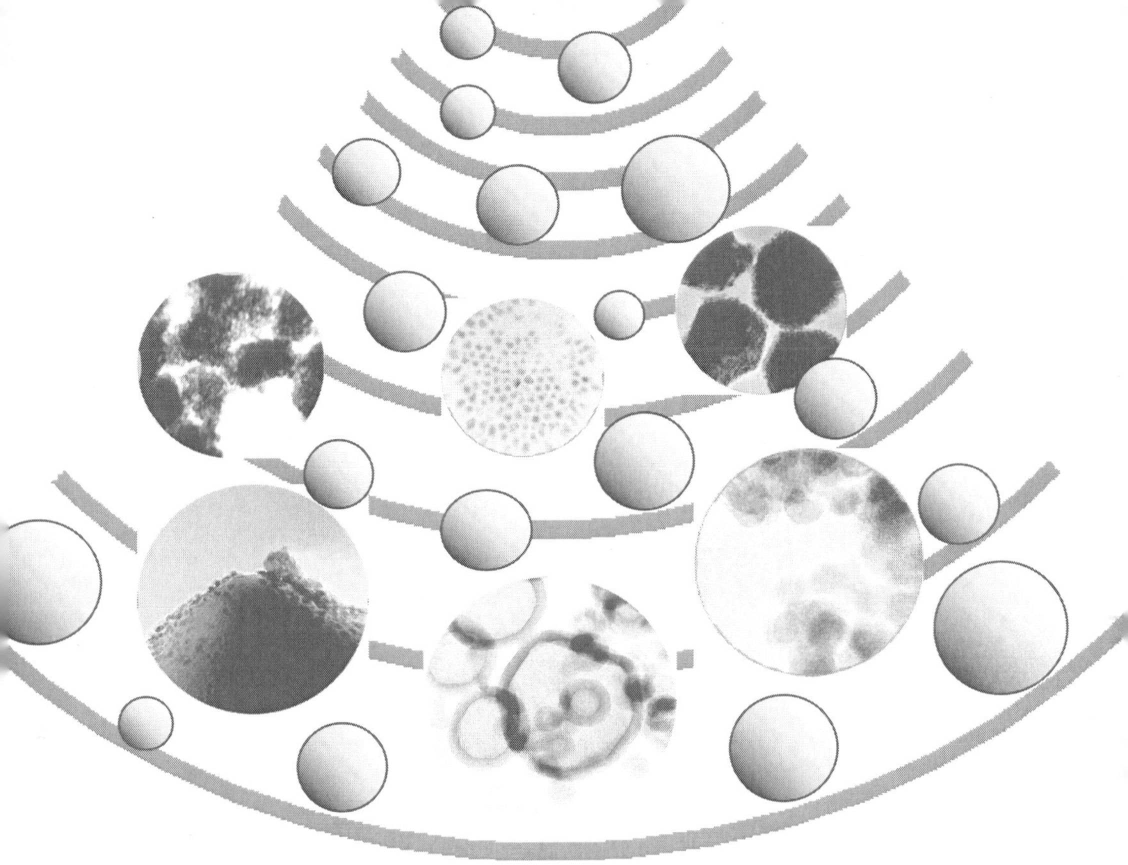

Cavitation

Cavitation

A Novel Energy-Efficient Technique for the Generation of Nanomaterials

edited by
Sivakumar Manickam
Muthupandian Ashokkumar

Pan Stanford Publishing

Published by

Pan Stanford Publishing Pte. Ltd.
Penthouse Level, Suntec Tower 3
8 Temasek Boulevard
Singapore 038988

Email: editorial@panstanford.com
Web: www.panstanford.com

British Library Cataloguing-in-Publication Data
A catalogue record for this book is available from the British Library.

Cavitation: A Novel Energy-Efficient Technique for the Generation of Nanomaterials

ISBN 978-981-4411-54-7 (Hardcover)
ISBN 978-981-4411-55-4 (eBook)

Printed in the USA

Contents

Foreword

It was an honor to be asked to provide the foreword for this book, and it is with great pleasure that I do so. *Cavitation: A Novel Energy-Efficient Technique for the Generation of Nanomaterials* is a timely survey of a field that is mature in parts but with new, ever more complex applications emerging at a consistent rate. It is edited by two colleagues, Professors Manickam and Ashokkumar, who have made their own seminal contributions but also collaborated with a wide range of others in pushing forward the field of "sonochemical materials science".

There is a constant demand for developing novel technologies to synthesize multifunctional materials and, perhaps more importantly, control their properties. Having used ultrasound in the synthesis of functional polymers and materials for the past 25 years or so, I have always appreciated the extreme, but useful reaction conditions generated by acoustic cavitation but it has been interesting to watch the many clever and inventive ways that have surfaced for producing a huge range of materials. Over recent years, sonochemistry has shown significant growth in a variety of fields, but none more so than in the preparation of functional bio- and inorganic materials covering length scales from the *nano-* to the *micro-*. Many of these are described in this book along with emerging methods that combine with great effect cavitation with other methods such as electrochemistry, photochemistry, or microwaves. In parallel, cavitation generated by hydrodynamic forces has also been shown to be useful.

The book contains chapters from a number of leaders in the field of application of cavitation to nanomaterial synthesis. These chapters will serve as an excellent introduction for those new to the area and a welcome review of applications for those more familiar with sonochemistry. It has something for everyone and will not only be beneficial to sonochemists and materials scientists, but also to many interdisciplinary scientists. It represents an excellent addition to the literature.

Gareth J. Price
Professor of Chemistry
University of Bath, UK

Preface

Nanomaterials and their end products have started occupying the pinnacle position of consumer markets. At this juncture it becomes vital to consider the processing means through which naomaterials are generated. Especially, energy efficacy is the foremost concern while dealing with the processing of nanomaterials regardless of the scale of operation. Conventionally a gamut of top-down and bottom-up techniques are explored to obtain these nanomaterials. One of the green chemistry principles underlines the need for unusual energy sources to generate nanomaterials. Utilizing the energy from the collapse of cavitation bubbles, generated either by ultrasound or hydrodynamic forces, for the generation of nanomaterials is a merit to consider in this "green chemical processing era".

The past decade has witnessed the development of a wide range of nanomaterials using cavitation. A step forward is the coupling of cavitation with other techniques such as microwave, photochemistry, and electrochemistry, which have seen innumerous advantages in the generation of nanomaterials. In many instances, nanomaterials are attained with unique morphologies, a reduction in size, and narrow size distribution. While a few currently available books deal with the fundamental aspects of cavitation and sonochemistry, there is no book devoted specifically to the technologically important nanomaterials obtained by cavitation. This stimulus made us think about it, and we ended up editing this book. The chapters have been contributed by leading researchers working on utilizing cavitation for the generation of nanomaterials. This book will be most useful to those who explore cavitation for the facile synthesis of diverse nanomaterials. Some fundamental aspects of cavitation have been discussed only to a certain extent, as the core theme of this book is to understand the nanomaterials generated by cavitation.

In Chapter 1, Prof. Manickam, Dr. Tang, and Prof. Ashokkumar have discussed newly emerging multifunctional nanomaterials and the preparative strategies using ultrasound, where a clear evidence for the proper selection of ultrasonic parameters for the formation of nanomaterials has been revealed. Dr. Okitsu in Chapter 2 has reviewed the role of sonochemical method in the synthesis of metal nanoparticles with precise control in size, shape, and structure. Dr. Anandan and Prof. Ashokkumar have focused in Chapter 3 on the ultrasonic synthesis of monometallic and bimetallic nanostructured materials with unique properties as well with their applications. Prof. Yu and coworkers have examined the recent advances in nanostructured metal oxides obtained using sonochemistry and ultrasonic spray pyrolysis methods in Chapter 4.

In Chapter 5, Prof. Pandit and coworkers have analyzed synthesizing nanomaterials using an alternative form of cavitation, that is, hydrodynamic cavitation. In this chapter, it is clearly shown that the control of the size of nanomaterials is possible using experimental parameters such as pump discharge pressure, liquid flow rate, and the design of the orifice plate employed. Moreover, they have shown the importance of this technique for obtaining hard, brittle, rubbery, and highly crystalline materials. Prof. Pollet and colleagues (Chapter 6) have discussed the concept of fabricating nanomaterials by combining power ultrasound and electrochemistry. More importantly, they have investigated the generation of intriguing nanostructures using this unique and combined methodology. Prof. Cravotto and Dr. Boffa (Chapter 7) have shed light on another hybrid technique, namely, a combination of ultrasound and microwaves and discussed the synergistic effects in the preparation of various nanomaterials.

Dr. Francesca's group (Chapter 8) has looked into the application of ultrasound technique in synthesizing and functionalizing micro-nanopolymeric materials for the microencapsulation of various bioactive agents. Prof. Gedanken and coworkers (Chapter 9) have discussed the principles behind the sonochemical process in embedding the nanoparticles onto textiles and discussed the advantages of this method compared with other coating techniques. Dr. Sonawane and coworkers (Chapter 10) discussed the generation of nanocomposites assisted by ultrasound via in situ emulsion polymerization. Dr. Ekaterina and Dr. Shchukin (Chapter 11) have provided the conceptual solutions for the controlled sonochemical fabrication of mesoporous surfaces and metal sponges. Following this, Dr. Yasui and Dr. Kato (Chapter 12) have provided a detailed discussion on the numerical simulations of nucleation and aggregation of $BaTiO_3$ crystals under ultrasound conditions. Finally, Prof. Ashokkumar and Prof. Manickam (Chapter 13) have discussed the issues on the development of large-scale ultrasound reactors besides future perspectives.

We sincerely thank all the contributors of the chapters as well the reviewers for their excellent contributions towards the successful completion of this book. We trust that this book will be extremely beneficial to researchers involved in the exploitation of cavitation for the generation of nanomaterials for different technological applications.

Sivakumar Manickam
Muthupandian Ashokkumar
Summer 2014

Chapter 1

Development of Multifunctional Nanomaterials by Cavitation

Sivakumar Manickam,[a] **Tang Siah Ying,**[a] **and Muthupandian Ashokkumar**[b,c]

[a]*Department of Chemical and Environmental Engineering, Faculty of Engineering, The University of Nottingham, Malaysia Campus, Jalan Broga, 43500 Semenyih, Selangor Darul Ehsan, Malaysia*

[b]*School of Chemistry, University of Melbourne, 1-100 Grattan Street, Parkville VIC 3010, Australia*

[c]*Chemistry Department, King Abdulaziz University, Abdullah Sulayman, Jeddah 22254, Saudi Arabia*

Sivakumar.Manickam@nottingham.edu.my

Synthesis of advanced nanomaterials using acoustic cavitation, often referred to as sonochemical route, has received considerable interest in the past few decades. In this chapter, a brief description of the preparation of some of the newly emerging nanomaterials has been presented. The sonochemical route is by far the most reliable and efficient preparative technique, primarily owing to its ability to concentrate the acoustic energy in a smaller volume. The implosion of cavitation microbubbles results in the formation of millions of short-lived, localized hot spots that have an extreme temperature of

Cavitation: A Novel Energy-Efficient Technique for the Generation of Nanomaterials
Edited by Sivakumar Manickam and Muthupandian Ashokkumar

ISBN 978-981-4411-54-7 (Hardcover), 978-981-4411-55-4 (eBook)
www.panstanford.com

thousands of Kelvin and pressures of few thousands of atmospheres. Consequently, the fabrication of an array of nanostructured materials that differ distinctly from their bulk counterparts has been promising with this powerful cavitation approach. There are also other forms of nanomaterials such as nanowires and nanospheres that have been successfully prepared by this sonochemical approach. A proper selection of ultrasonic irradiation parameters influences the satisfactory formation of nanomaterials. In short, this chapter will therefore be confined to discussing the recent developments of the ultrasound-mediated synthesis of various multifunctional nanomaterials and the involved preparative strategies.

1.1 Introduction

Recent advancements in the area of nanomaterials have led to many promising applications in the fields extending from solid oxide fuel, superconductors and lithium batteries, catalysis, cosmetics to pharmaceuticals. Materials in the nanometer scale often demonstrate distinctive properties that are different from their bulky counterparts, predominantly owing to their unusual and extremely large specific surface area and surface-to-volume ratio. The magnetic, electronic, optical, and chemical properties vary significantly once the size of materials becomes smaller and smaller, and reduced to nanosized regime. The morphology also differs since the shapes of the nanoparticles alter. As far as the impact of innovations in the nanotechnology is concerned, an exponential growth of research interest and endeavors in the design and preparation of novel nanomaterials with notable physical and chemical properties has been observed in the past decade. Nanostructured materials have been synthesized through a diverse known chemical methods, including sol–gel, micelle and inverse micelle, hydrothermal, solvothermal, chemical precipitation, direct oxidation, chemical vapor deposition, physical vapor deposition, electrodeposition, and microwave-assisted hydrothermal techniques. Nonetheless, the design and manufacture of nanomaterials together with tunable particle size and morphology toward a specific application yet remain an onerous and challenging task to most of the scientists working in the area of nanotechnology. Furthermore, it is the selection of an appropriate synthetic route that determines the success rate of nanomaterial synthesis, where the physical and chemical properties of nanostructured materials are strongly dependent upon their preparation techniques.

Among a wide range of approaches, the utilization of high intensity ultrasound for the synthesis of materials has been considerably investigated in the past decade and emerged to be one of the most efficient, powerful, and versatile synthetic tool for the nanosized compounds that are often not accessible by the conventional methods. The first application of ultrasound irradiation to nanomaterial synthesis dates back to 1966 where CdS thin film deposition as solar cell was reported (Chamberlin and Skarman, 1966). Nanostructured metals were first prepared by Suslick and his coworkers where they successfully produced amorphous nanostructured iron metal and colloidal iron nanoparticles with the size ranging from 10 to 20 nm by treating a nonaqueous solution of $Fe(CO)_5$ in the presence of ultrasound (Suslick et al., 1983). The synthesized iron nanoparticles showed narrow size distribution and were found to be superparamagnetic. Ever since rigorous efforts have been made with ultrasound, which has set off novel research prospects and agenda for the synthesis of myriad types of nanostructured materials, an array of induced effects of ultrasound on the nanomaterial synthesis can be ascribed to the acoustic cavitation. Cavitation is the growth of pre-existing gas nuclei to micron-sized bubbles and the implosive collapse of these bubbles in a liquid. Acoustic cavitation creates severe physical and chemical conditions within the collapsing bubble and serves as the impetus of a vast majority of sonochemical reactions in the liquid-liquid or liquid-solid systems. Cavitation induced by ultrasound waves imparts a unique molecular level interaction between energy and chemical species via the generation of millions of short-lived, localized hot spots, which have equivalent temperatures of approximately 5000 K, pressures of about 1000 bar, and heating and cooling rates above 10^{10} K/s, during the inertial bubble collapse phase. These extreme conditions can result in highly concentrated energy to accelerate and augment the chemical and biological reactions that are generally not viable to be achieved by the traditional methods, enabling the synthesis of a wide range of unusual nanomaterials (Suslick, 2001).

When the intensity of ultrasound is higher than the threshold intensity, the primary physical phenomena associated with the cavitational implosion of the bubbles that are relevant to material synthesis are shock waves and microstreaming. It is well recognized that the violent and implosive collapse provides a couple of strong physical effects outside the microbubbles: disruptive forces, liquid microjets, and shock waves that significantly lead to an enhanced mass transport, improved chemical reaction rates, decrease in the diffusion layer thickness and effective cleaning, and degassing action in the case of an electrode surface. In addition, the generation of free radicals from different species, and in particular the formation of free radicals by the sonolysis of water, has also been explored. This chapter strives to summarize the preparation techniques and recent developments in the applications of ultrasound in the synthesis of nanostructured materials. We focus on the recent progress in the synthesis, properties, modifications, and applications of nanomaterials. The synthesis of nanomaterials, including nanoparticles, nanorods, nanowires, and nanopowders, is primarily categorized with the synthetic routes. For detailed information, the readers are advised to refer to the corresponding literature.

1.2 Bimetallic Nanoparticles

1.2.1 Copper–Silver Core–Shell Nanoparticles

Cu–Ag core–shell nanoparticles were prepared by a sonoelectrochemical route followed by a displacement reaction (Mancier et al., 2010). In this case, the experimental setup essentially composed of a homemade galvanostat/potentiostat and an ultrasonic horn with tunable power from 7 W to 100 W. The generation of nanopowder was controlled by three electrodes electrochemical setup using the sonotrode as working electrode. The reference electrode was a saturated mercury sulfate electrode, while copper wire acted as a soluble anode. The reference electrode was placed in a separate beaker to avoid any deterioration by ultrasonic waves (Fig. 1.1). TEM observations revealed that the sonoelectrochemically produced powder consists mainly of isolated particles with different nanometric diameters but also with some small agglomerates (Fig. 1.2). However, these agglomerates were avoided by using surfactants.

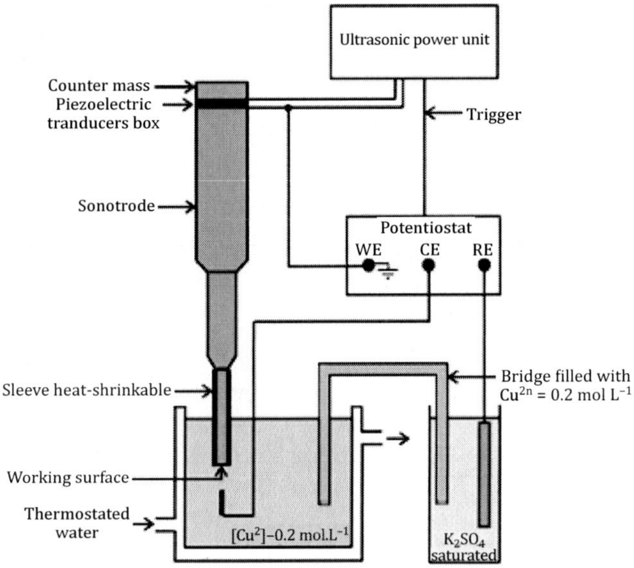

Figure 1.1 Schematic of pulsed sonoelectrochemical apparatus. WE = working electrode in titanium; RE = reference electrode (SSE); CE = counter electrode of copper. Reprinted from *Ultrason. Sonochem.*, 17, Mancier V., Rousse-Bertrand C., Dille J., Michel, J., and Frocoteaux, P., Sono and electrochemical synthesis and characterization of copper core–silver shell nanoparticles, 690–696, Copyright (2010), with permission from Elsevier.

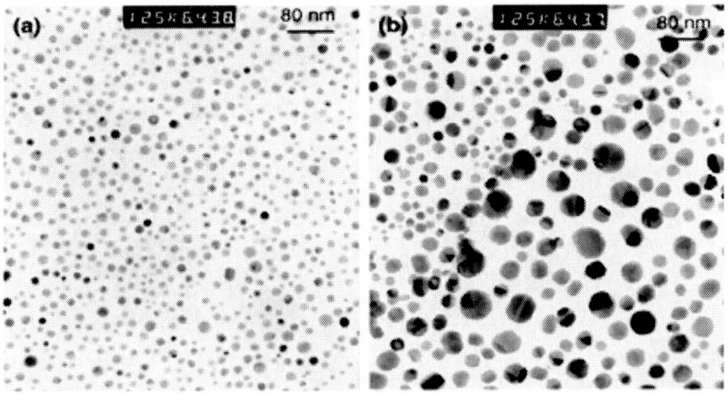

Figure 1.2 TEM of Cu-Ag nanoparticles. Reprinted from *Ultrason. Sonochem.*, 17, Mancier V., Rousse-Bertrand C., Dille J., Michel, J., and Frocoteaux, P., Sono and electrochemical synthesis and characterization of copper core–silver shell nanoparticles, 690–696, Copyright (2010), with permission from Elsevier.

1.2.2 Co-Sn-Graphite Nanoparticles

Preparation of Co-Sn-graphite electrode material using the high-intensity ultrasonication method has been recently reported by González et al. (2011). A composite electrode material containing amorphous Co-Sn nanoparticles and ultrathin layers of graphite has been successfully prepared by using the combination of polyol and sonochemical treatment for the first time (González et al., 2011). The utilization of ball-milled graphite, ultrasonication treatment, and lithium polyacrylate (LiPAA) permits accomplishing amorphous Co-Sn nanoparticles (Fig. 1.3) with improved electrochemical properties. The excellent electrochemical cycling behavior gained has been linked to the smaller particle size of Co-Sn, the low crystallinity of Co-Sn, and the dispersion of Co-Sn particles in an optimized carbon matrix.

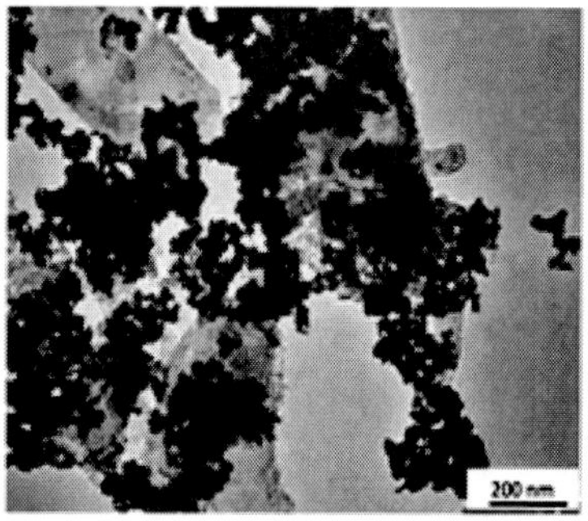

Figure 1.3 TEM of CO-Sn nanoparticles (40 h ball milling followed ultrasonication). Reprinted from *Ultrason. Sonochem.*, 15, Tai, G. and Guo, W., Sonochemistry-assisted microwave synthesis and optical study of single-crystalline CdS nanoflowers, 350–356, Copyright (2008), with permission from Elsevier.

1.3 Metal Oxide Nanoparticles

1.3.1 Zinc Oxide Nanoparticles

A sonochemical route was exploited by Banerjee et al. (2012) to synthesize zinc oxide nanoparticles from zinc acetate solution in aqueous methanol, ethanol, and *iso*propanol (5% by volume) in the absence of alkali and at neutral pH while maintaining the temperature at 65°C. Characterization with SEM (Fig. 1.4), XRD, AFM, and BET revealed that the as-prepared nanoparticles differed in size and shape and the yield was about 35% in all the cases. The sonochemically generated nanoparticles were then tested for the photocatalytic reduction of Cr(VI) and the efficiency was found to be nearly doubled as compared to commercially available microparticles.

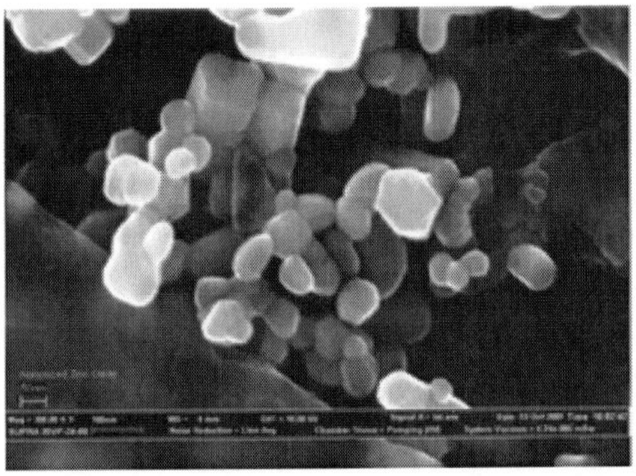

Figure 1.4 SEM of ZnO nanoparticles. Reprinted from *Ultrason. Sonochem.*, 19, Banerjee, P., Chakrabarti, S., Maitra, S., and Dutta, B. K., Zinc oxide nanoparticles: Sonochemical synthesis, characterization, and application for photo-remediation of heavy metal, 85–93, Copyright (2012), with permission from Elsevier.

1.3.2 Manganese Oxide Nanoparticles

Mn_3O_4 (hausmannite) nanoparticles were successfully prepared by using ultrasound in an aqueous and emulsion mediums under ambient conditions without employing any additives (Bastami and Entezari, 2010). Using water as a medium, the synthesis was conducted by direct sonication of manganese acetate in 30 mL of manganese acetate solution for 45 min; while in the case of emulsion, the sonication was performed in a mixture consisted of 4 mL of saturated solution of $Mn(OAc)_2$ and 30 mL of oil. The results showed that the size and shape of nanoparticles depend on the type of medium (Fig. 1.5). The XRD patterns confirmed that by using different media the same product, that is, Mn_3O_4 nanoparticles were obtained. Ultrasound itself generates intense forces to well disperse the oil phase into water, which resulted in the formation of fine droplets. These droplets act as reactors for the formation of product and affected the size and shape of the nanoparticles.

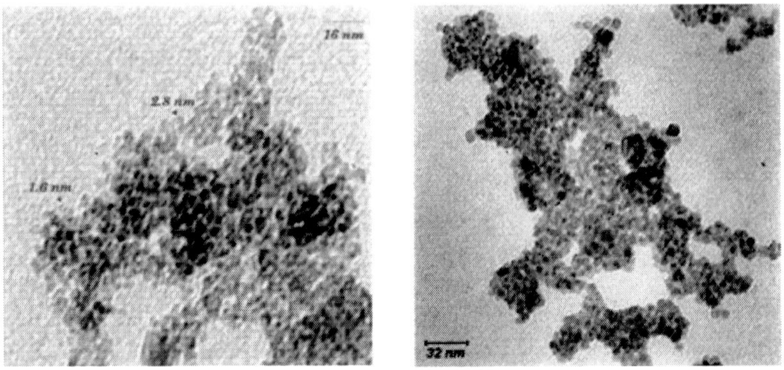

Figure 1.5 TEM images of hausmannite (Mn_3O_4) nanoparticles (aqueous medium (left) and olive oil medium (right)). Reprinted from *Chem. Eng. J.*, 164, Bastami, T. R., and Entezari M. H., Sono-synthesis of Mn_3O_4 nanoparticles in different media without additives, 261–266, Copyright (2010), with permission from Elsevier.

1.3.3 Tl_2O_3 Nanoparticles

Akhbari and Morsali (2012) employed slow evaporation technique coupled with sonochemical methods to synthesize a new thallium (I) supramolecular polymer, $[Tl_2(\mu_3\text{-TPOA}_c)(\mu_4\text{-TPOAc})]_n$ (Fig. 1.6). Besides, thallium (III) oxide with nanostructured surface was generated by calcinating thallium (I) supramolecular polymer at 873 K. This sonochemical approach for the generation of supramolecular polymers offers the following advantages over conventional methods: (1) shorter reaction time, (2) better yield, and (3) nanosize. Thus, ultrasound irradiation as an alternative synthetic approach provides new insights into metal-organic supramolecular assembly and nanochemistry (Akhbari and Morsali, 2012).

1.3.4 Copper (II) Oxide Nanoparticles

Nanosized Cu (II) supramolecule, $\{[Cu(L)_2(H_2O)_2]$, L = 1 H-1,2,4-triazole-3-carboxylate$\}$, was prepared using an aqueous solution of copper (II) chloride in the presence of a ligand, 1 H-1,2,4-triazole-

3-carboxylic acid, and sodium hydroxide via an ultrasound-assisted method (Safarifard and Morsali, 2012a). Reagents were employed with two different concentrations (0.01 and 0.05 M) to prepare the CuO nanoparticles. The as-prepared samples were characterized by SEM (Fig. 1.7), XRD, IR spectroscopy, and elemental analysis. Calcination of the nanosized Cu (II) supramolecule at 600°C under air atmosphere resulted into the generation of CuO nanoparticles.

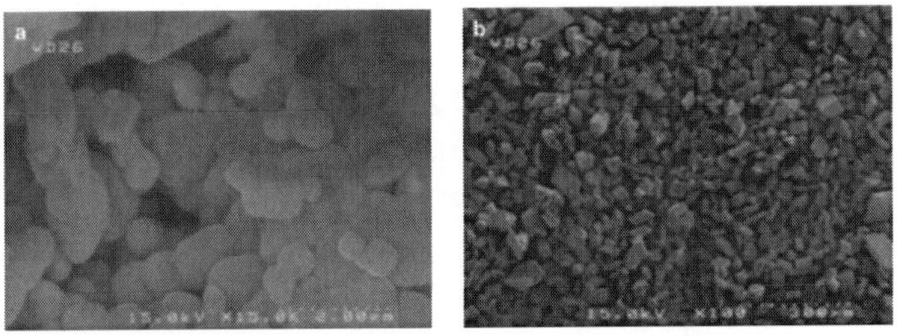

Figure 1.6 SEM images of thallium (I) supramolecular polymer obtained from (a) 0.025 M and (b) 0.05 M concentration of initial precursor using an ultrasonic bath. Reprinted from *J. Organomet. Chem.*, 700, Akhbari, S. and Morsali, A., Thallium(I) supramolecular polymer with Tl···π secondary interactions; preparation of nanostructures by sonochemical method as a new precursor for preparation of Ti_2O_3 with nano-structural surface, 125–131, Copyright (2012), with permission from Elsevier.

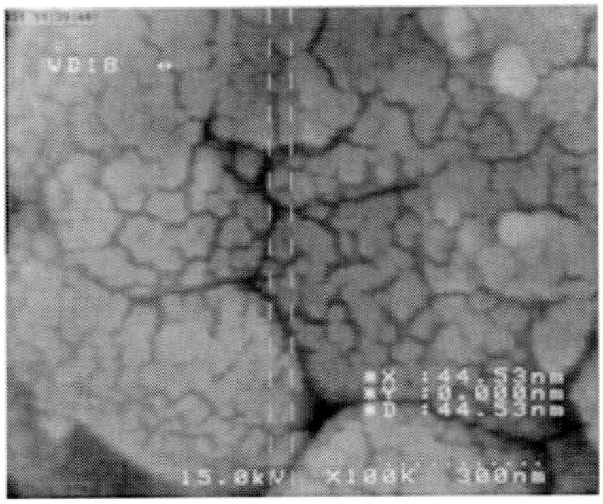

Figure 1.7 SEM images of agglomerated CuO nanoparticles. Reprinted from *Ultrason. Chem.*, 19, Safarifard, V., and Morsali, A., Sonochemical synthesis of nano-sized copper (II) supramolecule as a precursor for the synthesis of copper (II) oxide nanoparticles, 823–829, Copyright (2012), with permission from Elsevier.

Nanosized Pb (II) coordination polymer {[Pb(L)($\mu_{1,1}$-NCS)(H_2O)]$_n$, L = 1 H-1,2,4-triazole-3-carboxylate} has also been prepared via sonochemical irradiation (Safarifard and Morsali, 2012b). Effect of concentration of reagents employed and power of ultrasound irradiation on the size and morphology of nanoparticles was also evaluated and the authors have found that the power of ultrasound irradiation has a great influence on the resultant particle size and morphology of the nanoparticles. Calcination of the single crystals and nanosized Pb (II) coordination polymer at 400°C under ambient conditions gave a mixture of PbS and $Pb_2(SO_4)O$ nanoparticles. Results confirmed a decrease in the particle size with an increase in the power of ultrasound irradiation.

1.3.5 Synthesis of Pt-Graphene Oxide (GO)-TiO_2

Ultrasound-assisted hydrothermal method was reported for the synthesis of nanosized Pt-GO-TiO_2 photocatalyst (Neppolian et al., 2012). In this method, GO-TiO_2 was prepared by the sonochemical treatment of GO dispersion and crystalline TiO_2. Different wt% of Pt was then doped onto the above photocatalyst by a photochemical reduction method. The resultant Pt-GO-TiO_2 photocatalyst (Fig. 1.8) showed very high photocatalytic degradation efficiency toward an anionic surfactant, dodecylbenzenesulfonate (DBS), in the presence of both UV and visible light irradiation. In addition, the authors found that the formation of many intermediate products during the sonochemical process decreased the rate of sonochemical oxidation.

1.3.6 Manganese Oxide Nanocrystals

An ultrasound-based synthesis of manganese oxide nanocrystals was reported by Bastami and Entezari (2012). It is based on the sonication of magnetized precursor solution in the neutral as well as in the basic media. The sonication was accomplished using an ultrasonic bath (40 kHz) for 180 min at a temperature of 60°C. In the neutral medium, sonicating a magnetized solution for 24 h mainly led to a mixture of Mn_3O_4 (hausmannite) and MnO(OH) (manganite), whereas the sonication of unmagnetized solution led to pure Mn_3O_4

(Fig. 1.9). They observed the formation of rod-like structures in the presence of ultrasonic irradiation. Besides, the sonochemically prepared manganese oxide nanocrystals showed superparamagnetic behavior at room temperature without any saturation magnetization and hysteresis in the measured field strength.

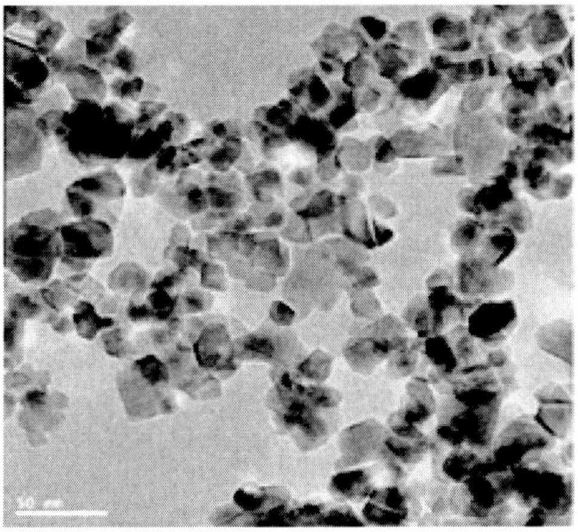

Figure 1.8 TEM images of Pt-GO-TiO_2 photocatalyst calcined at 650°C for 3 h. Reprinted from *Ultrason. Chem.*, 19, Neppolian, B., Bruno, A., Bianchi, C. L., Ashokkumar, M., Graphene oxide based Pt-TiO_2 photocatalyst: Ultrasound assisted synthesis, characterization and catalytic efficiency, 9–15, Copyright (2012), with permission from Elsevier.

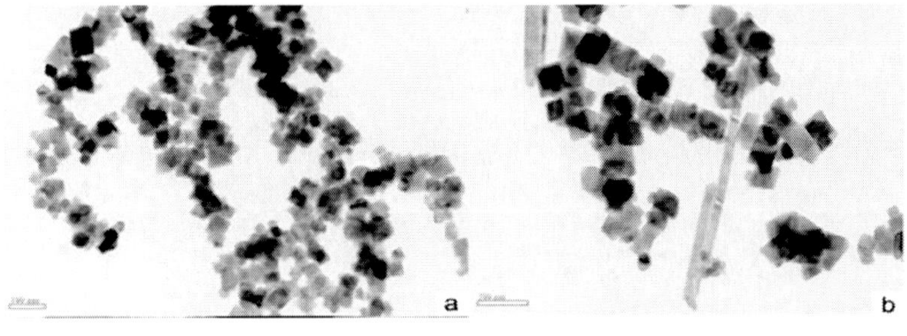

Figure 1.9 HRTEM of (a, b) manganese oxide nanocrystals using ultrasound irradiation in the neutral medium. Reprinted from *Ultrason. Sonochem.*, 19, Bastami, T. R., Entezari, M. H., Synthesis of manganese oxide nanocrystal by ultrasonic bath: Effect of external magnetic field, 830–840, Copyright (2012), with permission from Elsevier.

1.3.7 Nanocrystalline NiO Powder

Meybodi et al. (2012) reported a novel sonochemical technique to produce a wide band gap nanocrystalline nickel particles with the mean crystallite size of ~20 nm using $Ni(OH)_2$ precursor. Nanocrystalline NiO powder was produced via sonochemical route by the drop-wise addition of 0.1 M sodium hydroxide to 0.1 M nickel nitrate $Ni(NO_3)_2$ solution followed by ultrasound irradiation for 60 min. The obtained gel was then dried at 80°C for overnight. Dried precipitates were finally calcined at 320°C for 1 h to form the nanocrystalline NiO particles (Fig. 1.10). It has been mentioned that the cavitational implosion due to ultrasound irradiation led to the size reduction of the nickel oxide crystallites that imparted highly pure well-crystallized structure.

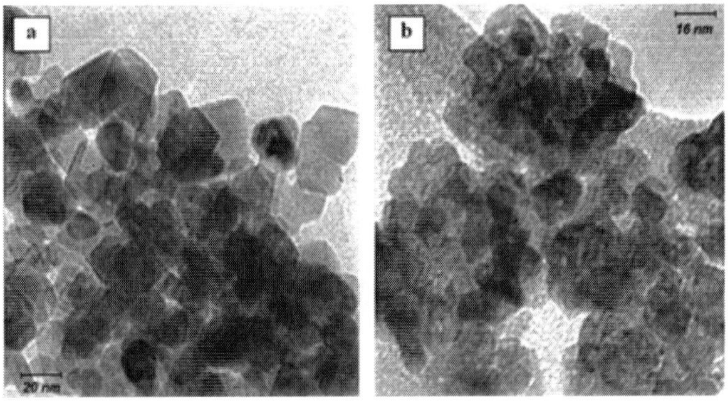

Figure 1.10 TEM images of (a) well-dispersed and (b) agglomerated nanocrystalline NiO powder. Reprinted from *Ultrason Sonochem.*, 84, Mohseni Meybodi S., Hosseini S. A., Rezaee M., Sadrnezhaad S. K., and Mohammadyani D., Synthesis of wide band gap nanocrystalline NiO powder via a sonochemical method, 841–845, Copyright (2012), with permission from Elsevier.

1.4 Magnetic Nanoparticles

1.4.1 Magnetite Nanoparticles

Magnetic hydroxyapatite (m-HAP) nanoparticles has potential applications in biomedicine particularly for the treatment of bone cancer. Functionalization of the hydroxyapatite (HAP) with magnetite nanoparticles (MNPs) through ultrasonic irradiation technique has been recently explored (Gopi et al., 2012). Ultrasonic irradiation was employed with two different frequencies of 28 kHz and 35 kHz at 150 W and 320 W, respectively, for the synthesis of m-HAP (Fig. 1.11). The ultrasound irradiation of 35 kHz at 320 W showed an enhanced diffusion of MNP to the HAP host matrix and led to the formation of m-HAP. This sonochemical method was rapid and cheaper as compared to other conventional methods as it does not require high temperature and stabilizer. This study also confirmed the formation of pure m-HAP nanocomposite with distinctive magnetic properties needed for biological applications.

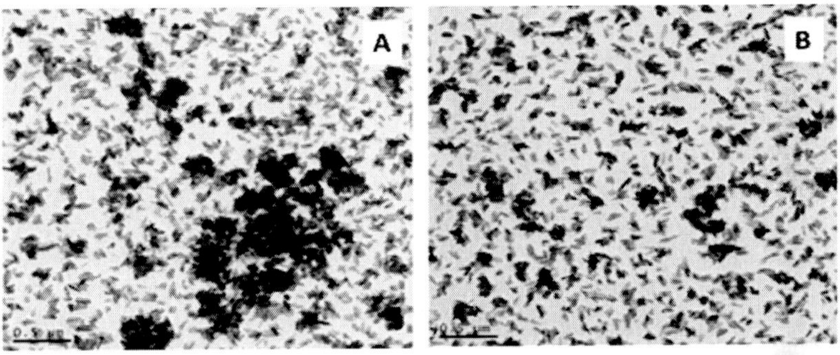

Figure 1.11 TEM images of (A) m-HAP synthesized at 28 kHz and at 150 W and (B) m-HAP synthesized at 35 kHz and at 320 W. Reprinted from *Spectrochim. Acta A: Mol. Biomol. Spectrosc.*, 87, Gopi, D., Ansari, M. T., Shinyjoy, E., and Kavitha, L., Synthesis and spectroscopic characterization of magnetic hydroxyapatite nanocomposite using ultrasonic irradiation, 245–250, Copyright (2012), with permission from Elsevier.

1.4.2 Peroxidase-Like Fe_3O_4 Magnetic Nanoparticles

Fe_3O_4 magnetic nanoparticles (Fe_3O_4 MNPs) with enhanced peroxidase-like activity were obtained by Wang et al. (2010) using an advanced reverse co-precipitation method with the assistance of ultrasound irradiation (US-RP). A significant improvement in the catalytic activity obtained was attributed to the smaller particle size (16.5 nm), greater BET surface area (82.5 m^2g^{-1}), and higher dispersibility of Fe_3O_4 MNPs in water (Wang et al., 2010).

It has been observed that at pH 5.4 and at a temperature of 40°C, the sonochemically prepared Fe_3O_4 MNPs were able to remove approximately 90% of Rhodamine B. The analysis also revealed that both ·OH and $O_2^{-\cdot}/H_2O^{\cdot}$ radicals are involved in the Fe_3O_4 participating catalytic reaction, and therefore the radicals are the main reactive species to degrade the organic pollutants.

1.5 Nanocomposites

1.5.1 CdS/TiO_2 Nanocomposites

A core–shell nanocomposite (CdS/TiO_2) with the particle size of ~11 nm was successfully synthesized by Ghows and Entezari (2012) at a relatively low temperature using a combination of ultrasound and new microemulsion (o/w) but without incorporating any surfactant. Application of ultrasound irradiation controlled the hydrolysis and condensation of titanium tetraisopropoxide (TTIP) and led to the formation of TiO_2 shell around the CdS core (Ghows and Entezari, 2012). The depth of coated TiO_2 onto CdS nanoparticles was also controlled in the nanoscale by using this method. Besides, this combined approach also avoids the tedious selection of different additives and obviates the need for calcination process. Synergistic action of sonochemical and microemulsion technique resulted into shorter reaction time, better crystallinity, better control, and higher speed of formation of core–shell structure with uniform shape. Formation of core–shell nanoparticles has been explained based on the cavitation process that drives the nanoparticles together at higher velocities. In addition, the intense local heating due to cavitation enhances the crystallinity of the generated nanoparticles (Fig. 1.12).

1.5.2 CdS Nanoparticles and CdS/TiO_2 Nanocomposites

Wang et al. (2012) have reported the preparation of CdS nanoparticles and CdS/TiO_2 nanocomposites under multibubble sonoluminescence (MBSL) conditions employing $CdCl_2$, $NaBH_4$, Se, and/or TiO_2. The mixture was sonicated with a high intensity ultrasonic horn (20 kHz, 55 W) at a constant temperature of 55°C. It has been noticed that ultrasonically produced nanoparticles were well crystallized (Fig. 1.13), and the suppression of crystal pattern transition as well as the control of CdS crystal growth could be implemented by coupling with TiO_2 semiconductor. The yield was about 95%. In comparison with other conventional preparation methods, this synthetic methodology has the advantages of brief reaction steps, convenient manipulation, and environmental friendliness.

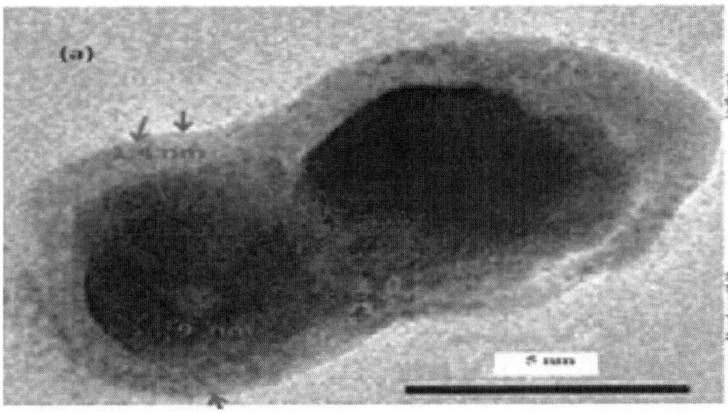

Figure 1.12 TEM images of CdS/TiO_2 nanoparticles prepared in the presence of surfactant with the molar ratio of 2.5:1. Reprinted from *Ultrason. Sonochem.*, 19, Ghows, N., and Entezari, M. H., Sono-synthesis of core–shell nanocrystal (CdS/TiO_2) without surfactant, 1070–1078, Copyright (2012), with permission from Elsevier.

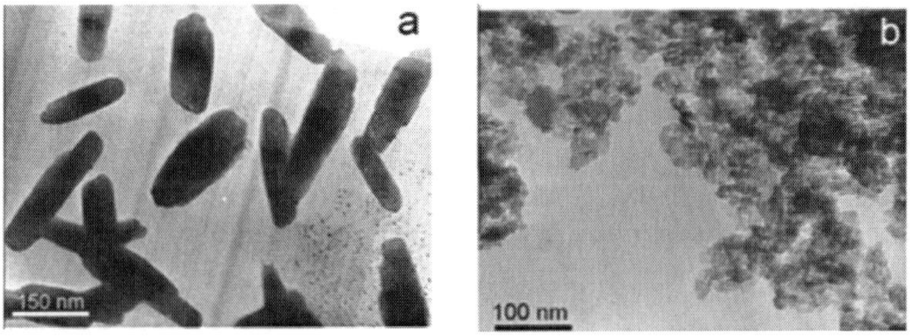

Figure 1.13 TEM images of the products synthesized under MBSL conditions (a) CdS nanorods and (b) CdS/TiO_2 nanocomposites. Reprinted Reprinted from *Ultrason. Sonochem.*, 19, Wang, J., Guo, W., Liu, S., and Li, D., Synthesis of CdSe and CdSe/TiO_2 nanoparticles under multibubble sonoluminescence condition, 464–468, Copyright (2012), with permission from Elsevier.

1.5.3 Exfoliated Poly(styrene-co-methyl methacrylate)/Montmorillonite [P(MMA-co-St)/O-MMT] Nanocomposites

Synthesis of exfoliated structure of P(MMA-co-St)/O-MMT nanocomposite with ultrasound-assisted in situ emulsion polymerization has been reported by Bhanvase et al. (2012). Such an ultrasound-assisted emulsion polymerization requires shorter reaction time (1 h) with improved polymerization rate as well as the dispersion of MMT in polymer latex compared to the conventional emulsion polymerization method, which requires 5 h (Bhanvase et al., 2012). Zeta potential and particle size analysis showed that nanocomposite latexes were electrostatically stable. The average particle size was in the range of 156.58–191.23 nm with narrow particle size distribution. In addition, the exfoliated P(MMA-co-St)/O-MMT nanocomposite exhibited a higher glass transition temperature and lower heat of reaction at 1% O-MMT loading as compared to the neat copolymer.

1.5.4 Au-Polypyrrole (PPy)/Prussian Blue (PB) Nanocomposites

The multicomponent nanocomposites have been synthesized successfully by Lu et al. (2012) in water via one-step sonochemical approach without adding any reductant and stabilizer. With the addition of pyrrole monomers into the solution mixture, the autopolymerization of pyrrole into PPy and $AuCl_4^-$ were reduced instantaneously to elemental Au along with PB (Lu et al., 2012). Then, sonochemically prepared Au-polypyrrole (PPy)/Prussian blue (PB) nanocomposites were immobilized on the glassy carbon electrode (GCE) to construct a H_2O_2 biosensor. Such a biosensor exhibited high sensitivity, fast response, and good stability. It is a green and easily scalable process and can be extended to prepare other multicomponent nanocomposites used for catalytic and other applications.

1.5.5 Chitosan/Gold Nanocomposites

Chitosan/gold (Ch/Au) nanocomposites were successfully prepared in aqueous solution by Ou et al. (2011) using electrochemical methods with the aid of ultrasound irradiation. The process primarily consisted of two steps: (1) preparation of Ch-capped Au by dissolving Au substrates via electrochemical treatments of oxidation-reduction cycles and (2) reduction of Ch-capped Au cations via sonoelectrochemical route to synthesize Ch/Au nanocomposites. The sonoelectrochemical method resulted in the particle size of prepared Au nanoparticles with the predominant (1 1 1) face on Ch is approximately 15 nm. Figure 1.14 shows the TEM image of the Au nanoparticles in the chitosan/Au nanocomposite. The catalysts of Ch-capped Au nanoparticles showed a high performance in decomposing the capped acetaldehydes and oxidizing CO into CO_2 which can protect people from poisoning in the mainstream smokes.

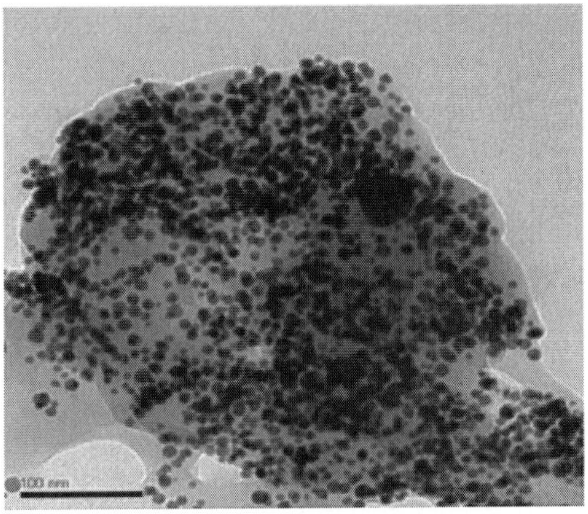

Figure 1.14 TEM image of Au nanoparticles in the chitosan/Au nanocomposite. Reprinted from *Mat. Res. Bull.*, 46, Ou, K.L., Yu, C.C., Liu, Y.C., Yang, K.H., Wang, C.C., Chen, Q.Y., Sonoelectrochemical synthesis of chitosan/gold nanocomposites for application in removing toxic materials in mainstream smokes, 2333–2337, Copyright (2011), with permission from Elsevier.

1.5.6 Magnesium–Aluminum-Layered Double Hydroxides Nanocomposite

A simple and effective ultrasound-assisted co-precipitation method in combination with a calcination treatment has been employed by Chang et al. (2011) to fabricate magnetic Mg–Al layered double hydroxides nanocomposite as an adsorbent material to remove fluoride ions from aqueous solutions. This sonochemical method promoted the formation of hydrotalcite-like phase and drastically shortened the preparation time. It has also been found that the use of ultrasound decreased the size of adsorption capacity and increased the specific surface area. Moreover, the sonochemically prepared nanocomposite exhibited a fairly high adsorption capacity for fluoride (47.7 mg g^{-1}), which was 60% higher than that of the composite prepared without using ultrasound but with the same aging time. Thus, the sonochemical methodology is a convenient synthesis process for calcined magnetic layered double hydroxides nanocomposite with high efficiency for fluoride removal.

1.6 Chalcogenides

1.6.1 Mercury Selenide Nanoparticles

Mercury selenide (HgSe) was successfully synthesized by Esmaeili-Zare et al. (2012) using a sonochemical method from $HgCl_2$ and $SeCl_4$ in the presence of hydrazine hydrate ($N_2H_4.H_2O$) and different capping agents. The synthesis was carried out with a multiwave ultrasonic generator (20 kHz, 60 W) for 30 min under controlled temperature. The effect of some of the parameters such as the type and amount of capping agent, reaction time, temperature, and power of sonication on the morphology and particle size of HgSe products were evaluated. Figure 1.15 shows the SEM images of HgSe obtained with the application of two different ultrasound power intensities, that is, 40 W/cm^2 and 50 W/cm^2. Application of ultrasonic irradiation in this case obviates the use of toxic chemicals and time-consuming conventional methods.

1.7 Metal Bromide and Sulfate Nanoparticles

1.7.1 AgBr Nanoparticles

Abbasi and Morsali (2012) synthesized nanostructured AgBr by the reaction between $AgNO_3$ and KBr using a molar ratio of 1:1

under ultrasound irradiation. The above reaction was carried out in a double jacketed ultrasonic bath. It has been observed that the particle size and morphology of the nanoparticles obtained were highly dependent on the temperature, sonication power, reaction time, and concentration of precursors. A decrease in the particle size was observed with an increase in the power of ultrasound. Obtained results also demonstrate that an increase in temperature, concentration, and reaction time led to a decrease in the particle size and hence the morphology of silver bromide nanoparticles.

Figure 1.15 SEM images of HgSe prepared in the presence of Tetraethylammonium (TEA) and hydrazine with different ultrasound power intensities (a) 40 W/cm^2 and (b) 50 W/cm^2. Reprinted from *Ultrason. Sonochem.*, 19, Esmaeili-Zare M, Salavati-Niasari M, Sobhani A., Simple sonochemical synthesis and characterization of HgSe nanoparticles, 1079–1086, Copyright (2012), with permission from Elsevier.

1.7.2 Barium Sulfate Nanoparticles

Two continuous flow sonochemical reactors were developed by Pohl et al. (2012) for the preparation of barium sulfate nanoparticles via precipitation. Both the reactors employed ultrasound as a mixing tool. The first geometry consisted of a conical chamber (10 mL), which was used to study the micromixing quality, the cavitation intensity, and the precipitation of barium sulfate nanoparticles. The second reactor was essentially a cavitation chamber (2.5 mL), which was an optimized conical reactor. Both reactors were compared with respect to the properties of the products obtained. This study showed that increasing ultrasonic output and a temperature rise had no influence on the precipitation. They also observed that at an optimal feed rate of 2.4 L/h and at a power output of 20–40 W in both the reactor geometries, no decoupling of the sonotrode from the liquid occurred and smaller nanoparticles were synthesized. At these ultrasonic outputs, particles with a size of 200 nm were produced using a much lower feed rate of 0.8 L/h as compared to 6 L/h in the absence of ultrasound. Figure 1.16 shows the SEM images of precipitated barium sulfate using ultrasound of 160 W at two different temperatures of 20°C and 70°C.

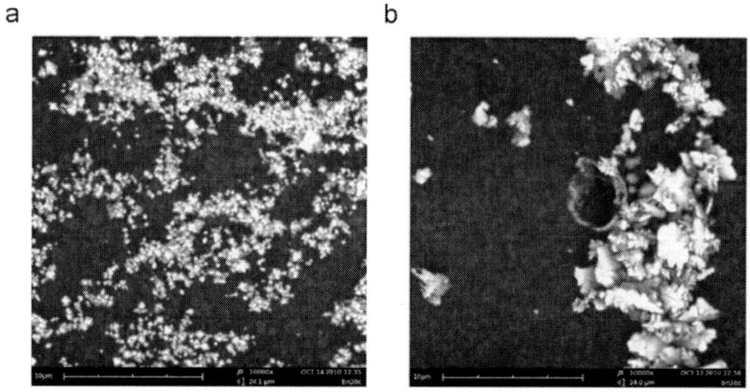

Figure 1.16 SEM images of precipitated barium sulfate using ultrasound with 160 W power at (a) 20°C and at (b) 70°C. Reprinted from *Chem. Eng. Sci.*, 69, Pohl. B., Jamshidi, R., Brenner, G., and Peuker, U. A., Experimental study of continuous ultrasonic reactors for mixing and precipitation of nanoparticles, 365–372, Copyright (2012), with permission from Elsevier.

1.8 Porous Materials

1.8.1 N-Doped Nanocrystalline Anatase TiO_2

A novel single-step sonochemical method has been established by Wang et al. (2011) to prepare N-doped nanocrystalline anatase TiO_2 at low temperature. The synthesis was performed by sonicating a solution of tetraisopropyl titanium and urea in water and isopropyl alcohol at 80°C for 150 min. The mean crystallite size of the N-doped anatase TiO_2 obtained was 10.8 nm. The obtained results also showed that N-doped nanocrystalline TiO_2 prepared via sonochemical route exhibit an excellent photocatalytic activity under UV light and simulated sunlight.

1.9 Other Nanomaterials

1.9.1 MoO_3 Nanospheres

With the assistance of ultrasonic irradiation, highly dispersed MoO_3 nanospheres were successfully synthesized by Du et al. (2008) using a silane agent, that is, 3-mercaptopropyltrimethoxysilane (HS-$(CH_2)_3Si(OCH_3)_3$(MPTS)) as figuration agent. The mixture was treated under mechanical agitation and ultrasound (40 kHz, 300 W) for 2 h at ambient conditions. The XRD and TEM results indicated that the obtained MoO_3 nanoparticles were spherical in shape with the particle size distribution of approximately 30–80 nm and 25–75 nm, respectively. The probable reason for the formation of highly dispersed MoO_3 nanospheres was mainly accounted based on the dispersion function of ultrasound irradiation and figuration of MPTS.

1.9.2 $LiMn_2O_4$ Nanospheres

Kim et al. (2012) successfully fabricated highly ordered mesoporous $LiMn_2O_4$ nanospheres through a chemical reaction in combination with ultrasonic waves. Using a direct immersion titanium alloy, sonicator operated at a periodic pulse mode (750 W, 20 kHz at 50% amplitude) resulted in the formation of hierarchically worm-like and stable $LiMn_2O_4$, imparting structural stability and high crystallinity with a large effective surface area to storage charge to the nanospheres. As a cathode material, $LiMn_2O_4$ not only had a similar value of theoretical discharge capacity but also showed an excellent rate capability and cyclability with a capacity retention of more than 98% up to 50 cycles at a 0.2 C rate. This has been explained due to the larger surface area of the active domains and high crystallinity of the sonochemically produced $LiMn_2O_4$ nanospheres. More importantly, it has been demonstrated that ultrasound is a simple, cost-effective, and powerful tool in the large-scale production of cathode materials with high electrochemical performance while being environmentally benign.

1.9.3 Chromium Dioxide Nanowires

Zhou et al. (2011) reported the fabrication of chromium dioxide (CrO_2) nanowires (Fig. 1.17) on a large scale by using CrO_2Cl_2 and homemade Cr_2O_3 hollow nanospheres employing ultrasound. This one-step sonochemical route is simpler as compared to that of multistep complicated solid-state reaction. This process also gave high-yield CrO_2 nanowires through a growth mechanism that has been suggested to be a complex eight-step approach. The experimental results showed that the nanowires with the controlled diameter of 100–200 nm possess a single crystal tetragonal structure and exhibit a colossal coercive field (1678 Oe) at room temperature, which could be beneficial for tuning and improving the magnetic properties of nanowire devices, particularly for spintronics, high density data storage, magneto-optical ones, and nanobiosensors.

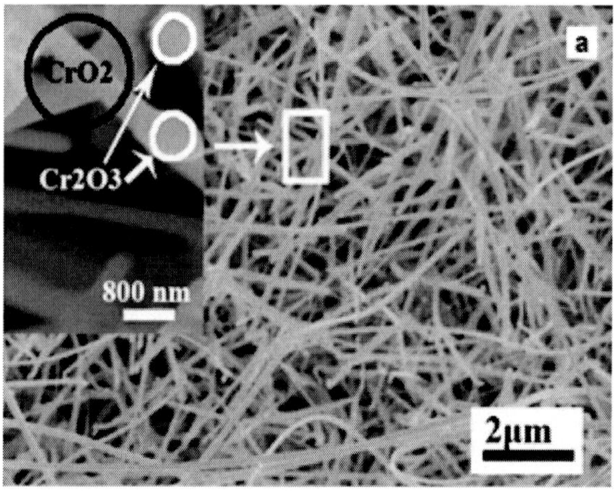

Figure 1.17 HRTEM images of CrO_2 nanowires. Reprinted from *Chem. Eng. J.*, 174, Zhou, S.-M., Zhu, G.-Y., Wang, Y.-Q., Lou, S.-Y. and Hao, Y.-M. Sonochemical preparation of massive CrO_2 nanowires, 432–435, Copyright (2011), with permission from Elsevier.

1.9.4 Nanoclay Bentonite-FeCo Nanocomposite Hybrid Hydrogel

Poly(acrylic acid)-bentonite-FeCo (PAA-B-FeCo) hydrogel nanocomposite was synthesized by Shirsath et al. (2011) through an ultrasound-assisted in situ emulsion polymerization of AA (acrylic acid) and then cross-linking with B-FeCo. The response of the in situ emulsion nanocomposite hydrogel obtained using this ultrasound cavitation technique was evaluated using cationic dye, crystal violet (CV), at different temperatures and pH. Addition of exfoliated bentonite clay platelets and FeCo was found to increase the strength and stability of hydrogel. The sonochemically cross-linked polymer hydrogels also showed good swelling behavior due to the presence of B-FeCo. Also, the combined effect of hydrogel and ultrasound showed a comparatively higher removal of the dye than the hydrogel alone. With hydrogel alone, 87% removal was achieved in 15 h, whereas with the combination of ultrasound 97% removal was achieved with 5 h only. This enhancement was mainly attributed to the mechanical agitation generated by the physical forces caused by acoustic cavitation. The color change observed in both the hydrogel and dye solutions has been shown in Fig. 1.18. The color of the hydrogel changed from brown to dark blue.

Figure 1.18 Color change of crystal violet dye and hydrogel before [(A), (C)] and after absorption test for dye removal [(B), (D)]. Reprinted from *Desalination*, 281, Shirsath. S. R., Hage, A. P., Zhou, M., Sonawane, S. H., and Ashokkumar, M., Ultrasound-assisted preparation of nanoclay Benonite-FeCo nanocomposite hybrid hydrogel: A potential responsive sorbent for removal of organic pollutant from water, 429–437, Copyright (2011), with permission from Elsevier.

1.9.5 Silicoaluminophosphate Molecular Sieve (SAPO)-34 Nanocrystals

Mesoporous silicoaluminophosphate molecular sieves (SAPOs) have considerable potential as acidic catalyst, for example, can act as a membrane or as an adsorbent in the sorption reactions and as a catalyst in the petrochemical reactions. Askari and Halladj (2012) have successfully produced SAPO-34 via a sonochemical method using tetraethylene ammonium hydroxide (TEAH) as a structure-directing agent (SDA). Ultrasound irradiation was performed by means of a direct immersion titanium sonotrode with 300 W/cm^2 output power. The resulting mean crystallite size was 50 nm which was much smaller than that of the synthesized one under hydrothermal conditions. The sonochemically prepared SAPO-34 nanocrystals possessed uniform morphology. Compared to hydrothermal synthesis, this technique appeared to be faster and more effective in the production of SAPO-34 nanocrystals. It has been proposed that the cavitational collapse of bubbles could have provided a huge density of energy for the crystallization and would have prevented the organization and agglomeration of particles that leads to crystals with smaller size (Fig. 1.19).

Figure 1.19 SEM images of Sono-SAPO-34. Reprinted from *Ultrason. Sonochem.*, 19, Askari, S., and Halladj, R., Ultrasonic pretreatment for hydrothermal synthesis, 554–559, Copyright (2012), with permission from Elsevier.

1.10 Conclusions

This chapter has outlined the recent progress concerned with making use of ultrasound in the fabrication, modification, and applications of various multifunctional materials including nanostructured metal oxides, nanocomposites, nanocrystals, nanorods, and nanopowders. It has been undoubtedly demonstrated that acoustic cavitation has the capability to be developed into a most useful and powerful synthetic tool in the nanoscience and nanotechnology arena if the underlying principles and key parameters are understood and controlled. With a simple modification in the reaction conditions, the utilization of sonochemical method has been yet extended to the generation of nanospheres, nanocomposites hydrogel, and nanowires. The synthesis of nanostructured materials via ultrasound emerges to be more flexible and efficient than the other conventional approaches as a means of generating a variety of nanostructured materials with different morphology and properties. The cavitation bubbles act as microreactors for accelerating the rate of myriad type of chemical reactions and thus allow researchers for the effective control over the bulk chemical composition and particle size of nanomaterials. The only issue with the sonochemical method is related to "scale-up" at least at this stage. Many sonochemical scientists and engineers have great difficulty with this as the cavitational bubble cloud and high shear intensity zone are only focused in a small region in a larger sonochemical reactor. Besides, to build and enable this large-scale production requires cooperative efforts among the mechanical, chemical, and electrical engineers and expertise from physicists and chemists. Overall, while the synthesis of nanostructured materials via ultrasound can be a generalized, versatile, and facile synthetic route nowadays, it has to be undertaken in conjunction with an even greater effort to control the reaction parameters. With the benefit of earlier findings gained through serious and intense studies, it behooves us to continue to understanding the physics of cavitation clouds chiefly in terms of their impact on the nanomaterial synthesis pathway to advance our knowledge in the area of sonochemistry.

References

Abbasi, A. R. and Morsali, A. (2012). Influence of solvents on the morphological properties of AgBr nanostructures prepared using ultrasound irradiation. *Ultrason. Sonochem.*, **19**(3), pp. 540–545.

Akhbari, K. and Morsali, A. (2012). Thallium (I) supramolecular polymer with Tl...π secondary interactions; preparation of nanostructures by sonochemical method as a new precursor for preparation of Tl_2O_3 with nanostructural surface. *J. Organomet. Chem.*, **700**, pp. 125–131.

Askari, S. and Halladj, R. (2012). Ultrasonic pretreatment for hydrothermal synthesis of SAPO-34 nanocrystals. *Ultrason. Sonochem.*, **19**(3), pp. 554–559.

Banerjee, P., Chakrabarti, S., Maitra, S. and Dutta, B. K. (2012). Zinc oxide nano-particles: Sonochemical synthesis, characterization and application for photo-remediation of heavy metal. *Ultrason. Sonochem.*, **19**(1), pp. 85–93.

Bastami, T. R. and Entezari, M. H. (2010). Sono-synthesis of Mn_3O_4 nanoparticles in different media without additives. *Chem. Eng. J.*, **164**(1), pp. 261–266.

Bastami, T. R. and Entezari, M. H. (2012). Synthesis of manganese oxide nanocrystal by ultrasonic bath: Effect of external magnetic field. *Ultrason. Sonochem.*, **19**(4), pp. 830–840.

Bhanvase, B. A., Pinjari, D. V., Gogate, P. R., Sonawane, S. H. and Pandit, A. B. (2012). Synthesis of exfoliated poly(styrene-co-methyl methacrylate)/ montmorillonite nanocomposite using ultrasound assisted in situ emulsion copolymerization. *Chem. Eng. J.*, **181–182**, pp. 770–778.

Chamberlin, R. R. and Skarman, J. S. (1966). Chemical spray deposition process for inorganic films. *J. Electrochem. Soc.*, **113**(1), pp. 86–89.

Chang, Q., Zhu, L., Luo, Z., Lei, M., Zhang, S. and Tang, H. (2011). Sono-assisted preparation of magnetic magnesium–aluminum layered double hydroxides and their application for removing fluoride. *Ultrason. Sonochem.*, **18**(2), pp. 553–561.

Du, K., Fu, W., Wei, R., Yang, H., Xu, J., Chang, L., Yu, Q. and Zou, G. (2008). Ultrasonic-assisted synthesis of highly dispersed MoO_3 nanospheres using 3-mercaptopropyltrimethoxysilane. *Ultrason. Sonochem.*, **15**(3), pp. 233–238.

Esmaeili-Zare, M., Salavati-Niasari, M. and Sobhani, A. (2012). Simple sonochemical synthesis and characterization of HgSe nanoparticles. *Ultrason. Sonochem.*, **19**(5), pp. 1079–1086.

Ghows, N. and Entezari, M. H. (2012). Sono-synthesis of core–shell nanocrystal (CdS/TiO_2) without surfactant. *Ultrason. Sonochem.*, **19**(5), pp. 1070–1078.

González, J. R., Alcántara, R., Nacimiento, F. and Tirado, J. L. (2011). CoSn-graphite electrode material prepared by using the polyol method and high-intensity ultrasonication. *Electrochim. Acta,* **56**(27), pp. 9808–9817.

Gopi, D., Ansari, M. T., Shinyjoy, E. and Kavitha, L. (2012). Synthesis and spectroscopic characterization of magnetic hydroxyapatite nanocomposite using ultrasonic irradiation. *Spectrochim Acta A: Mol. Biomol. Spectrosc.,* **87**, pp. 245–250.

Kim, J. M., Lee, G., Kim, B. H., Huh, Y. S., Lee, G.-W. and Kim, H. J. (2012). Ultrasound-assisted synthesis of Li-rich mesoporous $LiMn_2O_4$ nanospheres for enhancing the electrochemical performance in Li-ion secondary batteries. *Ultrason. Sonochem.,* **19**(3), pp. 627–631.

Lu, X., Li, Y., Zhang, X., Du, J., Zhou, X., Xue, Z. and Liu, X. (2012). A simple and an efficient strategy to synthesize multicomponent nanocomposites for biosensor applications. *Anal. Chim. Acta,* **711**, pp. 40–45.

Mancier, V., Rousse-Bertrand, C., Dille, J., Michel, J. and Fricoteaux, P. (2010). Sono and electrochemical synthesis and characterization of copper core–silver shell nanoparticles. *Ultrason. Sonochem.,* **17**(4), pp. 690–696.

Mohseni Meybodi, S., Hosseini, S. A., Rezaee, M., Sadrnezhaad, S. K. and Mohammadyani, D. (2012). Synthesis of wide band gap nanocrystalline NiO powder via a sonochemical method. *Ultrason. Sonochem.,* **19**(4), pp. 841–845.

Neppolian, B., Bruno, A., Bianchi, C. L. and Ashokkumar, M. (2012). Graphene oxide based Pt–TiO_2 photocatalyst: Ultrasound assisted synthesis, characterization and catalytic efficiency. *Ultrason. Sonochem.,* **19**(1), pp. 9–15.

Ou, K.-L., Yu, C.-C., Liu, Y.-C., Yang, K.-H., Wang, C.-C. and Chen, Q.-Y. (2011). Sonoelectrochemical synthesis of chitosan/gold nanocomposites for application in removing toxic materials in mainstream smokes. *Mat. Res. Bull.,* **46**(12), pp. 2333–2337.

Pohl, B., Jamshidi, R., Brenner, G. and Peuker, U. A. (2012). Experimental study of continuous ultrasonic reactors for mixing and precipitation of nanoparticles. *Chem. Eng. Sci.,* **69**(1), pp. 365–372.

Safarifard, V. and Morsali, A. (2012a). Sonochemical syntheses and characterization of nano-sized lead(II) coordination polymer with ligand 1H-1,2,4-triazole-3-carboxylate. *Ultrason. Sonochem.*, **19**(2), pp. 300–306.

Safarifard, V. and Morsali, A. (2012b). Sonochemical syntheses of a nano-sized copper(II) supramolecule as a precursor for the synthesis of copper(II) oxide nanoparticles. *Ultrason. Sonochem.*, **19**(4), pp. 823–829.

Shirsath, S. R., Hage, A. P., Zhou, M., Sonawane, S. H. and Ashokkumar, M. (2011). Ultrasound assisted preparation of nanoclay Bentonite-FeCo nanocomposite hybrid hydrogel: A potential responsive sorbent for removal of organic pollutant from water. *Desalination*, **281**, pp. 429–437.

Suslick, K. S. (2001). *Sonoluminescence and Sonochemistry: Encyclopedia of Physical Science and Technology*. Ed., R. A. Meyers. San Diego: Academic Press.

Suslick, K. S., Gawienowski, J. W., Schubert, P. F. and Wang, H. H. (1983). Alkane sonochemistry. *J. Phys. Chem.*, **87**, pp. 2299–2301.

Tai, G., Guo, W. (2008). Sonochemistry-assisted microwave synthesis and optical study of single-crystalline CdS nanoflowers. *Ultrason. Sonochem.*, **15**(4), pp. 350–356.

Wang, J., Guo, W., Liu, S., and Li, D. (2012). Synthesis of CdSe and CdSe/TiO_2 nanoparticles under multibubble sonoluminescence condition. *Ultrason. Sonochem.*, **19**(3), pp. 464–468.

Wang, N., Zhu, L., Wang, D., Wang, M., Lin, Z. and Tang, H. (2010). Sono-assisted preparation of highly-efficient peroxidase-like Fe_3O_4 magnetic nanoparticles for catalytic removal of organic pollutants with H_2O_2. *Ultrason. Sonochem.*, **17**(3), pp. 526–533.

Wang, X.-K., Wang, C., Guo, W.-L. and Wang, J.-G. (2011). A novel single-step synthesis of N-doped TiO_2 via a sonochemical method. *Mat. Res. Bull.*, **46**, pp. 2041–2044.

Zhou, S.-M., Zhu, G.-Y., Wang, Y.-Q., Lou, S.-Y. and Hao, Y.-M. (2011). Sonochemical preparation of massive CrO_2 nanowires. *Chem. Eng. J.*, **174**(1), pp. 432–435.

Chapter 2

Generation of Size, Structure, and Shape-Controlled Metal Nanoparticles Using Cavitation

Kenji Okitsu
Osaka Prefecture University, Osaka, Japan
okitsu@mtr.osakafu-u.ac.jp

Metal nanoparticles have a great potential for their application to industry and technology. To develop novel properties of metal nanoparticles, the size, structure and shape of metal nanoparticles must be controlled precisely. Although there are various methods to synthesize metal nanoparticles, a sonochemical method offers unique synthetic routes such as radical reactions and thermal reactions, which originate from hot cavitation bubbles during ultrasonic irradiation, where physical effects such as shock wave and microjet flow are simultaneously induced. In this chapter, the sonochemical reduction method to synthesize the size, structure, and shape-controlled metal nanoparticles is described.

Cavitation: A Novel Energy-Efficient Technique for the Generation of Nanomaterials
Edited by Sivakumar Manickam and Muthupandian Ashokkumar

ISBN 978-981-4411-54-7 (Hardcover), 978-981-4411-55-4 (eBook)
www.panstanford.com

2.1 Introduction

Metal nanoparticles have attracted considerable attention in various fields of industry and technology, which can contribute to comfortable human life in addition to the reduction of energy consumption and resources consumption. For example, Pd nanoparticles can be used as catalysts for cross-coupling reactions, hydrogenations and alcohol oxidations (Bianchini and Shen, 2009; Jia and Schuth, 2011). As the size and shape of metal nanoparticles strongly affects the surface property of metal nanoparticles, control in the size and shape is an important research area. In addition, in the case of bimetallic nanoparticles, their nanostructures such as core–shell structure, and a random alloy structure also expected to show novel and diverse physicochemical properties.

Although there are a number of methods for the synthesis of metal nanoparticles, the reduction processes of metal ions are often used to synthesize metal nanoparticles in an aqueous solution. Table 2.1 shows the representative approaches to synthesize metal nanoparticles in the reduction processes in a solution.

Table 2.1 Representative approaches to synthesize metal nanoparticles in the reduction processes in a solution

Approach 1 Selecting a suitable reductant and reduction process	**Addition of reductant** H_2, $NaBH_4$, citrate, hydrazine, CO, and alcohol in thermal chemistry (simple heating, micro-wave heating) **In situ generation of reductant** Electrochemistry, photochemistry, radiation-chemistry (radiation of γ–ray, X-ray, electron beam), sonochemistry
Approach 2 Selecting a suitable stabilizer	**Stabilizer** Citrate, polyvinyl pyrrolidone (PVP), sodium dodecyl sulfate (SDS), hexadecyltrimetylammonium bromide (CTAB), hexadecyltrimethylammonium chloride, polyethylene(40)glycol monostearate (PEG40MS), polyoxyethylene(20)sorbitan monolaurate (Tween20), triphenylphosphine, disulfide, dendrimer, sodium polyacrylate, sodium 3-mercaptopropionate, dodecanethiol

Selecting a suitable reductant is important to reduce metal ions in thermal chemistry. If one selected a strong reductant for metal ions, the reduction of metal ions would proceed quickly so that the nucleation behavior of metal nanoparticles is affected: if the number of the nucleation events was large, the size of the formed metal particles would become small.

On the contrary, if one selected an effective stabilizer that can adsorb on the surface of nuclei or small seed particles, the growth and/or aggregation of these particles would be suppressed, resulting in the formation of small metal particles and size-controlled metal particles. If the stabilizer adsorbed on the specific crystal surface of nuclei or small seed particles, the oriented growth of these particles would occur, resulting in the formation of shape-controlled metal particles. Therefore, to synthesize the size and shape-controlled metal nanoparticles, the balance of above two approaches has to be taken into account.

A thermal-chemical (simple heating) synthesis is widely studied, wherein the synthesis conditions with reductants and suitable stabilizers are explored to control the size, structure, and shape of metal nanoparticles precisely. Recently, microwave heating is used for the effective and homogeneous heating to proceed the reduction of metal ions and formation of metal nanoparticles.

As a controlled chemical method, electrochemical, photochemical, radiation-chemical, and sonochemical syntheses are also studied, because various types of reductants can be generated in situ in the scale from molecular size to micrometer under each experimental condition.

In particular, the sonochemical synthesis offers unique synthesis routes such as radical reactions and thermal reactions that originate from cavitation bubbles with local high temperature and high pressure, where physical effects such as shock wave and microjet flow are simultaneously induced in a solution. The size of cavitation bubbles is in the range of micrometers so that the reaction field induced by cavitation bubbles can be used as an effective microreactor. The specific chemical and physical effects of cavitation bubbles would enable us to synthesize novel nanoparticles with various physicochemical properties. In this chapter, the sonochemical synthesis of metal nanoparticles is reviewed in connection with the sonochemical reduction of metal ions in an aqueous solution. In particular, the results of the sonochemical reduction of Pd(II) and/or Au(III) to synthesize the size, structure, and shape-controlled metal nanoparticles are described in detail.

2.2 Size Control of Monometallic Nanoparticles

Monometallic nanoparticles can be prepared by the sonochemical reduction of the corresponding metal ions. In this chapter, the results obtained by employing a standing wave 200 kHz ultrasound irradiation system are mainly described. A representative irradiation setup is shown elsewhere (Okitsu et al., 2002). Figure 2.1 shows the changes in the absorption spectra of 0.5 mM $PdCl_2 \cdot 2NaCl$ (Pd(II)) and 8 mM SDS aqueous solution during ultrasonic irradiation under argon atmosphere (Okitsu et al., 1996). Before irradiation, the Pd(II) solution has small and trace absorption peaks at about 300 nm and 400 nm. These peaks gradually disappear with increasing irradiation time and the increase in the absorption spectra from the visible to the ultraviolet region is clearly observed. From the analysis of a light scattering experiment, X-ray diffraction (XRD), and transmission electron microscope (TEM), the formation of Pd nanoparticles with face-centered cubic structure was confirmed. The obtained spectrum in Fig. 2.1 is the typical spectrum of an aqueous solution containing Pd nanoparticles.

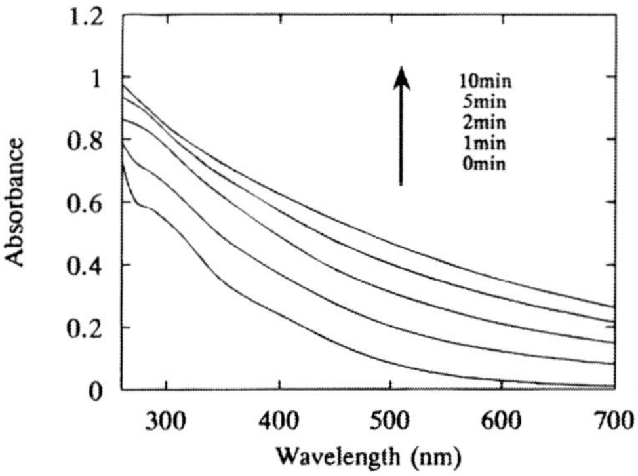

Figure 2.1 Changes in the absorption spectra of 0.5 mM Pd(II) and 8 mM SDS aqueous solution during 200 kHz ultrasonic irradiation under argon atmosphere (cell length 0.5 cm). Reprinted with permission from Okitsu et al. (1996). Copyright 1996 American Chemical Society.

Figure 2.2 shows changes in the absorption spectra of $NaAuCl_4$ (Au(III)) and SDS aqueous solution during ultrasonic irradiation under argon atmosphere (Nagata et al., 1996). Before irradiation, the Au(III) solution has a relatively large peak at around 300 nm. The peak gradually disappears with increasing irradiation time and the surface plasmon peak that is characteristic of Au nanoparticles (Creighton and Eadon, 1991; Haiss et al., 2007) is clearly observed at around 530 nm. This change in the absorption spectra is the typical result to show the formation of Au nanoparticles. From the analysis of a dynamic light scattering photometer and TEM, the size of the formed Au nanoparticles was measured to be 10 nm and 9 nm, respectively.

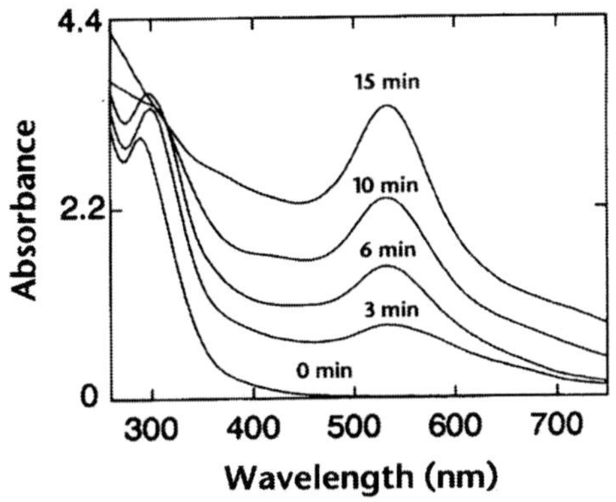

Figure 2.2 Changes in the absorption spectra of 1 mM Au(III) and 8 mM SDS aqueous solution during 200 kHz ultrasonic irradiation under argon atmosphere (cell length 1.0 cm). Reprinted with permission from Nagata et al. (1996). Copyright 1996 Radiation Research Society.

As described in the other chapters, it has been recognized that there are three different regions in the sonochemical reduction processes in an aqueous solution:

(1) The inside of the collapsing cavitation bubbles, where temperature of several thousand Kelvins and pressures of several hundred atmospheres are produced. Here, water vapor is pyrolyzed into •OH and •H radicals (Eq. 2.1). •H radicals have reducing potential for metal ions.

$$H_2O \rightarrow \bullet OH + \bullet H \quad (2.1)$$

(2) The interfacial region between the cavitation bubbles and the bulk solution, where the temperature is lower than that inside the cavitation bubbles, but still high enough for thermal decomposition of organic solutes to occur. Since the concentration of radicals is high, various radical reactions also take place.
(3) The bulk solution at ambient temperature, where the reactions of organic solutes with •OH or •H radicals, which escaped from the interfacial region, take place.

•OH and •H radicals can be recombined to form H_2O_2, H_2, and H_2O in the absence of a radical scavenger as represented in Eqs. 2.2–2.4.

$$2{\bullet}OH \rightarrow H_2O_2 \tag{2.2}$$

$$2{\bullet}H \rightarrow H_2 \tag{2.3}$$

$${\bullet}OH + {\bullet}H \rightarrow H_2O \tag{2.4}$$

As seen in Eqs. (2.1) and (2.3), the reductants of •H and H_2 are formed in the sonolysis of water. If ultrasonic irradiation was performed under air or O_2 atmosphere, it is probable that the oxidation of metal particles occurs. In addition, taking into account the efficiency of the sonochemical reduction, the use of monoatomic gases such as rare gases is better than that of diatomic or triatomic gases, taking into account the following equation (Neppiras, 1980):

$$T_{max} = T_0 P_{max} \left(\frac{\gamma - 1}{P} \right) \tag{2.5}$$

where T_{max} is the maximum temperature achievable in the collapsing bubble, T_0 is the temperature of the sample solution, P_{max} is the maximum pressure achievable in the collapsing bubble, γ is the ratio of specific heats, C_p/C_v, and P is the sum of the vapor pressure of the solvent and the atmospheric gas pressure. As seen from Eq. (2.5), the γ value is one of the most important factors influencing the attained temperature in the collapsing bubble: higher temperature can be attained with monoatomic gases (γ: approximately 1.6–1.7) than with diatomic (γ: approximately 1.4) or triatomic gases (γ: approximately 1.3). In addition, the maximum temperature in the collapsing bubble would decrease with increasing thermal conductivity of the gases. The solubility of the gases would also affect the number of cavitation bubbles.

Taking into account these factors, ultrasonic irradiation is often performed under rare gas atmosphere to reduce metal ions. In the presence of an organic additive (RH), the following chemical reactions (Eqs. 2.6–2.8) proceed under ultrasonic irradiation in an aqueous solution under rare gas atmosphere:

$$RH + \bullet OH\ (\bullet H) \rightarrow \bullet R + H_2O\ (H_2) \tag{2.6}$$

$$RH + H_2O \rightarrow \text{pyrolysis radicals, unstable products} \tag{2.7}$$

$$M^{n+} + \text{reductants} \rightarrow M^0 \tag{2.8}$$

where M^{n+} and M^0 correspond to a metal ion and metal atom, respectively. Equations (2.6) and (2.7) indicate the sonochemical formation of reductants: in Eq. (2.6), •R and H_2 are formed from the abstraction reaction of RH with •OH or •H, where •R acts as a reducing radical. In Eq. (2.7), pyrolysis radicals and unstable products are formed via pyrolysis of RH and water. As seen in the above equations, general organic stabilizers such as surfactants or water soluble polymers can act as a precursor for the formation of reductants under ultrasonic irradiation. Finally, the reduction of M^{n+} with reductants occurs to form M^0 as seen in Eq. (2.8).

In addition, nM^0 aggregates to form a small particle $(M^0)_n$ or the formed M^0 would deposit on $(M^0)_x$ to give $(M^0)_{x+1}$ as follows (Eqs. 2.9 and 2.10):

$$nM^0 \rightarrow (M^0)_n \tag{2.9}$$

$$M^0 + (M^0)_x \rightarrow (M^0)_{x+1} \tag{2.10}$$

To understand the effects of ultrasound on the reduction of metal ions, a colorimetric method is useful, because the concentration of unreduced metal ions can be determined (Nagata et al., 1996; Okitsu et al., 1996). Figure 2.3 shows the absorption spectra obtained from each of the sample solutions of Fig. 2.1, wherein an appropriate amount of a saturated NaI aqueous solution is added to the sample solutions and then filtered by a 0.2 μm pore size membrane filter before UV-vis measurements (Okitsu et al., 1996).

It is found that large spectral changes occur when NaI is added to the sample solution of Pd(II) as shown in Figs. 2.1 and 2.3. This is due to the fact that the formation of a Pd(II)-iodide complex with large absorbance (ε = 9600 M^{-1} cm^{-1} at wavelength of 408 nm) occurs as follows (Eq. 2.11):

$$Pd(II) + 4I^- \rightarrow [PdI_4]^{2-} \quad (2.11)$$

where Pd(II) corresponds to the mixture of $[PdCl_y(H_2O)_{4-y}]^{(y-2)-}$ or $[PdCl_y(OH)_{4-y}]^{2-}$ in an aqueous solution. As the structure of the complex is affected by the concentration of Pd(II) and solution pH, the spectrum of Pd(II) before NaI addition would be a broad and unclear spectrum.

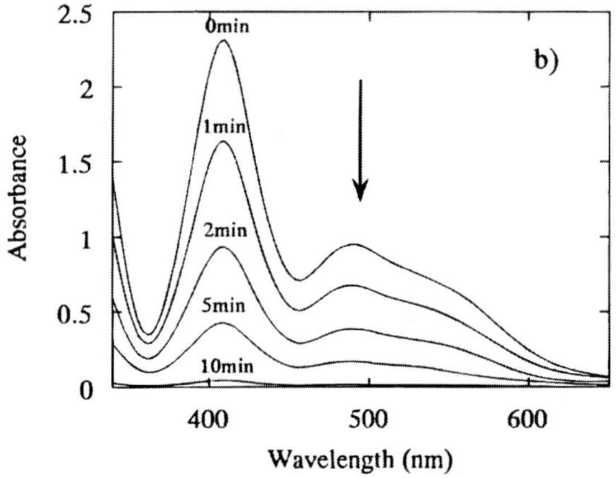

Figure 2.3 Absorption spectra after the addition of NaI to each sample solution of Fig. 2.1. Conditions: $PdCl_2 \bullet 2NaCl$ 0.5 mM, SDS 8 mM, cell length 0.5 cm. Reprinted with permission from Okitsu et al. (1996). Copyright 1996 American Chemical Society.

In addition to the spectral change, the aggregation of colloidal Pd nanoparticles occurs simultaneously after NaI addition so that the aggregated Pd nanoparticles can be removed by filtration with a membrane filter. Therefore, the concentration of Pd(II) in the sample solution can be determined without interference from colloidal Pd nanoparticles as seen in Fig. 2.3. However, the formation behavior of Pd(II) complex and/or the aggregation behavior of Pd nanoparticles may be changed depending on the types of the coexisting stabilizers and of the used Pd(II) complex in the sample solution (Okitsu et al., 1996, 2002). At that time, some kind of optimum conditions must be investigated to determine the concentration of Pd(II). From Fig. 2.3, the absorption peak attributed to $[PdI_4]^{2-}$ can be observed at 408 nm before ultrasonic irradiation and the peak gradually decreases with increasing irradiation time. It is clear that the reduction of Pd(II) proceeds and almost finishes at 10 min irradiation.

By using the colorimetric method, the rates of Pd(II) reduction can be determined. For example, in the absence and presence of stabilizers, the rates of Pd(II) reduction at an initial concentration of 1 mM Pd(II) were measured to be 6.9 μM/min in the absence of stabilizer, 87 μM/min at 1 g/L PVP, 130 μM/min at 8 mM SDS, 230 μM/min at 0.4 mM PEG40MS, and 400 μM/min at 5 g/L Tween20 (Okitsu et al., 1996). It should be noted that the rates are accelerated at least 10-fold by the presence of stabilizers. This is due to that stabilizers act as a precursor for the formation of reductants during ultrasonic irradiation. When the concentration of SDS was changed, the rates of Pd(II) reduction increased with increasing concentration of SDS as follows: 30 μM/min at 0.1 mM SDS < 100 μM/min at 1 mM SDS < 130 μM/min at 8 mM SDS. By the addition of alcohols or carboxylic acids, the rates of Pd(II) reduction can be changed. For example, the rates were 250 μM/min at 10 mM t-BuOH and 300 μM/min at 10 mM *n*-pentanoic acid.

The effects of types of alcohol on the rates of Pd(II) reduction are shown in Fig. 2.4, where ultrasonic irradiation is performed for Pd(II) aqueous solution containing 2.02 g/L Al_2O_3 powders to synthesize Pd nanoparticles supported on Al_2O_3 (Okitsu et al., 2000). From Fig. 2.4, even in a heterogeneous solution containing Al_2O_3 powders, it is found that hot cavitation bubbles are generated by ultrasonic irradiation so that the reduction of Pd(II) proceeds. It is also found that the rate increases in the order of methanol < ethanol < 1-propanol, although the same concentration of alcohol is added in the solution. This result is due to that the reductants are more efficiently formed from the sonolysis of higher hydrophobic alcohols, because higher hydrophobic molecules can more efficiently accumulate at the interface region of hot cavitation bubbles (Henglein and Kormann, 1985; Alegria et al., 1989; Okitsu et al., 2000).

In the case of the sonochemical reduction of Au(III), Pt(II), and Pt(IV), changes in the concentration of these metal ions can be determined by a similar colorimetric method. The absorption coefficient at the peak position of the formed metal complex after NaBr or NaI addition is as follows: $[AuBr_4]^-$ is 4100 M^{-1} cm^{-1} at wavelength of 382 nm (Nagata et al., 1996), $[PtI_4]^{2-}$ is 4600 M^{-1} cm^{-1} at wavelength of 388 nm and 7600 M^{-1} cm^{-1} at wavelength of 332 nm (Mizukoshi et al., 1999), and $[PtI_6]^{2-}$ is 9400 M^{-1} cm^{-1} at wavelength of 495 nm (Mizukoshi et al., 2001).

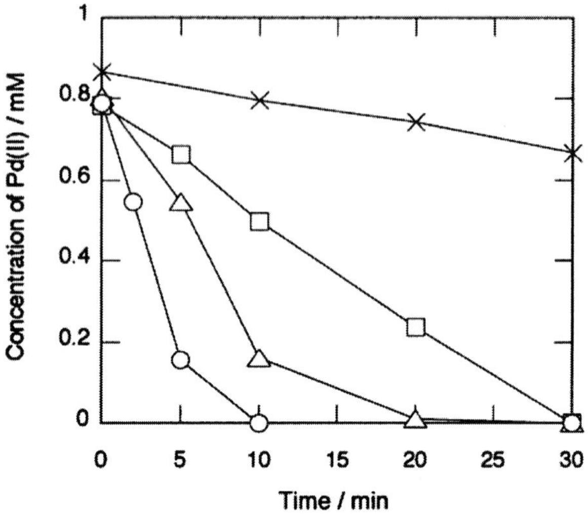

Figure 2.4 Changes in the concentration of Pd(II) during 200 kHz ultrasonic irradiation under argon in the presence and absence of alcohol additives. Conditions: 2.02 g/L Al_2O_3, 20 mM alcohol. (X) None, (□) methanol, (△) ethanol, (○) 1-propanol. Reprinted with permission from Okitsu et al. (2000). Copyright 2000 American Chemical Society.

The rate of reduction of metal ions is one of the most important factors to control the size of the formed metal nanoparticles. Figure 2.5 shows the TEM images of the products formed from the sonochemical reduction of Pd(II) in the presence of Al_2O_3 powders (Okitsu et al., 1999). It can be seen that the formed Pd nanoparticles are highly dispersed on the surface of Al_2O_3.

Table 2.2 shows the average size and standard deviation of the formed Pd nanoparticles on Al_2O_3 and the rate of Pd(II) reduction under different irradiation conditions (Okitsu et al., 1999, 2000). In the presence of 10.5 g/L Al_2O_3 (corresponding to the formation of 1 wt% Pd/Al_2O_3), the rate is 15 μM/min at 20 mM methanol, 52 μM/min at 20 mM ethanol, and 120 μM/min at 20 mM 1-propanol. In the presence of 2.02 g/L Al_2O_3 (corresponding to the formation of 5 wt% Pd/ Al_2O_3), the rate is 25 μM/min at 20 mM methanol, 53 μM/min at 20 mM ethanol, and 130 μM/min at 20 mM 1-propanol. In the same amount of Al_2O_3, smaller Pd particles are formed when the rate of Pd(II) reduction is faster. This is due to that the rate of Pd nucleation would be related to the rate of Pd(II) reduction.

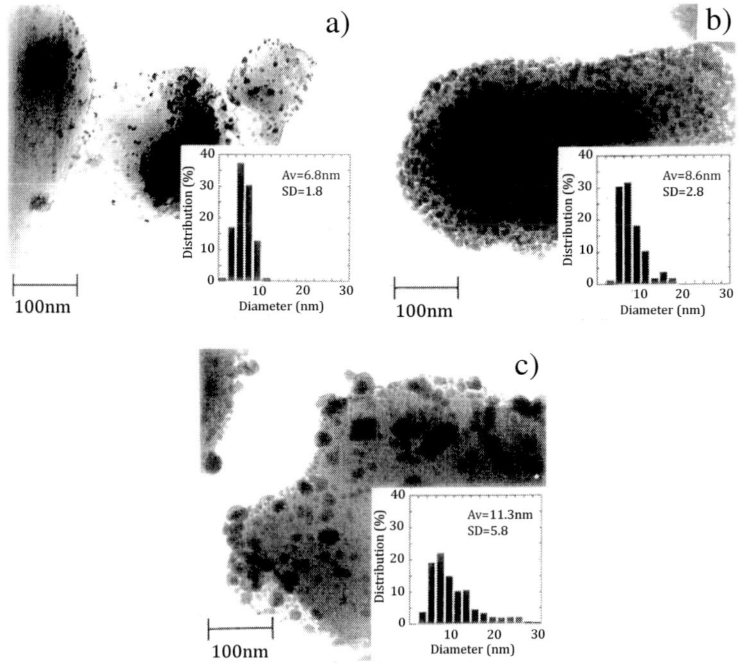

Figure 2.5 TEM images of the sonochemically prepared Pd nanoparticles on Al_2O_3 powders and size distribution of Pd nanoparticles formed by 30 min of irradiation. Conditions: 200 kHz, Ar, 20°C, 1 mM Pd(II), 20 mM alcohol. (a) 1 wt% Pd/Al_2O_3, 1-propanol, (b) 5 wt% Pd/Al_2O_3, 1-propanol, 5 wt% Pd/Al_2O_3, methanol. Reprinted with permission from Okitsu et al. (1999). Copyright 1999 The Chemical Society of Japan.

In addition, it can be seen from Table 2.2 that, at 20 mM 1-propanol, the size of the formed Pd nanoparticles and its standard deviation become smaller with increasing amount of Al_2O_3 [5 wt% Pd (Al_2O_3: 2.02 g/L) < 3 wt% Pd (Al_2O_3: 3.44 g/L) < 1 wt% Pd (Al_2O_3: 10.5 g/L)]. However, the rate of reduction was almost the same among them. Taking into account these results, the growth and agglomeration of Pd nanoparticles in the solution would be suppressed by its adsorption onto Al_2O_3 so that the size of the formed Pd nanoparticles was smaller as the amount of Al_2O_3 increased. It should be noted that the formed Pd nanoparticles are immobilized only on the surface of Al_2O_3 powders, suggesting that the effective application of Pd nanoparticles as a catalyst or sensor is possible.

Table 2.2 Average size and standard deviation of Pd nanoparticles formed on the Al_2O_3 surface and rate of Pd(II) reduction by ultrasonic irradiation[a]

Conditions[b] Pd/Al_2O_3, Alcohol	Average size (nm)	Standard deviation[c]	Rate[d] (μM/min)
1wt%, 1-propanol	6.8	1.8	120
1wt%, ethanol	7.4	2.5	52
1wt%, methanol	7.9	3.4	15
3wt%, 1-propanol	7.1	2.2	130
5wt%, 1-propanol	8.6	2.8	130
5wt%, ethanol	10.0	2.9	53
5wt%, methanol	11.3	5.8	25

[a]Irradiation time: 30 min of 200 kHz ultrasound, Pd(II): 1 mM, alcohol: 20 mM
[b]Weight ratio of Pd/Al_2O_3
[c]Standard deviation of size of Pd particles
[d]Average rate for initial 5 min irradiation

As described above, the type of added alcohol affects the rate of reduction of metal ions and size of the formed metal nanoparticles. In addition to this, we can control these with various irradiation parameters such as ultrasound intensity, distance between reaction vessel and the oscillator, solution temperature, and so on. For example, in the case of the sonochemical reduction of Au(III), the rates of Au(III) reduction and sizes of the formed Au particles can be controlled by such irradiation parameters, although identical starting materials and the same initial concentrations are used in each preparation (Okitsu et al., 2002). Figure 2.6 shows the relationship between the rate of Au(III) reduction and the size of the formed Au nanoparticles at 20 min of irradiation. As the irradiation time was fixed at 20 min, the reduction ratio of Au(III) was changed in the range of 55–100% (Okitsu et al., 2002). Although the reduction ratio was not the same among the samples, the size of the formed Au nanoparticles decreased with increasing rate of Au(III) reduction.

It is well known that stronger shock waves are generated by lower frequency ultrasonic irradiation. If shock waves affected the growth of Au nanoparticles, Au nanoparticles would become larger with increasing intensity of shock waves. However, Au nanoparticles became smaller with increasing intensity of ultrasound (which would be related to the intensity of shock waves) as seen in Fig. 2.6a–d. Therefore, the intensity of shock waves would be not so strong to affect the size of Au nanoparticles in the case of sonication at 200 kHz. Consequently, the size of the formed Au particles is mainly dependent on the number of the formed Au nuclei, which is related to the rate of Au(III) reduction.

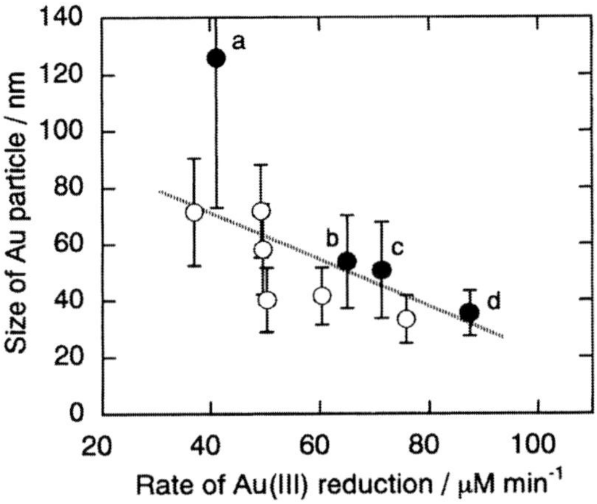

Figure 2.6 Relation between the rate of Au(III) reduction and the average size of the formed Au particles at various irradiation conditions. Each error bar corresponds to the standard deviation of the size of the formed Au particles. Closed circles correspond to the dependence of the ultrasonic intensities: (a) 0.53 W/cm^2, (b) 0.92 W/cm^2, (c) 1.20 W/cm^2, (d) 1.43 W/cm^2. Open circles are the results obtained by changing the distance between reaction vessel and oscillator, and changing the solution temperature. 200 kHz ultrasound, argon atmosphere, 20 mM 1-propanol. Reprinted with permission from Okitsu et al. (2002). Copyright 2002 The Chemical Society of Japan.

The rate of Au(III) reduction and size of the formed Au nanoparticles are also affected by ultrasound frequency. Figure 2.7 shows the relationship between the average size of the formed Au nanoparticles and the rate of Au(III) reduction (Okitsu et al., 2005). In the case of different frequency, different cavitation phenomena are created. For example, the changes in the temperature and pressure in the collapsing cavitation bubbles, the number and size distribution of bubbles, the size and lifetime of bubbles, the dynamics and symmetry (shape) of the bubble collapse, the effect of organic additive on bubble temperature, and the formation of primary and secondary radical, intermediate and product are induced by changing the ultrasound frequency. In addition, the sonication systems such as a horn-type sonication with a low frequency and a standing wave sonication system with a high frequency also affect the cavitation phenomena. Although the result in Fig. 2.7 was obtained by using different frequencies and different sonication system, it was confirmed that the rate of Au(III) reduction affected the size of the formed Au nanoparticles. Therefore, the rate of reduction related to the nucleation process is the important factor to determine the size of Au nanoparticles.

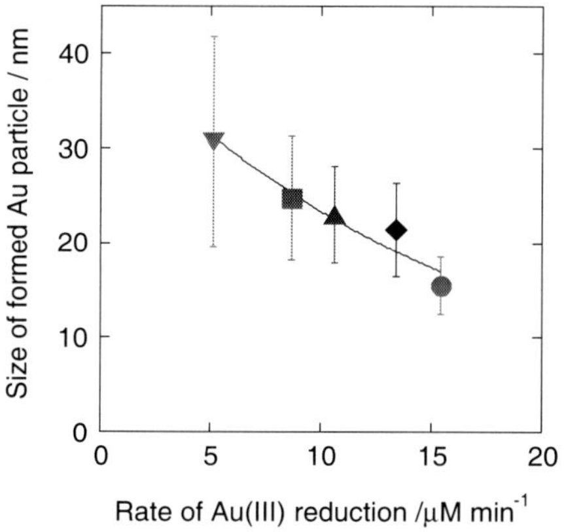

Figure 2.7 Average size of the formed Au nanoparticles as a function of the rate of Au(III) reduction, where ultrasound frequency is changed. (▼); 20 kHz, (●); 213 kHz, (◆); 358 kHz, (▲); 647 kHz, (■); 1062 kHz. Each error bar corresponds to the standard deviation of the size of the formed Au nanoparticles. Au(III): 0.2 mM, 1-propanol: 20 mM, atmosphere: Ar, ultrasonic power: 0.1 W/mL, irradiation time: 120 min. Reprinted with permission from Okitsu et al. (2005). Copyright 2005 American Chemical Society.

In general, the sizes of the sonochemically formed metal nanoparticles are dependent on the initial concentration of metal ions and the types of stabilizers (Okitsu et al., 1996) same as the conventional methods: smaller metal particles are formed in the presence of lower initial concentration of metal ions. If suitable stabilizers were selected, the size of metal nanoparticles could be changed (Mizukoshi et al., 2001; Nagata et al., 1992; Yeung et al., 1993; Okitsu et al., 1996). By using the sonochemical reduction method, various types of metal nanoparticles such as Ag, Au, Pd, Pt, and Ru (Nagata et al., 1992; Yeung et al., 1993; Grieser et al., 1996; Okitsu et al., 1996; Mizukoshi et al., 2001; He et al., 2006) as well as supported metal nanoparticles on metal oxides (Mizukoshi et al., 2005; Mizukoshi et al., 2008; Sivakumar et al., 2010) have been synthesized.

2.3 Structure Control of Bimetallic Nanoparticles

Structure-controlled bimetallic nanoparticles such as a core–shell structure and a random alloy structure have diverse catalytic activities. Bimetallic nanoparticles of Au/Pd are attractive catalysts for H_2O_2 production (Edwards et al., 2009), formic acid oxidation (Hong et al., 2011), alkene hydrogenation (Mizukoshi et al., 1997; Okitsu et al., 2000; Nakagawa et al., 2005), ethanol oxidation (Ksar et al., 2009), and trichloroethene hydrodechlorination (Nutt et al., 2005, 2006; Fang et al., 2011). It has been reported that bimetallic nanoparticles of Au/Pd can be prepared by the sonochemical reduction of Au(III) and Pd(II) in an aqueous solution (Mizukoshi et al., 1997; Okitsu et al., 2000; Nakagawa et al., 2005).

Figure 2.8a shows the changes in the absorption spectra of the sample solution containing 0.5 mM Au(III), 0.5 mM Pd(II), and 8 mM SDS during ultrasonic irradiation under Ar atmosphere (Mizukoshi et al., 1997). Before irradiation, no peak is observed in the region from 500 nm to 700 nm, but an absorption peak attributed to Au nanoparticles gradually emerges at around 520 nm with increasing irradiation time. After that, the peak becomes broader with further irradiation.

Figure 2.8b shows the changes in the concentrations of Au(III) and Pd(II) during ultrasonic irradiation, which are measured by colorimetric method (Mizukoshi et al., 1997). From Fig. 2.8b, it is found that the reduction of Au(III) and Pd(II) consecutively occurs: the reduction of Au(III) starts and, after completion of the reduction of Au(III), the reduction of Pd(II) starts. This is due to the different redox potential of Au(III) and Pd(II): if the formation of Pd(0) occurred in the presence of Au(III), the following reaction would occur (Eq. 2.12).

$$3Pd(0) + 2Au(III) \rightarrow 3Pd(II) + 2Au(0) \quad (2.12)$$

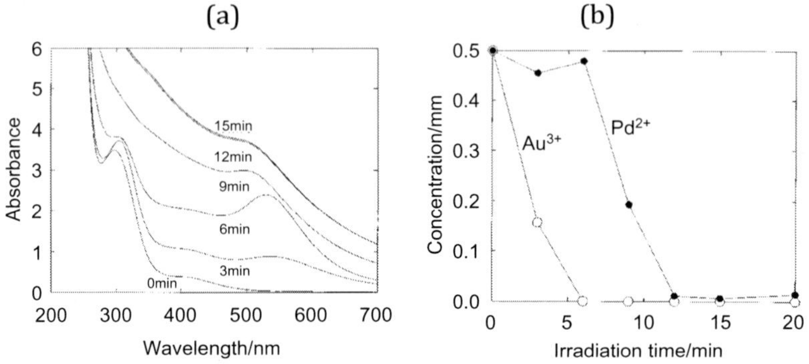

Figure 2.8 (a) Changes in the absorption spectra of sample solution containing 0.5 mM Au(III), 0.5 mM Pd(II), and 8 mM SDS during 200 kHz ultrasonic irradiation under Ar atmosphere. (b) Changes in the concentrations of Au(III) and Pd(II) as a function of the irradiation time. Reprinted with permission from Mizukoshi et al. (1997). Copyright 1997 American Chemical Society.

From TEM analysis (Fig. 2.9), the formation of Au core/Pd shell nanoparticles with the average diameter of approximately 8 nm was confirmed (Mizukoshi et al., 2000). Therefore, the reason why the surface plasmon peak at 6 min irradiation became broader after further irradiation as seen in Fig. 2.8a would be due to the fact that the sonochemically reduced Pd atoms are deposited on the already formed Au nanoparticles on continued irradiation. The sonochemically synthesized Au core/Pd shell bimetallic nanoparticles have a high catalytic activity for hydrogenation of olefin as described later.

Okitsu et al. (2000) reported that Au core/Pd shell nanoparticles immobilized on and inside porous silica can be prepared by successive processes composed of the sonochemical method and sol–gel method (Okitsu et al., 2000). That is, at first, Au core/Pd shell nanoparticles are synthesized by the sonochemical reduction of 0.5 mM Au(III) and 0.5 mM Pd(II) in the presence of 0.4 mM PEG40MS at 20°C under argon atmosphere and then the sol–gel processes with $Si(OC_2H_5)_4$ and NH_3 are performed to synthesize porous silica to support Au core/Pd shell nanoparticles. Owing to the silica matrix prepared by the sol–gel processes, the interparticle aggregation of Au core/Pd shell nanoparticles could be suppressed.

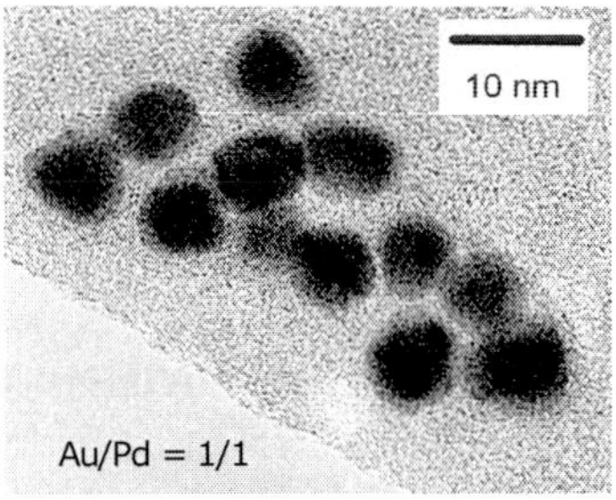

Figure 2.9 A representative TEM image of the sonochemically prepared Au core/Pd shell nanoparticles. The synthesis condition is the same as Fig. 2.8. Reprinted with permission from Mizukoshi et al. (2000). Copyright 2000 American Chemical Society.

The thermal stability of Au core/Pd shell nanostructure has been investigated by XRD. Figure 2.10 shows the effects of treatment temperature under H_2 on the XRD patterns of silica-supported Au/Pd nanoparticles.

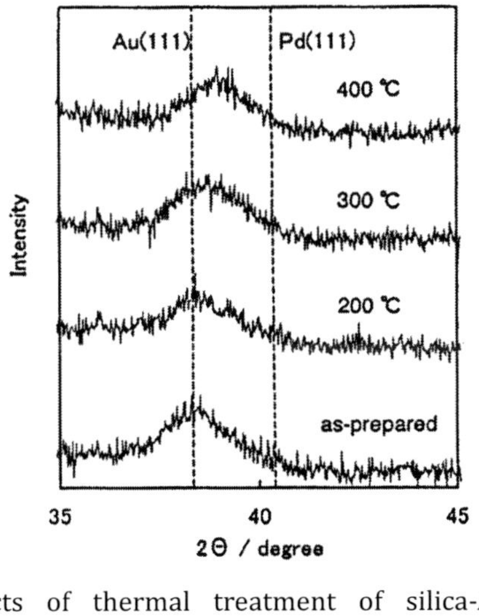

Figure 2.10 Effects of thermal treatment of silica-Au/Pd bimetallic nanoparticles on the XRD pattern (Cu Kα). Reprinted with permission from Okitsu et al. (2000). Copyright 2000 The Chemical Society of Japan.

It can be seen that the XRD pattern of the as-prepared sample has a broad peak at 38.2° assigned to Au(111) line. No peak assigned to Pd(111) was observed. This was due to the fact that Pd shell would be too thin for the diffraction. It was also found that the diffraction peak largely shifted to a higher angle with increasing temperature from 300°C to 400°C and was located at the middle of Au(111) and Pd(111) line at 400°C. This peak position was consistent with that of bulk alloy state (Maeland and Flanagan, 1965). The color of the sample also varied from reddish gray to dark gray at this temperature. These results indicated that the core–shell structure was destroyed at 300°C–400°C to form the random alloy structure. As the bimetallic nanoparticles were immobilized on and inside porous silica, the size of the bimetallic nanoparticles was hardly changed by the thermal treatment, as confirmed by the TEM observations.

The effect of treated temperature on the catalytic activity for the hydrogenation of cyclohexene has been investigated to discuss the role of core–shell structure in the catalytic activity. As Au nanoparticles (average size: 19.8 nm) exhibited no catalytic activity for the hydrogenation of cyclohexene at all, the catalytic activities for Au/Pd (atomic ratio 1:1, average size: 6.4 nm) and Pd (average size; 5.5 nm) nanoparticles were investigated as shown in Fig. 2.11 (Okitsu et al., 2000). Although the catalytic activity of Au/Pd nanoparticles was almost kept constant up to 300°C, the activity suddenly started to decrease with increasing temperature from 300°C to 400°C. The catalytic activity of Pd nanoparticles also started to decrease with increasing temperature in analogy with that of Au/Pd nanoparticles. However, the decrease in the ratio of the catalytic activity of Pd nanoparticles was smaller than that of Au/Pd nanoparticles. This decrease in the catalytic activity would be attributed not only to the decrease in the surface area of Au/Pd nanoparticles but also to the changes in other factors, because the size of the nanoparticles hardly changed by the thermal treatment.

Taking into account the results of Fig. 2.10, the extremely low activity of Au/Pd nanoparticles at 400°C should be due to the change in the Au/Pd nanostructure from the core–shell to the random alloy structure: the core–shell structure of Au/Pd would play an important role in the hydrogenation of cyclohexene.

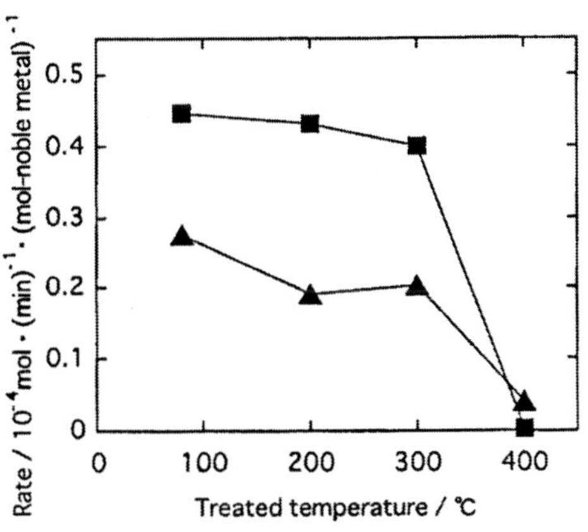

Figure 2.11 Effects of treated temperature under H_2 atmosphere on the catalytic activity for the hydrogenation of cyclohexene (■): silica-Au/Pd (Au/Pd=1/1), (▲): silica-Pd. Reprinted with permission from Okitsu et al. (2000). Copyright 2000 The Chemical Society of Japan.

On the basis of previous report (Toshima et al., 1992), a ligand effect derived from an electrostatic effect between Au core and Pd shell could affect the catalytic activity: a positively charged Pd shell deposited on Au core in the bimetallic particles is more active than noncharged monometallic Pd particles, as olefin having a double bond favors the electron-deficient surface. However, as the average size of the sonochemically synthesized Au/Pd nanoparticles was considerably larger than those of previous ones, there may be other reasons for the high catalytic activity of the sonochemically synthesized Au core/Pd shell nanoparticles. For example, as the thickness of Pd shell corresponds to several atomic layers, the ligand effect may be affected by any quantum size effects (Xu et al., 1997; Valden et al., 1998) with respect to the optimum size effect of Pd shell or Au core.

In addition to the above analysis, extended X-ray absorption fine structure (EXAFS) measurements are performed for the similar sample, wherein Au/Pd nanoparticles are prepared from the sonication of a 65 mL aqueous solution containing 5 mM Au(III), 5 mM Pd(II), and 4 mM PEG40MS (Nakagawa et al., 2005). As the concentration of Au(III) and Pd(II) is 10 times higher than the sample described above, some differences should be observed between these samples. However, EXAFS results also suggest the formation of core/shell nanoparticles of Au and Pd.

2.4 Shape Control of Metal Nanoparticles

Shape control of metal nanoparticles is an important research area, because the shape and crystal face of the nanoparticles affect the physicochemical property. In the case of Au nanorods, they have the longitudinal plasmon band at around 600–1000 nm, and therefore, they can be applied to sensors, surface-enhanced Raman spectroscopy, optical data storages, and photothermal therapy of cancer cells (Murphy et al., 2005; Huang et al., 2006; Zijlstra et al., 2009). For these applications, the size and aspect ratio of the nanorods must be controlled precisely. It is well known that CTAB acts as an excellent capping molecule to synthesize Au nanorods. A seed-mediated growth method is often used to synthesize Au nanorods, in which the following two steps are needed: a synthesis step of Au seeds with an appropriate particle size, and a growth step, wherein the synthesized seeds are used for the growth to nanorods. As this method consists of more than two steps and many interdependent parameters in the preparation of nanorods, it is desirable to develop a simple one-pot synthesis method. Here, the sonochemical method to synthesize Au nanorods is described.

Okitsu et al. (2009) reported that the reduction of Au(I) to Au(0) proceeds and the formation of Au nanorods occurs by ultrasonic irradiation of the sample solutions in the presence of Au(I), CTAB, $AgNO_3$, and ascorbic acid. Figure 2.12 shows the TEM images of the formed Au nanoparticles at pH 3.5, 5.0, 6.5, 7.7, and 9.8 from the sonochemical reduction of Au(I). It is found that the size and shape of Au nanoparticles change when the pH value of solutions changes.

Figure 2.12 (a–c) show that most of the formed nanoparticles have a rod shape in the size range of 10–50 nm. The average aspect ratio was in the order of 2.1 at pH 6.5 < 2.2 at pH 5.0 < 3.0 at pH 3.5, where particles with an aspect ratio of less than 1.5 were considered not to be nanorods and were therefore excluded from this calculation of aspect ratio. At pH 7.7 (Fig. 2.12d), most of the particles had an irregular shape. At pH 9.8 (Fig. 2.12e), most of the particles had a spherical shape with a smaller particle size than those formed in the lower pH solutions.

Figure 2.13 shows the extinction spectra of the sonochemically formed Au nanoparticles after 180 min of irradiation at pH 3.5, 5.0, 6.5, 7.7, and 9.8. It is noted that the extinction spectra of the sample solutions are quite different among different pH solutions. At pH 3.5, two extinction peaks at around 510 nm and 880 nm were clearly observed. These peaks corresponded with the surface plasmon bands of Au nanorods (Jana et al., 2001; Nikoobakht and El-Sayed, 2003; Perez-Juste et al., 2005). On the contrary, only one extinction peak was observed at pH 7.7 and 9.8. This was due to the absence in the formation of Au nanorods as seen in Fig. 2.12d,e. At pH less than 6.5, the peak wavelength of the longitudinal plasmon band increased with decreasing pH of the solution. This was due to the fact that the aspect ratio of Au nanorods increased with decreasing pH of the solution as seen in Fig. 2.12a–c.

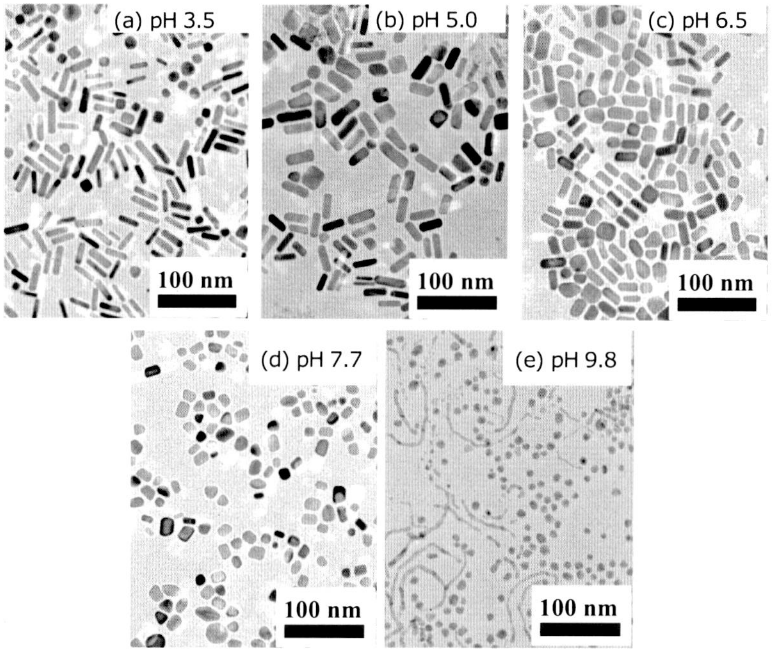

Figure 2.12 TEM images of Au nanorods and nanoparticles formed in different pH solutions of (a) pH 3.5, (b) pH 5.0, (c) pH 6.5, (d) pH 7.7, and (e) pH 9.8 after 180 min of irradiation under argon. Reprinted with permission from Okitsu et al. (2009). Copyright 2009 American Chemical Society.

In the sonochemical reduction process of Au(I) ions, some complex changes are induced by changing the pH of the solution: (i) the change in the redox potential of Au(I) ions, (ii) the change in the yield and reactivity of the sonochemically formed reductants, (iii) the change in the catalytic activity and surface property of the formed Au seeds, and (iv) the change in the adsorption property and reactivity of Au(I) ions on Au seeds. In addition to this, the change in pH affects the reducing ability of ascorbic acid (Wei et al., 2008). The mechanism of nanorod formation is not clear, but a schematic mechanism for one-pot sonochemical formation of Au nanorods and nanoparticles at pH 3.5 is proposed as follows: the number of Au seeds initiating the formation of nanorods is determined at the initial stage of Au seed formation. Once the formation of Au seeds occurs, the reduction of Au(I) ions with ascorbic acid should proceed catalytically on Au seeds as well as the sonochemical reduction of Au(I) ions. As CTAB and $AgNO_3$ acted as effective capping agents, the shape-controlled growth of Au seeds to Au nanorods succeeded even in the presence of ultrasound.

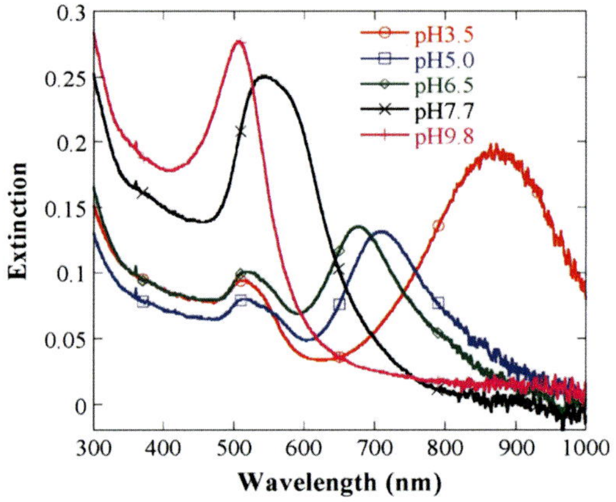

Figure 2.13 Extinction spectra of the sonochemically synthesized Au nanorods and nanoparticles at different pH solutions. Conditions: irradiation time; 180 min, atmosphere; argon, cell length: 0.5 cm. Reprinted with permission from Okitsu et al. (2009). Copyright 2009 American Chemical Society.

2.5 Concluding Remarks

Upon ultrasonic irradiation to an aqueous solution containing organic additives, various types of products such as •H, H_2, •R, pyrolysis radicals, and unstable products (intermediates) are generated in situ. As these products have different reactivity, they could be used as a reductant for the controlled reduction of metal ions and the formation of metal nanoparticles with controlled size, structure, and shape. The combination of an appropriate stabilizer with the sonochemical reduction method should provide us with more precisely controlled nanoparticles that have novel and attractive properties.

References

Alegria, A. E., Lion, Y., Kondo, T., Riesz, P. (1989). Sonolysis of aqueous surfactant solutions. Probing the interfacial region of cavitation bubbles by spin trapping, *J. Phys. Chem.* **93**, pp. 4908–4913.

Bianchini, C., Shen, P. K. (2009). Palladium-based electrocatalysts for alcohol oxidation in half cells and in direct alcohol fuel cells, *Chem. Rev.* **109**, pp. 4183–4206.

Creighton, J. A., Eadon, D. G. (1991). Ultraviolet–visible absorption spectra of the colloidal metallic elements, *J. Chem. Soc., Faraday Transactions* **87**, pp. 3881–3891.

Edwards, J. K., Solsona, B., Ntainjua, N. E., Carley, A. F., Herzing, A. A., Kiely, C. J., Hutchings, G. J. (2009). Switching off hydrogen peroxide hydrogenation in the direct synthesis process, *Science* **323**, pp. 1037–1041.

Fang, Y.-L., Heck, K. N., Alvarez, P. J. J., Wong, M. S. (2011). Kinetic analysis of palladium/gold nanoparticles as colloidal hydrodechlorination catalysts, *ACS Catalysis* **1**, pp. 128–138.

Grieser, F., Hobson, R., Sostaric, J., Mulvaney, P. (1996). Sonochemical reduction processes in aqueous colloidal systems, *Ultrasonics*. **34**, pp. 547–550.

Haiss, W., Thanh, N. T. K., Aveyard, J., Fernig, D. G. (2007). Determination of size and concentration of gold nanoparticles from UV-Vis spectra, *Anal. Chem.* **79**, pp. 4215–4221.

He, Y., Vinodgopal, K., Ashokkumar, M., Grieser, F. (2006). Sonochemical synthesis of ruthenium nanoparticles, *Res. Chem. Intermed.* **32**, pp. 709–715.

Henglein, A., Kormann, C. (1985). Scavenging of OH radicals produced in the sonolysis of water, *Int. J. Radiation Biol.*, **48**, pp. 251–258.

Hong, J. W., Kim, D., Lee, Y. W., Kim, M., Kang, S. W., Han, S. W. (2011). Atomic-distribution-dependent electrocatalytic activity of Au-Pd bimetallic nanocrystals, *Angew. Chem. Int. Ed.* **50**, pp. 8876–8880.

Huang, X., El-Sayed, I. H., Qian, W., El-Sayed, M. A. (2006). Cancer cell imaging and photothermal therapy in the near-infrared region by using gold nanorods, *J. Am. Chem. Soc.* **128**, pp. 2115–2120.

Jana, N. R., Gearheart, L., Murphy, C. J. (2001). Wet chemical synthesis of high aspect ratio cylindrical gold nanorods, *J. Phys. Chem. B* **105**, pp. 4065–4067.

Jia, C.-J., Schüth, F. (2011). Colloidal metal nanoparticles as a component of designed catalyst, *Phys. Chem. Chem. Phys.* **13**, pp. 2457–2487.

Ksar, F., Ramos, L., Keita, B., Nadjo, L., Beaunier, P., Remita, H. (2009). Bimetallic palladium-gold nanostructures: application in ethanol oxidation, *Chem. Mater.* **21**, pp. 3677–3683.

Maeland, A., Flanagan, T. B. (1965). X-ray and thermodynamic studies of the absorption of hydrogen by gold-palladium alloys, *J. Phys. Chem.* **69**, pp. 3575–3581.

Mizukoshi, Y., Fujimoto, T., Nagata, Y., Oshima, R., Maeda, Y. (2000). Characterization and catalytic activity of core-shell structured gold/palladium bimetallic nanoparticles synthesized by the sonochemical method, *J. Phys. Chem. B* **104**, pp. 6028–6032.

Mizukoshi, Y., Okitsu, K., Maeda, Y., Yamamoto, T. A., Oshima, R., Nagata, Y. (1997). Sonochemical preparation of bimetallic nanoparticles of gold/palladium in aqueous solution, *J. Phys. Chem. B* **101**, pp. 7033–7037.

Mizukoshi, Y., Oshima, R., Maeda, Y., Nagata, Y. (1999). Preparation of platinum nanoparticles by sonochemical reduction of the Pt(II) ion, *Langmuir* 1999, **15**, pp. 2733–2737.

Mizukoshi, Y., Sato, K., Konno, T.J., Masahashi, N., Tanabe, S. (2008). Magnetically retrievable palladium/maghemite nanocomposite catalysts prepared by sonochemical reduction method, *Chem. Lett.* **37**, pp. 922–923.

Mizukoshi, Y., Seino, S., Okitsu, K., Kinoshita, T., Otome, Y., Nakagawa, T., Yamamoto, T.A. (2005). Sonochemical preparation of composite nanoparticles of Au/γ-Fe_2O_3 and magnetic separation of glutathione, *Ultrason. Sonochem.* **12**, pp. 191–195.

Mizukoshi, Y., Takagi, E., Okuno, H., Oshima, R., Maeda, Y., Nagata, Y. (2001). Preparation of platinum nanoparticles by sonochemical reduction of the Pt(IV) ions: Role of surfactants, *Ultrason. Sonochem.* **8**, pp. 1–6.

Murphy, C. J., Sau, T. K., Gole, A. M., Orendorff, C. J., Gao, J., Gou, L., Hunyadi, S. E., Li, T. (2005). Anisotropic metal nanoparticles: Synthesis, assembly, and optical applications, *J. Phys. Chem. B* **109**, pp. 13857–13870.

Nagata, Y., Mizukoshi, Y., Okitsu, K., Maeda, Y. (1996). Sonochemical formation of gold particles in aqueous solution, *Radiat Res.*, **146**, pp. 333–338.

Nagata, Y., Watanabe, Y., Fujita, S., Dohmaru, T., Taniguchi, S. (1992). Formation of colloidal silver in water by ultrasonic irradiation, *J. Chem. Soc. Chem. Commun.* pp. 1620–1622.

Nakagawa, T., Nitani, H., Tanabe, S., Okitsu, K., Seino, S., Mizukoshi, Y., Yamamoto, T. A. (2005). Structural analysis of sonochemically prepared Au/Pd nanoparticles dispersed in porous silica matrix, *Ultrason. Sonochem.* **12**, pp. 249–254.

Neppiras, E. A. (1980). Acoustic cavitation, *Phys. Rep.*, 61, pp. 159–251.

Nikoobakht, B., El-Sayed, M. A. (2003). Preparation and growth mechanism of gold nanorods (NRs) using seed-mediated growth method, *Chem. Mater.* **15**, pp. 1957–1962.

Nutt, M. O., Heck, K. N., Alvarez, P., Wong, M. S. (2006). Improved Pd-on-Au bimetallic nanoparticle catalysts for aqueous-phase trichloroethene hydrodechlorination, *Appl. Catal. B: Environ.* **69**, pp. 115–125.

Nutt, M. O., Hughes, J. B., Wong, M. S. (2005). Designing Pd-on-Au bimetallic nanoparticle catalysts for trichloroethene hydrodechlorination, *Environ. Sci. Technol.* **39**, pp. 1346–1353.

Okitsu, K., Ashokkumar, M., Grieser, F. (2005). Sonochemical synthesis of gold nanoparticles: Effects of ultrasound frequency, *J. Phys. Chem. B* **109**, pp. 20673–20675.

Okitsu, K., Bandow, H., Maeda, Y., Nagata, Y. (1996). Sonochemical preparation of ultrafine palladium particles, *Chem. Mater.* **8**, pp. 315–317.

Okitsu, K., Mizukoshi, Y., Bandow, H., Maeda, Y., Yamamoto, T., Nagata, Y. (1996). Formation of noble metal particles by ultrasonic irradiation, *Ultrason. Sonochem.* **3**, pp. S249–S251.

Okitsu, K., Murakami, M., Tanabe, S., Matsumoto, H. (2000). Catalytic behavior of Au core/Pd shell bimetallic nanoparticles on silica prepared by sonochemical and sol–gel processes, *Chem. Lett.* **29**, pp. 1336–1337.

Okitsu, K., Nagaoka, S., Tanabe, S., Matsumoto, H., Mizukoshi, Y., Nagata, Y. (1999). Sonochemical preparation of size-controlled palladium nanoparticles on alumina surface, *Chem. Lett.* **28**, pp. 271–272.

Okitsu, K., Sharyo, K., Nishimura, R. (2009). One-pot synthesis of gold nanorods by ultrasonic irradiation: The effect of pH on the shape of the gold nanorods and nanoparticles, *Langmuir* **25**, pp. 7786–7790.

Okitsu, K., Yue, A., Tanabe, S., Matsumoto, H. (2000). Sonochemical preparation and catalytic behavior of highly dispersed palladium nanoparticles on alumina, *Chem. Mater.* **12**, pp. 3006–3011.

Okitsu, K., Yue, A., Tanabe, S., Matsumoto, H. (2002). Formation of palladium nanoclusters on Y-zeolite via a sonochemical process and conventional methods, *Bull. Chem. Soc. Jpn.* **75**, pp. 449–455.

Okitsu, K., Yue, A., Tanabe, S., Matsumoto, H., Yobiko, Y., Yoo, Y. (2002). Sonolytic control of rate of gold(III) reduction and size of formed gold nanoparticles: Relation between reduction rates and sizes of formed nanoparticles, *Bull. Chem. Soc. Jpn.* **75**, pp. 2289–2296.

Pérez-Juste, J., Pastoriza-Santos, I., Liz-Marzán, L. M., Mulvaney, P. (2005). Gold nanorods: Synthesis, characterization and applications, *Coordin. Chem. Rev.* **249**, pp. 1870–1901.

Sivakumar, M., Towata, A., Yasui, K., Tuziuti, T., Kozuka, T., Tsujimoto, M., Zhong, Z., Iida, Y. (2010). Fabrication of nanosized Pt on rutile TiO_2 using a standing wave sonochemical reactor (SWSR) – Observation of an enhanced catalytic oxidation of CO, *Ultrason. Sonochem.* **17**, pp. 213–218.

Toshima, N., Harada, M., Yamazaki, Y., Asakura, K. (1992). Catalytic activity and structural analysis of polymer-protected Au-Pd bimetallic clusters prepared by the simultaneous reduction of $HAuCl_4$ and $PdCl_2$, *J. Phys. Chem.* **96**, pp. 9927–9933.

Valden, M., Lai, X., Goodman, D. W. (1998). Onset of catalytic activity of gold clusters on titania with the appearance of nonmetallic properties, *Science*, **281**, pp. 1647–1650.

Wei, Q., Ji, J., Shen, J. (2008). pH controlled synthesis of high aspect-ratio gold nanorods, J. *Nanosci. Nanotechnol.* **8**, pp. 5708–5714.

Xu, C., Lai, X. Zajac, G. W., Goodman, D. W. (1997). Scanning tunneling microscopy studies of the TiO_2(110) surface: Structure and the nucleation growth of Pd, *Phys. Rev. B*, **56**, pp. 13464–13482.

Yeung, S.A, Hobson, R., Biggs, S., Grieser, F. (1993). Formation of gold sols using ultrasound, *J. Chem. Soc. Chem. Commun.* pp. 378–379.

Zijlstra, P., Chon, J. W. M., Gu, M. (2009). Five dimensional optical recording mediated by surface plasmons in gold nanorods, *Nature* **459**, pp. 410–413.

Chapter 3

Sonochemical Synthesis of Noble Monometallic and Bimetallic Nanoparticles for Catalytic Applications

Sambandam Anandan[a] and Muthupandian Ashokkumar[b,c]
[a]*Nanomaterials & Solar Energy Conversion Lab, Department of Chemistry, National Institute of Technology, Trichy, India*
[b]*School of Chemistry, University of Melbourne, 1-100 Grattan Street, Parkville VIC 3010, Australia*
[c]*Chemistry Department, King Abdulaziz University, Abdullah Sulayman, Jeddah 22254, Saudi Arabia*
masho@unimelb.edu.au

The synthesis of metal nanoparticles with unique properties has been intensively investigated through the process of acoustic cavitation. Acoustic cavitation induced "sonochemistry" provides a unique interaction between energy and matter. Extraordinary conditions, namely, hot spots of 5000 K, pressures of nearly 1000 bar, and heating and cooling rates of more than 10^{10} K/s, generated during acoustic cavitation provide access to a range of chemical reaction conditions. In this review, we summarize the sonochemical synthesis of monometallic and bimetallic nanostructured materials and their applications.

Cavitation: A Novel Energy-Efficient Technique for the Generation of Nanomaterials
Edited by Sivakumar Manickam and Muthupandian Ashokkumar

ISBN 978-981-4411-54-7 (Hardcover), 978-981-4411-55-4 (eBook)
www.panstanford.com

3.1 Introduction

Nanomaterials are currently the subject of intense research. The unusual properties of the nanomaterials result from their ultrafine structure (i.e., a grain size of <50 nm), which can be classified into two categories: properties that are relative to the bulk and properties that are relative to the surface. Among all nanostructured materials, novel metal nanoparticles have attracted particular interest due to their stability, biocompatibility, surface plasmon resonance effect, and unique catalytic activities (Elghanian et al., 1997; Storhoff et al., 1998; West and Halas, 2000; Haruta and Date, 2001; Park et al., 2002; Schmid and Corain, 2003; Alivisatos, 2004; Daniel and Astruc, 2004; Daté et al., 2004a, 2004b; Liu and Lu, 2004; Molina et al., 2004; Storhoff et al., 2004; Zhu et al., 2004; Astruc et al., 2005a, 2005b; Dadosh et al., 2005; Hasobe et al., 2005; Sönnichsen et al., 2005; Tseng et al., 2005; Liu et al., 2006; Pradhan et al., 2006). A number of approaches have been developed for the synthesis of plasmonic metallic nanoparticles with controlled morphology. These include wet chemical synthesis, sonochemical synthesis, photochemical synthesis, microwave synthesis, sputter deposition, electroless plating, and electrochemical synthesis (Papavassiliou, 1976; Sato et al., 1991; Mulvaney et al., 1993; Toshima et al., 1993; Liz-Marzan and Philipse, 1995; Freeman et al., 1996; Remita et al., 1996; Silvert et al., 1996; Mizukoshi et al., 1997, 2001; Treguer et al., 1998; Link et al., 1999; Hodak et al., 2000; Chen and Yeh, 2001; Yonezawa and Toshima, 2005; Vinodgopal et al., 2006). Among these synthetic methods, the sonochemical reduction has drawn particular attention due to the generation of nanoparticles with a narrow size distribution and higher surface area because of the unique reaction routes originating from the transient high temperatures and pressures induced by the acoustic cavitation phenomenon (Suslick, 1988).

Acoustic cavitation is generated in liquids at high and moderate intensities of ultrasonic irradiation. Dissolved gas nuclei in a liquid are forced to grow by rectified diffusion and coalescence pathways in an acoustic field. Bubbles continue to grow until they reach a critical size, which depends on the ultrasound frequency. The bubbles then grow to a maximum size and collapse within one acoustic cycle. The collapse of bubbles is very rapid, and due to the adiabatic nature of the collapse, localized hot spots are generated. Thus, cavitation can create extreme physical and chemical effects within the bubbles and at the bubble/solution interface despite the normal temperature conditions in the bulk solution. Hence, sonochemistry offers specific experimental conditions to perform synthetic chemical processes either within the bubble or at the bubble interface with extreme and non-equilibrium conditions, which could result in the formation of new materials with unique properties. Thus, the advent of promising sonochemical synthesis of nanomaterials meets the demand of nanotechnology as well as other physical sciences. A number of reviews and monographs devoted to various aspects of sonication have been reported in the last decade (Bang and Suslick, 2010). The main focus of this chapter is to review the sonochemical synthesis of robust noble monometallic and bimetallic with a particular attention to the resulting structures and properties of these nanomaterials.

3.2 Preparation, Characterization, and Applications of Monometallic Nanoparticles

Since the days of Michael Faraday, who first scientifically elucidated the preparative method for aqueous dispersions of metal nanoparticles (Faraday, 1857), these metal nanoparticles have often been produced by chemical reduction of the corresponding metal salts in solutions in the presence of suitable stabilizers (Kerker, 1986). Faraday gold nanoparticle dispersions can be considered to be dispersed without any stabilizer, but in fact, citrate ions used as the reductant were adsorbed on the surface of Au nanoparticles and stabilized them by the electrostatic repulsive forces. Thus, the citrate ions can reduce Au(III) ions, control the growth of Au atoms to metal nanoparticles, and stabilize the obtained Au nanoparticles. By adopting a similar citrate-reduction procedure, gold seed particles of size approximately 15 nm were prepared by Frens (1972) in 1973. Later, Murphy and his coworkers (Gole and Murphy, 2004; Sau and Murphy, 2004) reported a seed growth method in the presence of cationic surfactants for controlling the size and shape of gold nanoparticles. They have used conventional reductants, such as $NaBH_4$ and glucose. Electrochemical (Yu et al., 1997; Mohamed et al., 1998) and laser/photo irradiation techniques (Kim et al., 2003; Niidome et al., 2003) have also been used for the synthesis of shape-controlled particles with various aspect ratios (ratio between the length and width of the particles). From these reports, it has been concluded that the size and shape of the particles produced depend upon the preparation method: the nature of reductants and stabilizers, purification of seed and grown particles, and experimental conditions such as temperature and solute concentration. Although gold nanoparticles can be prepared by various methods (Daniel and Astruc, 2004), the sonochemical route (Yeung et al., 1993; Nagata et al., 1996; Caruso et al., 2002; Okitsu et al., 2002) offers some advantages over other synthetic processes. The reduction of Au(III) ion can proceed smoothly upon ultrasound irradiation of the solution (Okitsu et al., 2005). Figure 3.1a shows the changes in the absorption spectrum of the sonicated solution during ultrasound irradiation. An absorption band centered at 217 nm that corresponds to gold chloride complex ($AuCl_4^-$) was observed to decrease with irradiation time, whereas a surface plasmon band attributed to gold nanoparticles emerged at nearly 540 nm. These sonochemically produced gold seed particles were very stable in a colloidal state because citric acid acts as a good stabilizer. Not only ligands (citric acid) even polymers (polyvinylpyrrolidone, PVP) have some affinity for metals and are often used as stabilizers of metal nanoparticles (Hirai et al., 1976). These substances also control both the reduction rate of metal ions and the aggregation process of metal atoms. Figure 3.1(b) shows the transmission electron microscope (TEM) photograph of gold seed particles prepared sonochemically. The average size of the gold particles was nearly 17 nm.

The sonochemical method involves radical and/or thermal reactions, which originate from the high temperatures and pressures generated within cavitation bubbles (Misik and Riesz, 1996; Didenko et al., 1999; Nagata et al., 2001; Yasui et al., 2003; Rae et al., 2005). In a sonicated aqueous medium, Reactions 3.1–3.4 can be expected to occur (Yeung et al., 1993; Nagata et al., 1996; Caruso et al., 2002; Okitsu et al., 2002). Homolysis (thermolysis) of water molecules (reaction 1) occurs within the cavitation bubbles due to the high temperatures generated during the collapse of these bubbles. The recombination of these primary radicals (Reactions 3.2–3.4) generated in Reaction 3.1 can occur within these bubbles, at the bubble/solution interface, or in the bulk solution.

$$H_2O +))))) \rightarrow {}^{\bullet}H + {}^{\bullet}OH \tag{3.1}$$

$$2\ {}^{\bullet}OH \rightarrow H_2O_2 \quad (3.2)$$

$$2\ {}^{\bullet}H \rightarrow H_2 \quad (3.3)$$

$${}^{\bullet}H + {}^{\bullet}OH \rightarrow H_2O \quad (3.4)$$

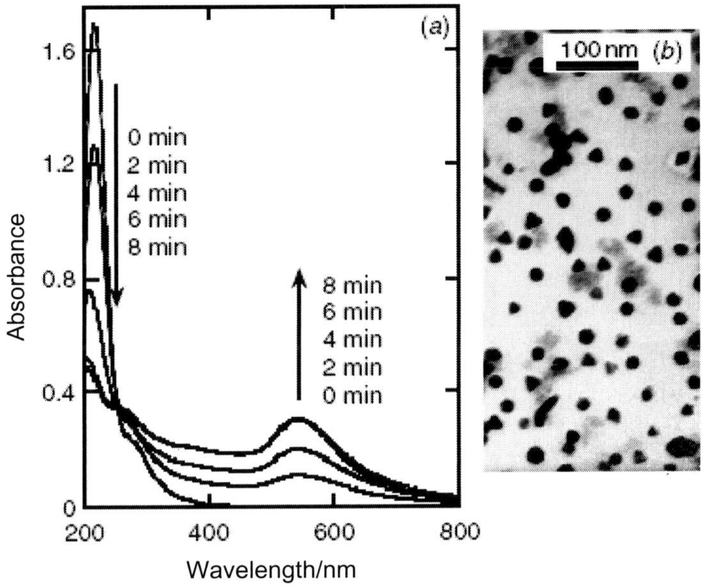

Figure 3.1 (a) Change in the absorption spectrum of Au (III) solution during irradiation under Ar. Conditions: 0.1 mM HAuCl4, 0.2 mM citric acid, 100 mM 2-propanol. (b) TEM image of gold seed particles prepared by 10 min irradiation (Okitsu et al., 2005).

Further, sonochemical reduction has been reported to be enhanced by the addition of organic additives with appropriate hydrophobic properties by using alcohols and surfactants (Yeung et al., 1993; Nagata et al., 1996; Okitsu et al., 2002). In aqueous solutions, 2-propanol (RHOH) adsorbs at the interface of the cavitation bubbles and reacts with primary radicals (Reaction 3.5),

$$RHOH + {}^{\bullet}H\ ({}^{\bullet}OH) \rightarrow {}^{\bullet}ROH + H_2O(H_2) \quad (3.5)$$

where secondary reducing radicals, ${}^{\bullet}ROH$, are formed by hydrogen abstraction from 2-propanol by ${}^{\bullet}OH$ radicals and ${}^{\bullet}H$ atoms. Thus, the oxidizing ${}^{\bullet}OH$ radicals are advantageously scavenged by 2-propanol (Reaction 3.5) and converted into highly reducing radicals, ${}^{\bullet}ROH$. In addition, 2-propanol molecules would be pyrolyzed at the interface and within the bubbles (Reaction 3.6),

$$\text{RHOH} + \text{sonication }))))) \rightarrow \text{Pyrolysis intermediates and products} \quad (3.6)$$

Some of the pyrolysis intermediates and products can also act as reductants (Nagata et al., 1996; Okitsu et al., 2002). Several types of reducing species subsequently react with Au(III) to form gold atoms (Au^0, Reaction 3.7), which can agglomerate to produce gold seed particles (Reaction 3.8).

$$Au^{3+} + {}^{\bullet}H/{}^{\bullet}ROH/\text{other reducing species} \rightarrow Au^0 + 3H^+ + \text{products} \quad (3.7)$$

$$nAu^0 \rightarrow (Au^0)_n \quad (3.8)$$

The actual mechanism involved in the reduction of Au(III) ions to form Au(0) is a complex process (Caruso et al., 2002). Au(II) ions might be formed as an intermediate in the reduction processes, but these are very unstable so that the following disproportionation reaction would also occur (Kurihara et al., 1983).

$$2Au^{2+} \rightarrow Au^+ + Au^{3+} \quad (3.9)$$

In continuation, the size and distribution of the Au nanoparticles produced can also be correlated with the rate of Au(III) reduction, which in turn is influenced by the applied frequency (e.g., 20 kHz, 213 kHz, 358 kHz, 647 kHz, and 1062 kHz) (Okitsu et al., 2005). Okitsu et al. (2005) performed systematic studies by varying the ultrasound frequency and suggested that maximum amount of reducing radicals [maximum rate of Au (III) reduction] were generated at around 213 kHz in the presence of 1-propanol. They observed that the ultrasound frequency significantly affected the size and size distribution of Au particles, indicating that the frequency is an important parameter that controls the size of the formed particles. In addition, the particle size can also be found to be inversely related to the rate of Au(III) reduction: the size of the particles decreased with an increase in the rate of reduction.

Zhang et al. (2006) produced gold nanobelts by ultrasonically irradiating an aqueous solution of $HAuCl_4$ containing α-D-glucose (Fig. 3.2). Zhang et al. (2006) speculated the reason for such nanobelt formation on the basis of Ostwald ripening processes at the expense of the small gold nanoparticles formed at the initial irradiation period (Roosen and Carter, 1998; Sun et al., 2002; Jin et al., 2001). That is,

(i) gold nuclei form and grow into nanoparticles; (ii) the gold nanoparticles directionally aggregate, melt, and then bind to each other (actually a recrystallization process); and (iii) growth along the Au[111] direction and further crystallization result in the single crystalline nanobelt structure. Likewise, Au nanoparticles with different shapes and diameters were ultrasonically formed by many researchers (Qiu et al., 2003; Li et al., 2006) owing to their potential applications in a range of areas, such as catalysis, biology, and nanotechnology.

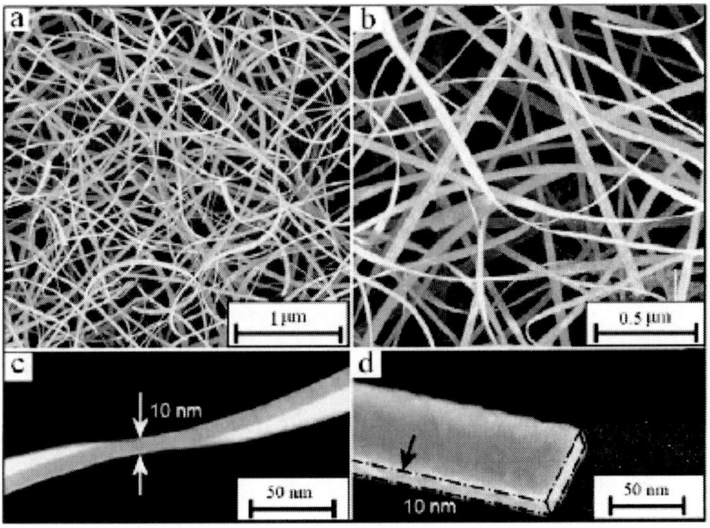

Figure 3.2 (a,b) SEM images and (c,d) high-magnification SEM images of as-synthesized gold nanobelts (Zhang et al., 2006).

Similarly, nanoparticles of other noble metals have also been prepared by the sonochemical technique. Nagata et al. (1992) prepared stable colloidal dispersions of silver by the ultrasonic irradiation of aqueous $AgClO_4$ or $AgNO_3$ solutions in the presence of surface active species [i.e., SDS, PEG40-MS, polyoxyethylene(20)sorbitan monolaurate (Tween 20)] with an ultrasonic wave of 200 kHz under air or Ar atmosphere. In continuation, Salkar et al. (1999) prepared amorphous Ag nanoparticles with an average size of 20 nm by the sonochemical reduction of an aqueous silver nitrate solution in an Ar–H_2 atmosphere. Anandan et al. (2008) prepared monometallic pure silver nanoparticles from an aqueous solution of $AgNO_3$ (0.2 mM) containing polyethylene glycol (0.1 wt%) and ethylene glycol (0.1 M) at room temperature under argon atmosphere. The solution containing silver nanoparticles was yellow in color and the absorption spectrum showed a band (λ_{max} = 437 nm) in the visible region (Fig. 3.3). TEM observations for the as-prepared sample indicate that the silver nanoparticles were approximately 15 nm in size (Fig. 3.4).

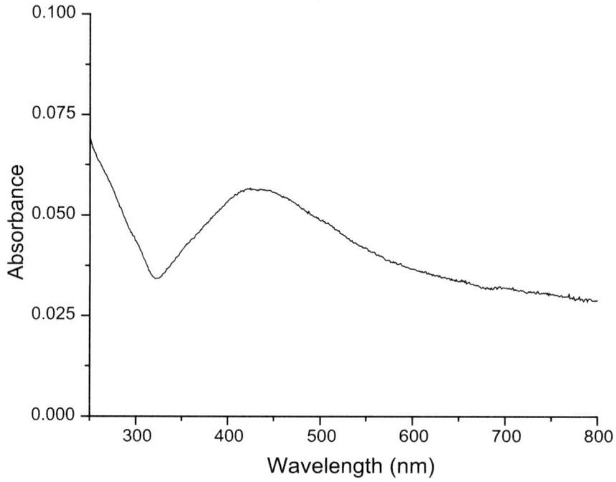

Figure 3.3 Absorption spectrum of silver nanoparticles prepared from an aqueous solution (70 mL) of $AgNO_3$ (0.2 mM) containing polyethylene glycol (0.1 wt%) and ethylene glycol (0.1 M), which was sonicated at room temperature under argon atmosphere for 30 min (Anandan et al., 2008).

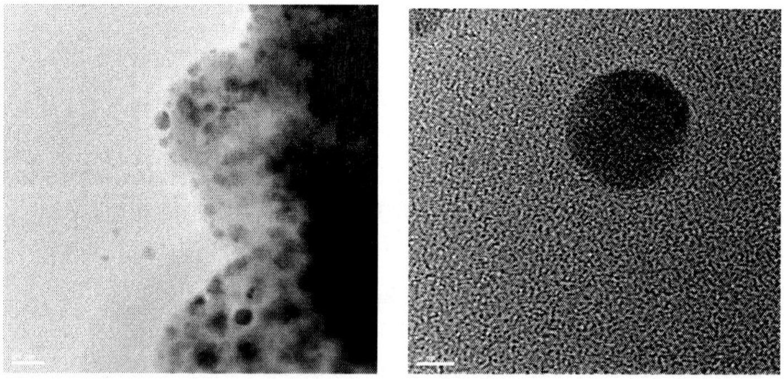

Figure 3.4 TEM images of monometallic pure silver nanoparticles prepared from an aqueous solution (70 mL) of $AgNO_3$ (0.2 mM) containing polyethylene glycol (0.1 wt%) and ethylene glycol (0.1 M), which was sonicated at room temperature under argon atmosphere for 30 min (Anandan et al., 2008).

Using high-intensity ultrasound (200 kHz, 6 W/cm^2), first Mizukoshi et al. (1999, 2001) reduced Pt(IV) ions in aqueous solution and soon after, Caruso et al. (2000) prepared colloidal platinum of various sizes using low-frequency ultrasound (20 kHz, calorimetric power = 10 W/cm^2) in the presence of different types of aliphatic alcohols (i.e., ethanol, propanol, and pentanol). Figure 3.5 shows the change in the absorption spectra (UV region gradually decreased and afterward the absorbance in the longer wavelength region grew) containing 1 mM of H_2PtCl_6 and alcohol), indicating the formation of platinum nanoparticles (Mizukoshi et al. 2001). Similarly, Vinodgopal et al. (2006) applied ultrasound to the tetrachloroplatinate solution containing surfactant and alcohol (8 mM SDS, 200 mM propanol, and 0.1M $HClO_4$) to produce colloidal platinum. TEM observations for the as-prepared, platinum nanoparticles are approximately 15 nm in size (Fig. 3.6).

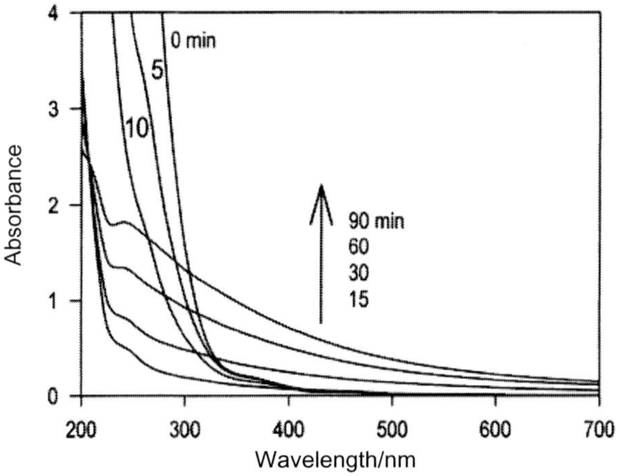

Figure 3.5 Change in the absorption spectra of an aqueous solution containing 1 mM of H_2PtCl_6 and 8 mM of SDS with the duration of the sonication (Mizukoshi et al., 2001).

Nemamcha et al. (2006) have used a 50 kHz ultrasonic generator and prepared stable palladium nanoparticles in different liquid media (aqueous, polyols) starting from palladium (II) nitrate. After ultrasonic irradiation for 180 min, the color of the palladium (II) nitrate solutions turned from the initial pale yellow to a dark brown indicating the formation of palladium nanoparticles. The synthesis of ruthenium nanoparticles in the presence of PVP stabilizer took a longer time for its formation (13 h) than the formation of other metal nanoparticles. This is due to the fact that reduction of Ru(III) is a very complex process and the complete reduction of Ru(III) is often difficult to achieve. Upon ultrasound irradiation, the Ru^{3+} species undergo sequential one-electron reduction to produce Ru^0 (He et al., 2006), which is identified by the disappearance of absorption peaks centered at 560, 450, and 400 nm (Fig. 3.7). The TEM image of the prepared Ru^0 nanoparticles using PVP as the stabilizer is shown in Fig. 3.8. If PEG is used as a stabilizer instead of PVP, the reduction is achieved in 7 h, because of the dual role played by PEG: as a reducing agent and as a stabilizer (Sathish Kumar et al., 2010).

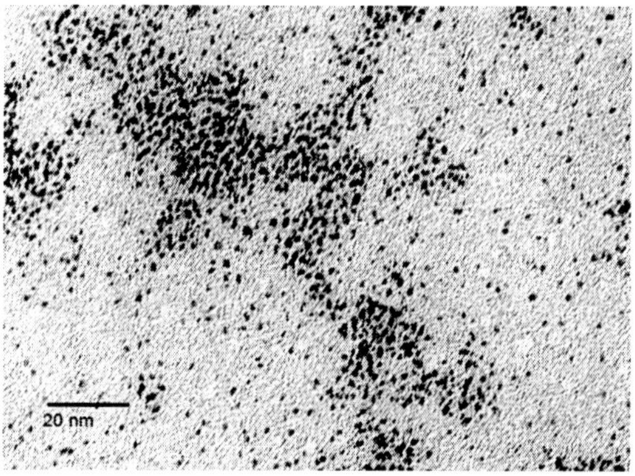

Figure 3.6 TEM image of Pt nanoparticles synthesized following 3 h of simultaneous sonication of 1 mM Pt(II) at 213 kHz. The accelerating TEM voltage was 100 kV (Vinodgopal et al., 2006).

Among the various metallic nanoparticles mentioned above, Au nanoparticles in particular have been widely studied with applications in a range of areas, such as catalysis, biology, and nanotechnology (Kusumi et al., 1993; Tanahashi et al., 1996; Haruta and Date, 2001; Olson et al., 2001; Zhong et al., 2005). Gold nanoparticles enhance the in vitro activities of antibacterial drugs, such as vancomycin (Gu et al., 2003) and ciprofloxacin (Tom et al., 2004). In addition, gold nanoparticles show antimicrobial activities due to their cytotoxicity toward various bacterial cells (Zhang et al., 2008). For the in vivo treatment of diseases, drugs have to be disseminated to the targeted tissues through the circulatory system.

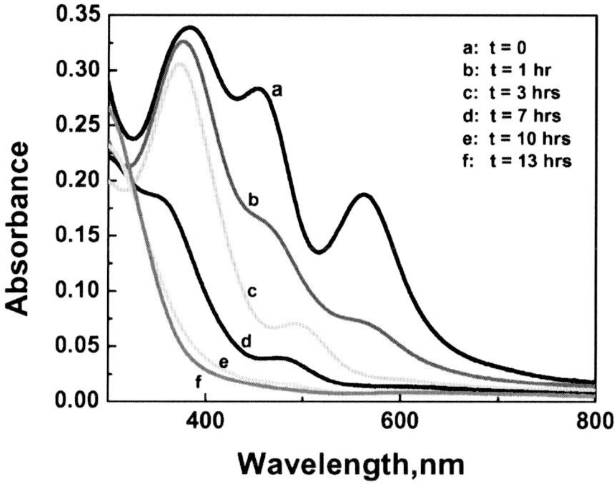

Figure 3.7 Changes in the absorption spectra of an aqueous solution of 1 mM $RuCl_3$ containing 0.1 M $HClO_4$, 8 mM SDS, and 0.2 M propanol as a function of sonication time: (a) t = 0, (b) t = 1 h, (c) t = 3 h, (d) t = 7 h, (e) t = 10 h, and (f) t = 13 h. The ultrasound frequency used was 213 kHz (He et al., 2006).

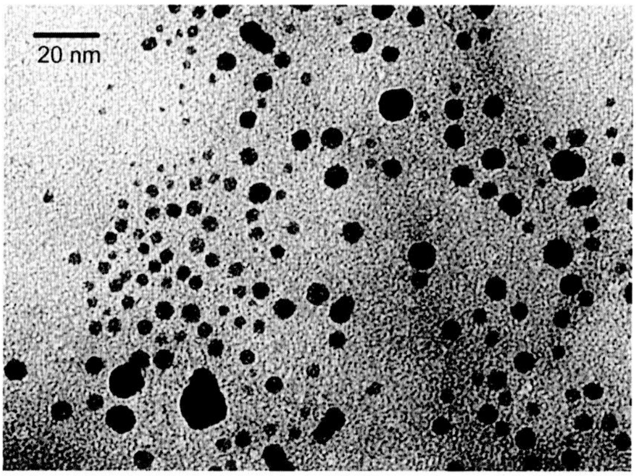

Figure 3.8 TEM image of Ru nanoparticles synthesized following 13 h of sonication at 213 kHz (He et al., 2006).

Serum albumins, the most abundant and major soluble protein constituent present in the blood stream of vertebrates, serve as a transport carrier of drugs due to their greater ability of binding reversibly with a large variety of endogenous and exogenous ligands present in blood (Gao et al., 2004). Hence, binding studies involving gold nanoparticles are biologically significant. Naveenraj et al. (2010) studied the interaction of globular fluorescent protein serum albumins, bovine serum albumin (BSA), and human serum albumin (HSA) with sonochemically synthesized gold nanoparticles using absorption and fluorescence spectroscopic techniques (Fig. 3.9). Experimental results suggested that quenching of the fluorescence emission from serum albumins by gold nanoparticles follows a static quenching mechanism and gold nanoparticles bind to serum albumins with high affinity. The binding constants obtained from modified Stern–Volmer plots and the apparent association constants from Benesi–Hildebrand plots suggested that HSA has more affinity toward sonochemically synthesized gold nanoparticles than that of BSA, whereas in the case of gold nanoparticles synthesized by $NaBH_4$ reduction method, BSA showed more affinity than that of HSA (Gao et al., 2005). Also, they determined the distance between the serum albumins and gold nanoparticles using Forster resonance energy transfer theory (FRET).

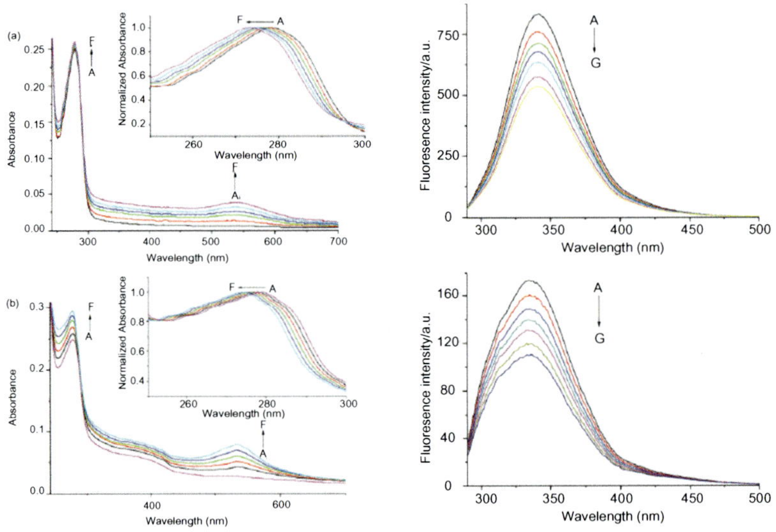

Figure 3.9 Absorption and fluorescence spectra of serum albumin (4 μM) [(a) BSA and (b) HSA] in the absence and presence of gold nanoparticles in the concentration range of 0–8 μM. Inset shows that their absorption spectra normalized to the 270–280 nm peak of a spectrum (Naveenraj et al., 2010).

Further, renewed interest in luminescence from metals came with a number of discoveries related to noble metals with roughened surfaces. Boyd et al. (1986) investigated the excitation energy dependence and the effect of surface roughness on the photoluminescence (PL) of noble metallic films and found that local field enhancement due to surface plasmon resonance is a prerequisite for PL. After the discovery of size-dependent PL of nanosized semiconductor materials by Mooradian (1969), the visible PL of sufficiently small gold nanoclusters has also been observed (Akiyama et al., 2008). Anandan et al. (2010) studied the spectral properties of gold nanoparticles prepared by the sonochemical method and the inherent PL images of the Au nanoparticles incorporated with single DNA molecules using a laser scanning, confocal, microscope-coupled, PL detection system.

Sonochemically synthesized Au nanoparticles exhibited three PL emission bands at 380, 415, and 670 nm (Fig. 3.10) when excited at 300 nm. The main reason for such PL emission is due to excitation of the surface plasmon coherent electronic motion. Upon comparison of gold nanoparticles prepared by other methods (Jian et al., 2004a, 2004b, 2005) with capping agents, sonochemically synthesized Au nanoparticles exhibit a new peak at 670 nm. This might be due to the generation of smaller sized Au nanoparticles with high surface area in the absence of any capping agents by the sonochemical method. Also by excitation of Au nanoparticles at 690 nm, a new PL feature was observed (Fig. 3.10) in terms of a weak PL emission at 780 nm that was remarkably different from that reported in literature (Jian et al., 2005). According to theoretical calculation of PL properties, gold nanoparticles prepared by other methods exhibit two emission bands at 741 nm and 793 nm attributed to the transverse and longitudinal surface plasmon resonances, respectively (Jian et al., 2004a, 2004b, 2005; Burda et al., 2005; Li et al., 2005), and Anandan et al. (2010) have suggested that longitudinal surface plasmon resonance is dominating for the sonochemically prepared Au nanoparticles.

To further understand the nature of electronic interactions between the nanoparticles and DNA, the inherent PL image of Au nanoparticles in the presence and absence of DNA was recorded by the laser confocal scanning microscope (LCSM)-coupled PL spectrophotometer equipped with an argon ion laser (excitation wavelength: 300 nm). Pure Au nanoparticles show bright PL image

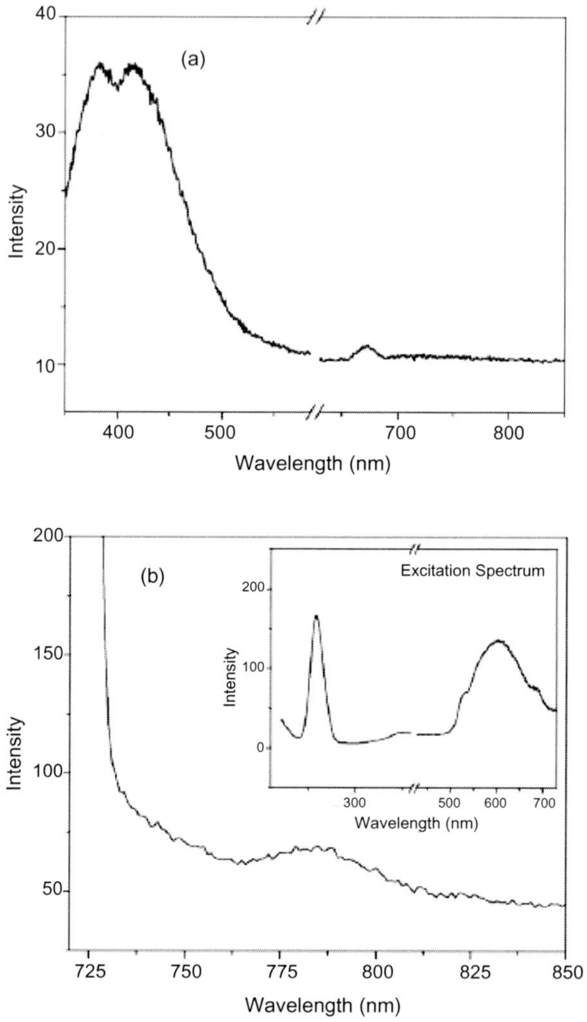

Figure 3.10 Photoluminescence spectra of Au nanoparticles prepared by sonication method: (a) Excitation at 300 nm, (b) excitation at 690 nm. Inset of b shows the excitation spectrum of Au nanoparticles detected with the emission wavelength of 780 nm (Anandan et al., 2010).

with strong emission at 510 nm (Fig. 3.11a). However, the observed PL spectrum is somewhat different in the case of adenine thymine (AT)-DNA (580 nm) and guanine cytosine (GC)-DNA (545 nm) molecules. This may be due to the difference in the electronic interactions

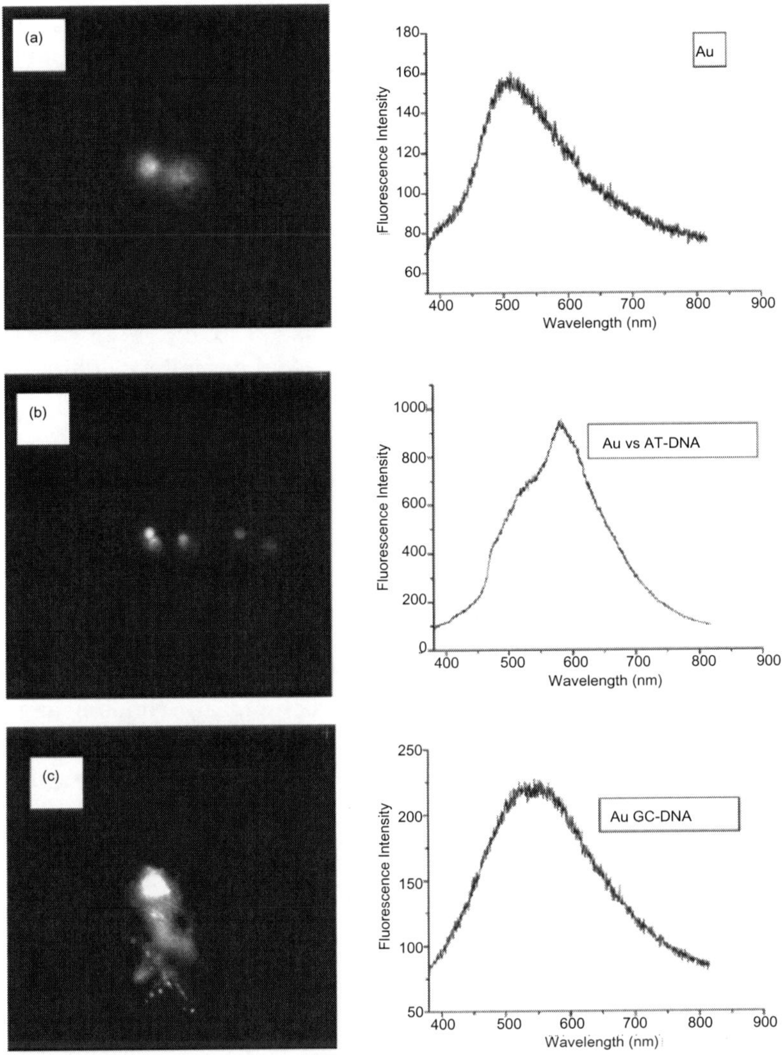

Figure 3.11 Confocal photoluminescence image and its emission spectrum for Au nanoparticle without any DNA (a), with AT-DNA (b), and with GC-DNA (c). The image dimensions are 2.5 μm × 2.5 μm (Anandan et al., 2010).

between AT-DNA and GC-DNA molecules with Au nanoparticles (Fig. 3.11b, c). The interaction between Au nanoparticles and DNA bases is nothing but Lewis acid (Au)–Lewis base (N- or O-groups of DNA) type of interaction (Sharma et al., 2007). Au nanoparticles prefer to interact more strongly through bare nitrogen atoms of the DNA bases instead of oxygen atoms. However, the absence of bare nitrogen atoms in thymine base DNA forces the Au nanoparticles to interact through O-atoms, and hence, the interaction between AT-DNA and Au nanoparticles is significantly lower than GC-DNA and Au nanoparticles. Apart from direct N–Au binding interaction, the presence of more hydrogen-bonding interaction also occurs in GC-DNA. As expected, such different shapes of DNAs were clearly observed in the confocal PL images (Fig. 3.11b, c). These results suggest that the Au nanoparticles exhibit greater binding affinity for GC-DNA than for AT-DNA.

3.3 Preparation, Characterization, and Applications of Bimetallic Nanoparticles

Bimetallic nanoparticles are being investigated in many fields of science and industries owing to their interesting physicochemical properties. Further, bimetallic nanoparticles have different internal structures and the three common structures are random alloy, core–shell, and cluster-in-cluster structures as shown in Fig. 3.12 (Harada et al., 1993; Major et al., 2009). Generally, the preparation of bimetallic nanoparticles from metal salts can be divided into two categories: coreduction (simultaneous) and successive (sequential) reduction of two metal salts. Coreduction is the simplest preparative method of bimetallic nanoparticles, that is, the same as that of monometallic nanoparticles and the only difference is that system originally contains both metal sources, whereas the coreduction method anticipated to generate alloy particles generally yielded core–shell particles. Successive reduction is usually carried out to prepare core–shell structured bimetallic nanoparticles. In contrast, successive reductions did not give core–shell structured products but, instead, gave "cluster-in-cluster" structured products or the mixtures of monometallic particles (Harada et al., 1993; Major et al., 2009). In order to confirm the above contrast, the morphologies of gold/palladium bimetallic particles have been studied since the early 1970s (Okamoto et al., 1992). This is because the combination of gold and palladium is one of the most popular examples that are miscible at any ratio as can be seen in their phase diagram (Turkevich and Kim, 1970). Also, nanoparticles with an Au core and Pd shell (Fig. 3.13) show higher catalytic activity in hydrogenation of 4-pentenoic acid than the mixtures of monometallic nanoparticles with a similar Au:Pd ratio (Mizukoshi et al., 2000). Further, the reason for the interest in the core–shell bimetallic nanoparticles from both fundamental scientific and technological points of view is that one of the metals (the shell) determines the surface properties of the particles while the other (the core) may be responsible for specific functional (optical, catalytic, magnetic, and so on) properties of the system.

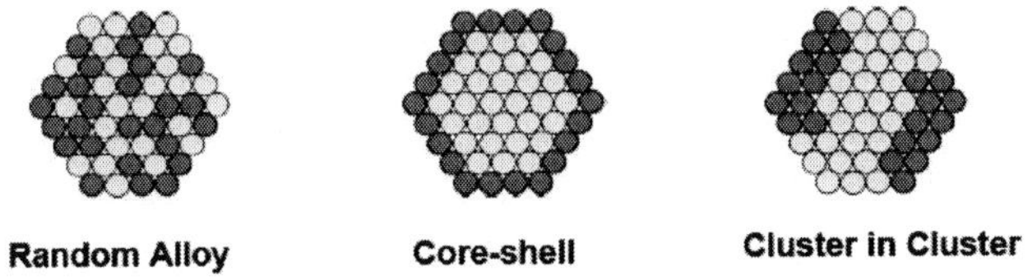

Figure 3.12 Several structures of bimetallic nanoparticles.

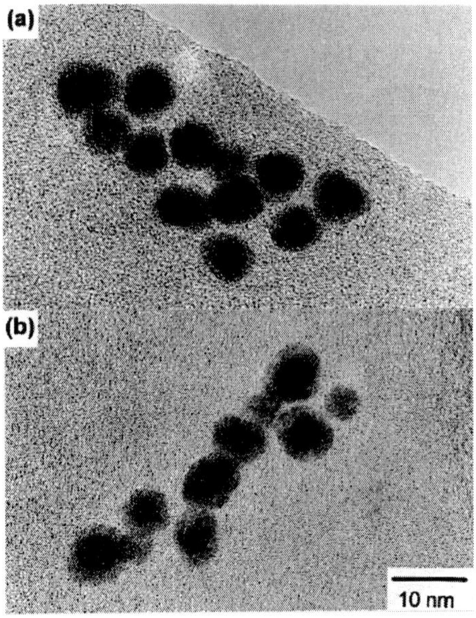

Figure 3.13 TEM images of sonochemically prepared gold/palladium: (a) (1:1) and (b) (1:4) bimetallic nanoparticles (Mizukoshi et al., 2000).

In this context, Mizukoshi et al. (1997, 2000) have synthesized bimetallic particles of gold and palladium with a narrow size distribution by the sonochemical reduction of an aqueous solution containing both Au(III) and Pd(II) ions. They reported core–shell morphology for the resulting nanoparticles with a Pd shell over the Au core. Kan et al. (2003) synthesized bimetallic particles with Au core and Pd shell by successive sonochemical reduction of the precursors $Pd(NO_3)_2$ and $HAuCl_4$ in ethylene glycol. The successive reduction of two metal salts can be considered as one of the most suitable methods to prepare core–shell structured bimetallic particles compared with coreduction/simultaneous reduction of the precursor ions. However, Harada et al. (1993) achieved Pd core–Au shell nanoparticles by the coreduction method instead of successive reduction procedure. The difference in the structure was argued to be due to the difference in the reduction potentials of Pd and Au ions. When Au ions were added in the presence of Pd nanoparticles, some Pd^0 atoms of the nanoparticles were oxidized and reduced Au^{III} ions; the oxidized Pd ions were reduced again by the reductants such as alcohols. This process led to the formation of particles with core–shell structure in the coreduction method.

Belloni et al. (1998) have described in their excellent review that the kinetics of competitive reduction in mixed systems has a dominant control on the final form of the mixed metal particles. Similarly, several preparation methods are also available for bimetallic nanoparticles, and for the particular case of Au–Ag nanoparticles, a number of proposals have been put forward to explain the nanoparticle formation processes (Bright et al., 1996; Rivas et al., 2000; Lu et al., 2002; Liu and Guyot-Sionnest, 2004; Raveendran et al., 2006). Sato et al. (1991) have prepared bilayered Ag–Au composites, with the core being composed of gold and with a silver-rich shell layer, by a photochemical method. Papavassiliou (1976) prepared gold-silver alloy nanoparticles by evaporation and condensation of the alloys. Silver particles with a gold layer were prepared by Henglein using γ-radiolysis (1993). In this context, Anandan et al. (2008) reported the synthesis of gold–silver bimetallic nanoparticles with core–shell morphology by the sonochemical coreduction of Au(III) and Ag(I) ions using 20 kHz ultrasound in aqueous solutions with polyethylene glycol as a stabilizer under argon atmosphere. TEM image (Fig. 3.14) is provided in order to support the formation of core–shell nanoparticles and gold core with silver shell upon simultaneous sonochemical irradiation of gold and silver ions. Further, to understand the core–shell structure formation process, Anandan et al. (2008) monitored the growth rates of individual metal nanoparticles under similar sonochemical irradiation conditions. The growth rate of Au could be followed by monitoring the absorption changes at 530 nm (Au plasmon absorption band) and at 265 nm ($AuCl_4^-$ absorption band). The rate of reduction of Ag^+ was monitored at 303 nm. A comparison of the rate of formation of individual metal clusters indicates that the rate of formation of gold nanoparticles is higher than that of silver nanoparticles. What these results suggest is that the formation of core–shell morphology might be due to the difference in the reduction rates of the individual metal ions. As gold ions are easily reduced under the sonochemical conditions, it may be suggested that gold particles are first generated followed by the reduction of Ag ions on the surface of the gold particles. In addition, the structure of Au–Ag nanoparticles was also controlled by the redox potentials of Au and Ag (i.e., $Au^{3+} \rightarrow Au$: 1.02 V, $Ag^+ \rightarrow Ag$: 0.80 V).

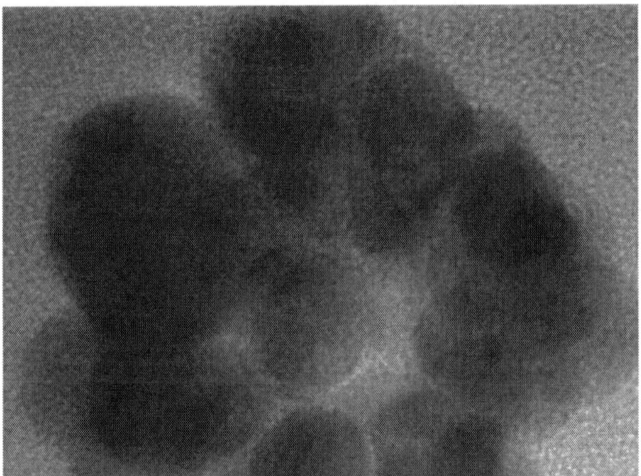

Figure 3.14 TEM image of sonochemically coreduced Au–Ag core–shell bimetallic nanoparticles (Anandan et al., 2008).

Although the ultrasound-assisted synthesis of platinum, silver, or gold-containing bimetallic nanoparticles has been well established, not much has been reported on the synthesis of ruthenium-containing bimetallic particles. This is presumably due to the difficulties associated with ruthenium reduction as compared with the reduction of other precious metals. Sathish Kumar et al. (2010) studied the sonochemical synthesis of gold/ruthenium bimetallic nanoparticles by simultaneous and sequential reduction of Au(III) and Ru(III) ions at three different molar ratios (Au^{3+}:Ru^{3+}; 1:1, 1:3, and 1:5) in the presence of PEG. A significant change in the absorption as a function of sonication time is observed for Au-Ru bimetallic particles (Fig. 3.15), which indicates the simultaneous reduction of both Au^{3+} and Ru^{3+} ions. For 1:1 molar ratio, the absorption maximum at 536 nm corresponds to the plasmon absorption of gold particles. It can be noticed in Fig. 3.15(a) that the absorption at 536 nm increases for the first 30 min and then slightly decreases. The increase in absorption is due to the initial formation of gold nanoparticles and the consecutive decrease in absorption is ascribed to the formation of a ruthenium layer on the surface of gold nanoparticles. The reason for the formation of Au particles prior to Ru particles is most probably due to the ease of reduction of gold ions compared with that of Ru ions. The Au particles may also catalyze the reduction of Ru ions leading to the formation of a Ru shell on the surface of the gold particles. Similar absorption characteristics were also observed with 1:3 molar ratio sample. However, the absorption characteristics were significantly different for the 1:5 molar ratio sample (Fig. 3.15b). This is due to the higher ruthenium ion concentration. Further, bimetallic nanoparticles containing gold as one substituent also have enhanced activity and selectivity for desired products for oxidation reactions. Hence in this regard, basic electrochemical characterizations were performed for the as-prepared bimetallic Au–Ru nanoparticles in order to evaluate their usefulness as a catalyst in direct methanol fuel cells. The cyclic voltammetric behavior of Au, Ru, and Au:Ru (1:1) nanoparticles modified glassy carbon (GC) electrodes was recorded in 0.5 M H_2SO_4 at a scan rate of 50 mV/s. In Fig. 3.16, a sharp oxidation peak around 1.2 V is observed with a corresponding reduction peak at 0.4 V for the GC–Au nanoelectrode. This electrochemical response corresponds to the formation of gold oxide and its reduction (Hoogvliet et al.,

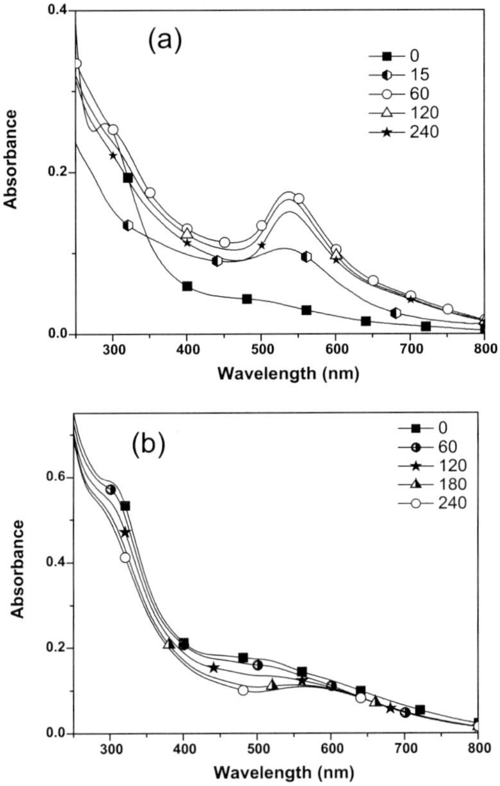

Figure 3.15 UV-vis absorption spectra of gold–ruthenium bimetallic nanoparticles prepared by the sonochemical coreduction method using (a) 1:1 and (b) 1:5 gold–ruthenium compositions (Sathish Kumar et al., 2010).

2000). This observation confirms that the gold nanoparticles are dispersed on the electrode surface and are in electrical contact with GC electrode. Further, it is well known that Ru nanoparticles undergo dissolution during the positive potential cycle of a cyclic voltammetry (CV) scan (Basnayake et al., 2006) with an increase in the double layer current. To avoid this, the anodic potential should be limited to about 0.8 V. Also, they recorded the cyclic voltammograms beyond that potential intentionally to notice irreversible oxidative dissolution of the Ru nanoparticles. A sharp peak obtained above 1.2 V indicates the formation of unstable ruthenium oxide. This provides evidence for the presence of Ru nanoparticles on the electrode.

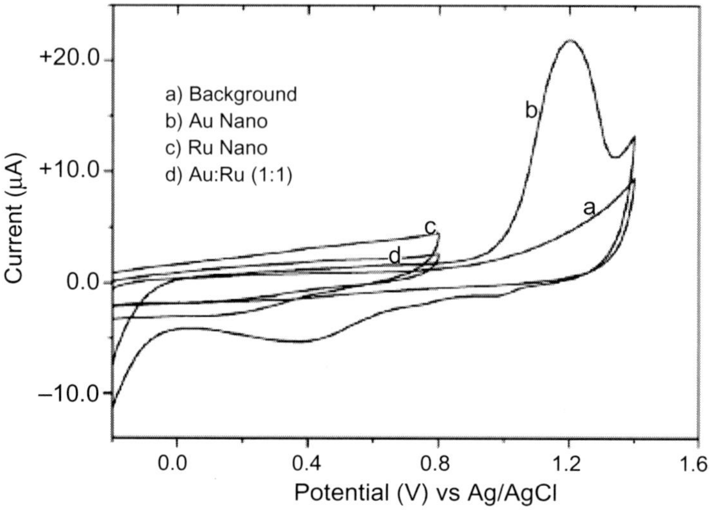

Figure 3.16 Cyclic voltammograms of pure Au, pure Ru, and Au–Ru bimetallic in 0.5 M H_2SO_4 at a scan rate of 50 mV/s (Sathish Kumar et al., 2010).

A similar trend is observed for Au:Ru bimetallic nanoparticles; however, when comparing the CVs of the bimetallic nanoparticles with pure Ru nanoparticles, a decrease in the double layer current is noticed. Thus, the incorporation of Au increases the oxidation potential of ruthenium, which in turn reduces the dissolution of Ru, critical factor responsible for the stability of the catalyst in fuel cell applications (Liang et al., 2008).

A similar core–shell structure was demonstrated by simultaneous sonochemical synthesis of Au and Pt in the presence of SDS and polyethylene glycol monostearate (PEG-MS) at 200 kHz with an input power of 4.2 W/cm^2 (Takatani et al., 2003). When using SDS and PEG-MS, the surface plasmon absorption of Au nanoparticles (about 530 nm) appeared shortly after the beginning of the sonication. Consequently, the surface plasmon absorption of Pt nanoparticles emerged, and that of Au nanoparticles remained for the last time. The spectrum change indicates that Au^{3+} ions are reduced at first, and then Pt^{4+} ions are reduced to Pt^{2+}. Because of the overlapping of the surface plasmon absorption of Au and Pt, Au and Pt nanoparticles may coexist individually. There are two kinds of particles with sizes of about 1 nm and about 6.5 nm, which were seen in TEM and HRTEM micrographs (Fig. 3.17). The selected area electron diffraction of the larger particles included not only Au diffraction lined but also a portion of Pt diffraction lines. The EDX measurement showed that the particles had a core–shell-like structure composed of Au-core and Pt-shell, in which the Pt content of edge region is higher than that of center. The smaller particles were confirmed to be Pt nanoparticles.

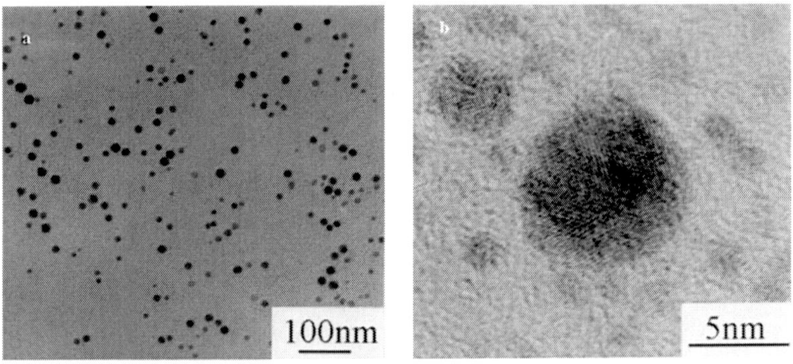

Figure 3.17 TEM and HRTEM images of sonochemically prepared Au/Pt (Takatani et al., 2003).

Further, among the bimetallic systems, bimetallic materials of Pt and Ru have a long history of applications in electrochemistry, primarily for the promotion of CO and CH_3OH oxidation in support of low-temperature fuel cell development (Watanabe and Motoo, 1975a,b). Through the sonochemical method, nanometer scale Pt–Ru bimetallic electrocatalysts rich in Ru and possessing alloy properties were produced by Basnayake et al. (2006) in order to achieve high CH_3OH oxidation rates. Vinodgopal et al. (2006) performed sequential sonolysis (platinum was first reduced followed by ruthenium reduction) to yield core–shell structure, that is, a platinum core with a ruthenium shell around it. However, when sonicated simultaneously with Pt(II), the reduction rate for the ruthenium ions is quite fast. Within 4 h, nearly all of the absorption bands of Pt and Ru (Fig. 3.18) have disappeared, and a clear brown solution is obtained. The accelerated reduction rate of Ru(III) is most likely due to the excess electrons on the platinum surface donated by the reducing radicals produced by sonolysis. The change in the absorption spectrum also indicates that reduction of

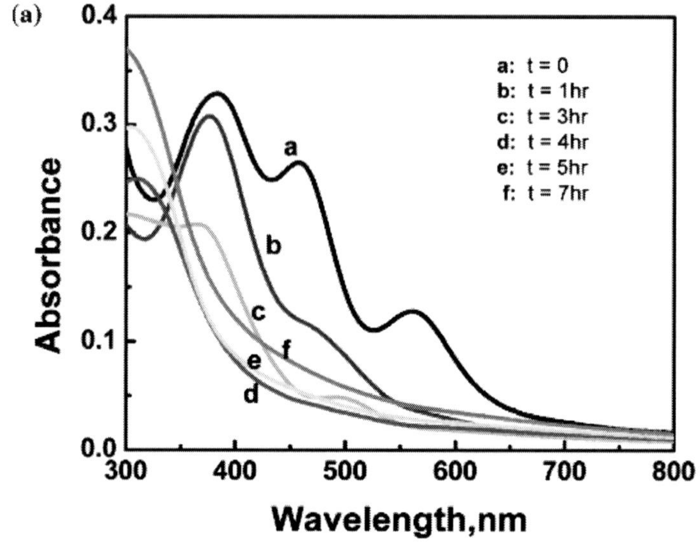

Figure 3.18 Absorption spectra showing the simultaneous sonochemical reduction of an aqueous solution of 1 mM $RuCl_3$ and 1 mM K_2PtCl_4 containing 0.1 M $HClO_4$, 8 mM SDS, and 0.2 M propanol under argon atmosphere; (a) t = 0. (b) t = 1 h. (c) t = 3 h. (d) t = 4 h. (e) t = 5 h. (f) t = 7 h. The ultrasound frequency used was 213 kHz (Vinodgopal et al., 2006).

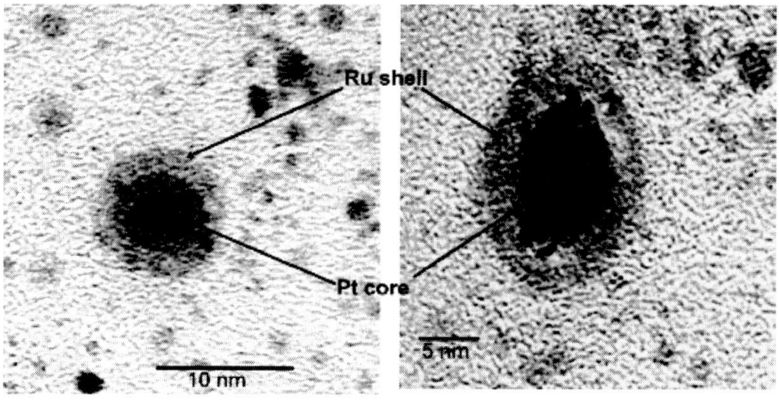

Figure 3.19 TEM images obtained at an accelerating voltage of 200 kV of Pt–Ru nanoparticles synthesized following sequential sonication of 1 mM Pt(II) and 1 mM Ru(III) at 213 kHz. Two representative particles are shown at different magnifications (Vinodgopal et al., 2006).

the Ru(III) occurs simultaneously with the reduction of the Pt (II). TEM image (Fig. 3.19) shows that particles with different sizes (the larger ones presumably ruthenium particles and the ultrasmaller ones represent platinum) support the above reduction processes. Thus, the sonochemical reduction technique has an advantage over other preparation methods and this approach can also be extended to produce nanoparticles of other bimetallic compounds.

3.4 Conclusion

The emerging sonochemical technique is unique in that it enables the preparation of new structures under nonequilibrium conditions. The preparation, characterization, and application of monometallic and bimetallic noble nanoparticles formed in an ultrasonic field have been summarized in this review. The morphological difference of the obtained products seemed to originate from the chemical and physical effects of ultrasound-induced cavitation phenomenon. This is mainly due to the vigorous microstirring and microjets generated during acoustic cavitation. The shock waves generated during the collapse of the cavities might promote the interaction between preformed metal core and post-reduced second metal to develop core–shell structures. The continued progress in the sonochemical synthesis of metal nanoparticles will help to improve the controlled synthesis of nanomaterials and broaden the range of applications of sonochemically prepared metal nanoparticles.

Acknowledgment

The research described herein was supported by the Department of Innovation, Industry, Science and Research (DIISR), Australia, and Department of Science and Technology (DST), India.

References

Akiyama, T., Kawahara, T., Arakawa, T. and Yamada, S. (2008). Fabrication of densely packed gold nanoparticle films and their fluorescence enhancement effect, *Jpn. J. Appl. Phys.*, **47**, pp. 3063–3066.

Alivisatos, P. (2004). The use of nanocrystals in biological detection, *Nat. Biotechnol.*, **22**, pp. 47–52.

Anandan, S., Oh, S., Yoon, M. and Ashokkumar, M. (2010). Photoluminescence properties of sonochemically synthesized gold nanoparticles for DNA biosensing, *Spectrochim. Acta A: Mol. Biomol. Spec.*, **76**, pp. 191–196.

Anandan, S., Ashokkumar, M. and Grieser, F. (2008). Sonochemical synthesis of Au–Ag core-shell bimetallic nanoparticles, *J. Phys. Chem. C*, **112**, pp. 15102–15105.

Astruc, D., Lu, F. and Aranzaes, J. R. (2005a). Nanopartikel als regenerierbare katalysatoren: an der nahtstelle zwischen homogener und heterogener katalyse, *Angew. Chem.*, **117**, pp. 8062–8083.

Astruc, D., Lu, F. and Aranzaes, J. R. (2005b). Nanoparticles as recyclable catalysts: the frontier between homogeneous and heterogeneous catalysis, *Angew. Chem. Int. Ed.*, **44,** pp. 7852–7872.

Bang, J. H. and Suslick, K. S. (2010). Applications of ultrasound to the synthesis of nanostructured materials, *Adv. Mater.*, **22**, pp. 1039–1059 and the references cited therein.

Basnayake, R., Li, Z., Katar, S., Zhou, W., Rivera, H., Smotkin, E. S., Casadonte Jr., D. J. and Korzeniewski, C. (2006). PtRu nanoparticle electrocatalyst with bulk alloy properties prepared through a sonochemical method, *Langmuir*, **22,** pp. 10446–10450.

Belloni, J., Mostafavi, M., Remita, H., Marignier, J. L. and Delcourt, M. O. (1998). Radiation-induced synthesis of mono- and multi-metallic clusters and nanocolloids, *New J. Chem.*, **22**, pp. 1239.

Boyd, G.T., Yu, Z. H. and Shen, Y.R. (1986). Photoinduced luminescence from the noble metals and its enhancement on roughened surfaces, *Phys. Rev. B*, **33**, pp. 7923–7936.

Bright, R. M., Walter, D. G., Musick, M. D., Jackson, M. A., Allison, K. J. and Natan, M. J. (1996). Chemical and electrochemical Ag deposition onto preformed Au colloid monolayers: approaches to uniformly-sized surface features with Ag-like optical properties, *Langmuir*, **12**, pp. 810–817.

Burda, C., Chen, X., Narayanan, R. and El-Sayed, M. A. (2005). Chemistry and properties of nanocrystals of different shapes, *Chem. Rev.*, **105,** pp. 1025–1102.

Caruso, R. A., Ashokkumar, M. and Grieser, F. (2001). Sonochemical formation of gold sols, *Langmuir*, **18**, pp. 7831–7836.

Caruso, R. A., Ashokkumar, M. and Grieser, F. (2000). Sonochemical formation of colloidal platinum, *Colloids Surf. A*, **169,** pp. 219–225.

Chen, Y. H. and Yeh, C. S. (2001). A new approach for the formation of alloy nanoparticles: laser synthesis of gold–silver alloy from gold–silver colloidal mixtures. *Chem. Commun.*, pp. 371–372.

Dadosh, T., Gordin, Y., Krahne, R., Khivrichm, I., Mahalu, D., Frydman, V., Sperling, J., Yacoby, A. and Bar-Joseph I. (2005). Measurement of the conductance of single conjugated molecules, *Nature,* **436,** pp. 677–680.

Daniel, M. C. and Astruc, D. (2004). Gold nanoparticles: assembly, supramolecular chemistry, quantum-size-related properties, and applications toward biology, catalysis, and nanotechnology, *Chem. Rev.,* **104,** pp. 293–346.

Daté, M., Okumura, M., Tsubota, S. and Haruta, M. (2004a). Vital role of moisture in the catalytic activity of supported gold nanoparticles. *Angew. Chem.,* **116,** pp. 2181–2184.

Daté, M., Okumura, M., Tsubota, S. and Haruta, M. (2004b). Vital role of moisture in the catalytic activity of supported gold nanoparticles, *Angew. Chem. Int. Ed.,* **43,** pp. 2129–2132.

Didenko, Y. T., McNamara, W. B. and Suslick, K. S. (1999). Hot spot conditions during cavitation in water. *J. Am. Chem. Soc.,* **121**, pp. 5817–5818.

Elghanian, R., Storhoff, J. J., Mucic, R. C., Letsinger, R. L. and Mirkin, C. A. (1997) Selective colorimetric detection of polynucleotides based on the distance-dependent optical properties of gold nanoparticles, *Science,* **277,** pp. 1078–1081.

El-Sayed, I. H., Huang, X. H. and El-Sayed, M. A. (2005). Surface plasmon resonance scattering and absorption of anti-EGFR antibody conjugated gold nanoparticles in cancer diagnostics: applications in oral cancer, *Nano Lett.,* **5**, pp. 829–834.

Faraday, M. (1857). The bakerian lecture: Experimental relations of gold (and other metals) to light, *Philos. Trans. R. Soc. London,* **147**, pp. 145–181.

Freeman, R. G., Hommer, M. B., Grabar, K. C., Jackson, M. A. and Natan, M. J. (1996). Ag- clad Au nanoparticles: Novel aggregation, optical, and surface-enhanced raman scattering properties, *J. Phys. Chem.,* **100**, pp. 718–724.

Frens, G. (1972). Controlled nucleation for the regulation of the particle size in monodisperse gold suspensions, *Nature,* **241**, pp. 20–22.

Gao, H., Lei, L., Liu, J., Kong, Q., Chen, X. and Hu, Z. (2004). The study on the interaction between human serum albumin and a new reagent with antitumour activity by spectrophotometric methods, *J. Photochem. Photobiol. A: Chem.,* **167**, pp. 213–221.

Gao, D., Tian, Y., Bi, S., Chen, Y., Yu. A. and Zhang, H. (2005). Studies on the interaction of colloidal gold and serum albumins by spectral methods, *Spectrochim. Acta A: Mol. Biomol. Spectrosc.,* **62**, pp. 1203–1208.

Gole, A. and Murphy, C. J. (2004). Seed-mediated synthesis of gold nanorods: role of the size and nature of the seed, *Chem. Mater.,* **16**, pp. 3633–3640.

Gu, H., Ho, P. L., Tong, E., Wang, L. and Xu, B. (2003). Presenting vancomycin on nanoparticles to enhance antimicrobial activities, *Nano Lett.,* **3**, pp. 1261–1263.

Harada, M., Asakura, K. and Toshima, N. (1993). Catalytic activity and structural analysis of polymer-protected gold/palladium bimetallic clusters prepared by the successive reduction of hydrogen tetrachloroaurate(III) and palladium dichloride, *J. Phys. Chem.,* **97**, pp. 5103–5114.

Hasobe, T., Imahori, H., Kamat, P. V., Ahn, T. K., Kim, S. K., Kim, D., Fujimoto, A., Hirakawa, T., and Fukuzumi, S. (2005). Photovoltaic cells using composite nanoclusters of porphyrins and fullerenes with gold nanoparticles, *J. Am. Chem. Soc.,* **127**, pp. 1216–1228.

Haruta, M. and Date M. (2001). Advances in the catalysis of Au nanoparticles, *Appl. Catal. A,* **222**, pp. 427–437.

He, Y., Vinodgopal, K., Ashokkumar, M. and Grieser, F. (2006). Sonochemical synthesis of ruthenium nanoparticles, *Res. Chem. Intermed.,* **32**, pp. 709–715.

Henglein, A. (1993). Physicochemical properties of small metal particles in solution: "microelectrode" reactions, chemisorption, composite metal particles, and the atom-to-metal transition, *J. Phys. Chem.,* **97**, pp. 5457–5471.

Hirai, H., Nakao, Y., Toshima, N. and Adachi, K. (1976). Colloidal rhodium in polyvinyl alcohol as hydrogenation catalyst of olefins, *Chem. Lett.,* **5,** pp. 905–910.

Hodak, J. H., Henglein, A., Giersig, M. and Hartland, G. V. (2000). Laser-induced inter-diffusion in AuAg core-shell nanoparticles, *J. Phys. Chem. B,* **104**, pp. 11708–11718.

Hoogvliet, J.C., Dijksma, M., Kamp, B. and van Bennekom, W. P. (2000). Electrochemical pretreatment of polycrystalline gold electrodes to produce a reproducible surface roughness for self-assembly: a study in phosphate buffer pH 7.4, *Anal. Chem.,* **72,** pp. 2016–2021.

Jian, Z. (2005). Shape dependent full width at half maximum of the absorption band in gold nanorods, *Phys. Lett. A,* **339**, pp. 466–471.

Jian, Z., Liqing, H., Yongchang, W. and Yimin, L. (2004a). Fluorescence spectrum properties of gold nanochains, *Physica E,* **25**, pp. 114–118.

Jian, Z., Yongchang, W. and Shinong, Y. (2004b). Fluorescence spectrum characteristics of gold nanorods, *Chin. Phys. Lett.,* **21**, pp. 559–561.

Jin, R., Cao, Y., Mirkin, C. A., Kelly, K. L., Schatz, G. C. and Zheng, J. G. (2001). Photoinduced conversion of silver nanospheres to nanoprisms, *Science,* **294**, pp. 1901–1903.

Kan, C., Cai,W., Li, C., Zhang, L. and Hofmeister, H. (2003). Ultrasonic synthesis and optical properties of Au/Pd bimetallic nanoparticles in ethylene glycol, *J. Phys. D: Appl. Phys.,* **36,** pp. 1609–1614.

Kerker, M. (1986). Classics and classicists of colloid and interface science: 1. Michael faraday, *J. Colloid. Interface Sci.,* **112**, pp. 302–305.

Kim, F., Song, J. H. and Yang, P. (2002). Photochemical synthesis of gold nanorods, *J. Am. Chem. Soc.,* **124**, pp. 14316–14317.

Kurihara, K., Kizling, J., Stenius, P. and Fendler, J. H. (1983). Laser and pulse radiolytically induced colloidal gold formation in water and in water-in-oil microemulsions, *J. Am. Chem. Soc.,* **105**, pp. 2574–2579.

Kusumi, A., Sako, Y. and Yamamoto, M. (1993). Confined lateral diffusion of membrane receptors as studied by single particle tracking (nanovid microscopy). Effects of calcium-induced differentiation in cultured epithelial cells, *Biophys. J.,* **65**, pp. 2021– 2040.

Li, C., Cai, W., Li, Y., Hu, J. and Liu, P. (2006). Ultrasonically induced Au nanoprisms and their size manipulation based on aging, *J. Phys. Chem. B,* **110**, pp. 1546–1552.

Li, C. Z., Male, K. B., Hrapovic, S. and Luong, J. H. T. (2005). Fluorescence properties of gold nanorods and their application for DNA biosensing, *Chem. Commun.,* pp. 3924–3926.

Liang, Z. X., Zhao, T. S. and Xu, J. B. (2008). Stabilization of the platinum-ruthenium electrocatalyst against the dissolution of ruthenium with the incorporation of gold, *J. Power Sources,* **185**, pp. 166–170.

Link, S., Wang, Z. L. and El-Sayed, M. A. (1999). Alloy formation of gold–silver nanoparticles and the dependence of the plasmon absorption on their composition, *J. Phys. Chem. B,* **103**, pp. 3529–3533.

Liu, G. L., Yin, Y., Kunchakarra, S., Mukherjee, B., Gerion, D., Jett, S. D., Bear, D. G., Gray, J. W., Alivisatos, A. P., Lee, L. P. and Chen, F. Q. F. (2006). A nanoplasmonic molecular ruler for measuring nuclease activity and DNA footprinting, *Nat. Nanotechnol.,* **1,** pp. 47–52.

Liu, J. and Lu, Y. (2004). Adenosine-dependent assembly of aptazyme-functionalized gold nanoparticles and its application as a colorimetric biosensor, *Anal. Chem.,* **76,** pp. 1627–1632.

Liu, M. Z. and Guyot-Sionnest, P. (2004). Synthesis and optical characterization of Au/Ag core/shell nanorods, *J. Phys. Chem. B,* **108**, pp. 5882–5888.

Liz-Marzan, L. M. and Philipse, A. P. (1995). Stable hydrosols of metallic and bimetallic nanoparticles immobilized on imogolite fibers, *J. Phys. Chem.*, **99**, pp. 15120–15128.

Lu, L., Wang, H., Zhou, Y., Xi, S., Zhang, H., Hu, J. and Zhao, B. (2002). Seed-mediated growth of large, monodisperse core–shell gold–silver nanoparticles with Ag-like optical properties, *Chem. Commun.*, pp. 144–145.

Major, K. J., De, C. and Obare, S. O. (2009). Recent advances in the synthesis of plasmonic bimetallic nanoparticles, *Plasmonics*, **4**, pp. 61–78.

Misik, V. Riesz, P. (1996). Recent applications of EPR and spin trapping to sonochemical studies of organic liquids and aqueous solutions, *Ultrasonics Sonochem.*, **3**, pp. S173–S186.

Mizukoshi, Y., Takagi, E., Okuno, H., Maeda, Y. and Nagata, Y. (2001). Preparation of platinum nanoparticles by sonochemical reduction of the Pt(IV) ions: role of surfactants, *Ultrason. Sonochem.*, **8**, pp. 1–6.

Mizukoshi, Y., Fujimoto, T., Nagata, Y., Oshima, R., Maeda, Y. (2000). Characterization and catalytic activity of core-shell structured gold/palladium bimetallic nanoparticles synthesized by the sonochemical method, *J. Phys. Chem. B*, **104**, 6028–6032.

Mizukoshi, Y., Oshima, R., Maeda, Y. and Nagata, Y. (1999). Preparation of platinum nanoparticles by sonochemical reduction of the Pt (II) ion, *Langmuir*, **15**, pp. 2733–2737.

Mizukoshi, Y., Okitsu, K., Maeda, Y., Yamamoto, T. A., Oshima, R. and Nagata, Y. (1997). Sonochemical preparation of bimetallic nanoparticles of gold/palladium in aqueous solution, *J. Phys. Chem. B*, **101**, pp. 7033–7037.

Mohamed, M. B., Ismail, K. Z., Link, S. and El-Sayed, M. A. (1998). Thermal reshaping of gold nanorods in micelles, *J. Phys. Chem. B*, **102**, pp. 9370–9374.

Molina, L. M., Rasmussen, M. D. and Hammer, B. (2004). Adsorption of O_2 and oxidation of CO at Au nanoparticles supported by $TiO_2(110)$, *J. Chem. Phys.*, **120**, pp. 7673–7680.

Mooradian, A. (1969). Photoluminescence of metals, *Phys. Rev. Lett.*, **22**, pp. 185–187.

Mulvaney, P., Giersig, M. and Henglein, A. (1993). Electrochemistry of multilayer colloids: preparation and absorption spectrum of gold-coated silver particles, *J. Phys. Chem.*, **97**, pp. 7061–7064.

Nagata, Y., Watanabe, Y., Fujita, S., Dohmaru, T. and Taniguchi, S. (1992). Formation of colloidal silver in water by ultrasonic irradiation, *J. Chem. Soc., Chem. Commun.*, pp. 1620–1622.

Nagata, Y., Okuno, H., Mizukoshi,Y.and Maeda, Y. (2001). Acoustic cavitation in water under rare gas atmosphere, *Chem. Lett.,* **30**, pp. 142–143.

Nagata,Y., Mizukoshi, Y., Okitsu, K. and Maeda, Y. (1996). Sonochemical formation of gold particles in aqueous solution, *Radiat. Res.,* **146**, pp. 333–338.

Naveenraj, S., Anandan, S., Kathiravan, A., Renganathan, R. and Ashokkumar, M. (2010). The interaction of sonochemically synthesized gold nanoparticles with serum albumins, *J. Pharma. Biomed. Anal.*, **53,** pp. 804–810.

Nemamcha, A., Rehspringer, J. L. and Kharmi, D. (2006). Synthesis of palladium nanoparticles by sonochemical reduction of palladium(II) nitrate in aqueous solution, *J. Phys. Chem. B,* **110**, pp. 383–387.

Niidome, Y., Nishioka, K., Kawasaki, H. and Yamada, S. (2003). Rapid synthesis of gold nanorods by the combination of chemical reduction and photoirradiation processes; morphological changes depending on the growing processes, *Chem. Commun.,* pp. 2376–2377.

Okamoto, H., Subramanian, R. P. and Kacpraz, L. (1992) Eds. *Binary Alloy Phase Diagrams,* 2nd ed.; **Vol. 1,** ASM International.

Okitsu, K., Teo, B. M., Ashokkumar, M. and Grieser, F. (2005). Controlled growth of sonochemically synthesized gold seed particles in aqueous solutions containing surfactants, *Aust. J. Chem.,* **58**, pp. 667–670.

Okitsu, K., Ashokkumar, M. and Grieser, F. (2005). Sonochemical synthesis of gold nanoparticles: effects of ultrasound frequency, *J. Phys. Chem. B,* **109,** pp. 20673–20675.

Okitsu, K., Yue, A., Tanabe, S., Matsumoto, H., Yobiko, Y. and Yoo, Y. (2002). Sonolytic control of rate of gold (III) reduction and size of formed gold nanoparticles: relation between reduction rates and sizes of formed nanoparticles, *Bull. Chem. Soc. Jpn.,* **75**, pp. 2289–2296.

Olson, L. G., Lo, Y. S., Beebe Jr. T. P. and Harris, J. M. (2001). Characterization of silane-modified immobilized gold colloids as a substrate for surface-enhanced Raman spectroscopy, *Anal. Chem.,* **73**, pp. 4268–4276.

Papavassiliou, G. C. (1976). Surface plasmons in small Au-Ag alloy particles, *J. Phys. F: Met. Phys.,* **6**, L103–L105.

Park, S. J., Taton, T. A. and Mirkin, C. A. (2002). Array-based electrical detection of DNA with nanoparticle probes, *Science,* **295**, pp. 1503–1506.

Pradhan, S., Sun, J., Deng, F. J. and Chen, S. (2006*)*. Single-electron transfer in nanoparticle solids, *Adv. Mater.,* **18,** pp. 3279–3283.

Qiu, X. F., Zhu, J. J. and Chen, H. Y. (2003). Controllable synthesis of nanocrystalline gold assembled whiskery structures via sonochemical route, *J. Crystal Growth,* **257,** pp. 378–383.

Rae, J., Ashokkumar, M., Eulaerts, O., von Sonntag, C., Reisse, J. and Grieser, F. (2005). Estimation of ultrasound induced cavitation bubble temperatures in aqueous solutions, *Ultrason. Sonochem.*, **12**, pp. 325–329.

Raveendran, P., Fu, J. and Wallen, S. L. (2006). A simple and "green" method for the synthesis of Au, Ag, and Au–Ag alloy nanoparticles, *Green Chem.*, **8**, pp. 34–38.

Remita, H., Mostafavi, M. and Delcourt, M. O. (1996). Bimetallic Ag-Pt and Au-Pt aggregates synthesized by radiolysis, *Radiat. Phys. Chem.*, **47**, pp. 275–279.

Rivas, L., Sanchez-Cortes, S., Garcia-Ramos, J. V. and Morcillo, G. (2000). Mixed silver/gold colloids: a study of their formation, morphology, and surface-enhanced Raman activity, *Langmuir*, **16**, pp. 9722–9728.

Roosen, A. R. and Carter, W. C. (1998). Simulations of microstructural evolution: anisotropic growth and coarsening, *Physica A*, **261**, pp. 232–247.

Salkar, R. A., Jeevanandam, P., Aruna, S. T., Koltypin, Y. and Gedanken A. (1999). The sonochemical preparation of amorphous silver nanoparticles, *J. Mater. Chem.*, **9**, pp. 1333–1335.

Sathish Kumar, P., Manivel, A., Anandan, S., Zhou, M., Grieser, F. and Ashokkumar, M. (2010). Sonochemical synthesis and characterization of gold-ruthenium bimetallic nanoparticles, *Colloids & Surf. A*, **356**, pp. 240–244.

Sato, T., Kuroda, S., Takami, A., Yonezawa, Y.and Hada, H. (1991). Photochemical formation of silver-gold (Ag-Au) composite colloids in solutions containing sodium alginate, *Appl. Organomet. Chem.*, **5**, pp. 261–268.

Sau, T. K. and Murphy, C. J. (2004). Seeded high yield synthesis of short Au nanorods in aqueous solution, *Langmuir*, **20**, pp. 6414–6420.

Schmid, G. and Corain, B. (2003). Nanoparticulated gold: syntheses, structures, electronics, and reactivities, *Eur. J. Inorg. Chem.*, **2003**, pp. 3081–3098.

Sharma, P., Singh, H., Sharma, S. and Singh, H. (2007). Binding of gold nanoclusters with size-expanded DNA bases: a computational study of structural and electronic properties, *J. Chem. Theory Comput.*, **3**, pp. 2301–2311.

Silvert, P. Y., Vijayakrishnan, V., Vibert, P., Herrara-Urbina, R. and Elhsissen, K. T. (1996). Synthesis and characterization of nanoscale Ag-Pd alloy particles, *Nanostruct. Mater.*, **7**, pp. 611–618.

Sönnichsen, C., Reinhard, B. M., Liphardt, J. and Alivisatos, A. P. (2005). A molecular ruler based on plasmon coupling of single gold and silver nanoparticles, *Nat. Biotechnol.*, **23**, pp. 741–745.

Storhoff, J. J., Lucas, A. D., Garimella, V., Bao, Y. P. and Muller, U. R. (2004). Homogeneous detection of unamplified genomic DNA sequences based on colorimetric scatter of gold nanoparticle probes, *Nat. Biotechnol.*, **22**, pp. 883–887.

Storhoff, J. J., Elghanian, R., Mucic, R. C., Mirkin, C. A. and Let-singer, R. L. (1998). One-pot colorimetric differentiation of polynucleotides with single base imperfections using gold nanoparticle probes, *J. Am. Chem. Soc.*, **120,** pp. 1959–1964.

Sun, Y., Yin, Y., Mayers, B., Herricks, T. and Xia, Y. (2002). Uniform silver nanowires synthesis by reducing $AgNO_3$ with ethylene glycol in the presence of seeds and poly(vinyl pyrrolidone), *Chem. Mater.*, **14**, pp. 4736–4745.

Suslick, K. S. (1995). Applications of ultrasound to materials chemistry, *MRS Bull.*, **20**, 29–34.

Suslick, K. S. (1988) *Ultrasound: Its Chemical, Physical and Biological Effects*; Suslick, K. S., Ed.; VCH: Weinheim, Germany.

Takatani, H., Kago, H., Nakanishi, M., Kobayashi, Y. and Hori F. (2003). Characterization of noble metal alloy nanoparticles prepared by ultrasound irradiation, *Rev. Adv. Mater. Sci.*, **5,** pp. 232–238.

Tanahashi, I., Manabe, Y., Tohda, T., Sasaki, S. and Nakamura, A. (1996). Optical nonlinearities of Au/SiO_2 composite thin films prepared by a sputtering method, *J. Appl. Phys.*, **79**, pp. 1244–1249.

Tom, R. T., Suryanarayanan, V., Reddy, P. G., Baskaran, S. and Pradeep, T. (2004). Ciprofloxacin protected gold nanoparticles, *Langmuir*, **20**, pp. 1909–1914.

Toshima, N., Yonezawa, T. and Kushihashi, K. (1993). Polymer-protected palladium–platinum bimetallic clusters: preparation, catalytic properties and structural considerations, *J. Chem. Soc., Faraday Trans.*, **89**, pp. 2537–2543.

Treguer, M., de Cointet, C., Remita, H., Khatouri, J., Mostafavi, M., Amblard, J., Belloni, J. and de Keyzer, R. (1998). Dose rate effects on radiolytic synthesis of gold–silver bimetallic clusters in solution, *J. Phys. Chem. B*, **102**, pp. 4310–4321.

Tseng, R. J., Huang, J. X., Ouyang, J., Kaner, R. B. and Yang, Y. (2005). Polyaniline nanofiber/gold nanoparticle nonvolatile memory, *Nano Lett.*, **5,** pp. 1077–1080.

Turkevich, J. and Kim, G. (1970). Palladium: preparation and catalytic properties of particles of uniform size, *Science,* **169**, pp. 873–879.

Vinodgopal, K., He, Y., Ashokkumar, M. and Grieser, F. (2006). Sonochemically prepared platinum-ruthenium bimetallic nanoparticles, *J. Phys. Chem. B,* **110,** pp. 3849–3844.

Watanabe, M. and Motoo, S. (1975a). Electrocatalysis by ad-atoms. Part III. Enhancement of the oxidation of carbon monoxide on platinum by ruthenium ad-atoms, *J. Electroanal. Chem.,* **60,** pp. 275–283.

Watanabe, M. and Motoo, S. (1975b). Electrocatalysis by ad-atoms. Part II. Enhancement of the oxidation of methanol on platinum by ruthenium ad-atoms, *J. Electroanal. Chem.,* **60**, pp. 267–273.

West, J. L. and Halas, N.J. (2000). Applications of nanotechnology to biotechnology: Commentary, *Curr. Opin. Biotechnol.,* **11**, pp. 215–217.

Yasui, K., Tuziuti, T., Iida, Y. and Mitome, H. (2003). Theoretical study of the ambient- pressure dependence of sonochemical reactions, *J. Chem. Phys.,* **119**, pp. 346–357.

Yeung, S. A., Hobson, R. Biggs, S. and Grieser, F. (1993). Formation of gold sols using ultrasound, *J. Chem. Soc. Chem. Commun.,* pp. 378–379.

Yonezawa, T. and Toshima, N. (1995). Mechanistic consideration of formation of polymer- protected nanoscopic bimetallic clusters, *J. Chem. Soc., Faraday Trans.,* **91**, pp. 4111–4119.

Yu, Y. Y., Chang, S. S., Lee, C. L. and Wang, C. R. C. (1997). Gold nanorods: electrochemical synthesis and optical properties, *J. Phys. Chem. B,* **101**, pp. 6661–6664.

Zhang, Y., Peng, H., Huang, W., Zhou, Y. and Yan, D. (2008). Facile preparation and characterization of highly antimicrobial colloid Ag or Au nanoparticles, *J. Colloid Interface Sci.,* **325**, pp. 371–376.

Zhang, J., Du, J., Han, B., Liu, Z., Jiang, T. and Zhang, Z. (2006). Sonochemical formation of single-crystalline gold nanobelts, *Angew. Chem. Int. Ed.,* **45**, pp. 1116–1119.

Zhong, X., Yuan, R., Chai, Y., Liu, Y., Dai, J. and Tang, D. (2005). Glucose biosensor based on self-assembled gold nanoparticles and double-layer 2d-network (3-mercaptopropyl)- trimethoxysilane polymer onto gold substrate, *Sens. Actuators B,* **104**, pp. 191–198.

Zhu, M. Q., Wang, L. Q., Exarhos, G. J. and Li, A. D. Q. (2004). Thermosensitive gold nanoparticles, *J. Am. Chem. Soc.,* **126**, pp. 2656–2657.

Chapter 4

Ultrasound-Assisted Synthesis of Metal Oxide Nanomaterials

Chuanhao Li, Donald Ka Long Chan, and Jimmy C. Yu
Department of Chemistry and Institute of Environment, Energy and Sustainability, The Chinese University of Hong Kong, Shatin, New Territories, Hong Kong, China
jimyu@cuhk.edu.hk

Great progress has been achieved in ultrasound-assisted synthesis of a variety of nanomaterials. This chapter reviews the recent advances in nanostructured metal oxides. Two different ultrasound-assisted synthetic methods, namely sonochemistry and ultrasonic spray pyrolysis, are discussed. The principle and mechanisms are briefly introduced and detailed synthesis, characterization, and applications are described. Perspectives for both methods are also presented.

4.1 Introduction

Nanostructured metal oxides with unique optical, electrical, and magnetic properties are promising candidates for diverse applications in electronics, catalysis, sensors, environmental remediation,

Cavitation: A Novel Energy-Efficient Technique for the Generation of Nanomaterials
Edited by Sivakumar Manickam and Muthupandian Ashokkumar

ISBN 978-981-4411-54-7 (Hardcover), 978-981-4411-55-4 (eBook)
www.panstanford.com

magnetic storage, and biotechnology (Sue et al., 2006). These properties are dependent not only on the composition but also on the particle size, morphology, surface area, and crystallinity (Zhou et al., 2006). Various approaches have been developed for the controlled synthesis of metal oxide nanoparticles. These include sol–gel, mechanochemical, chemical vapor deposition, electrochemical, hydrothermal/solvothermal, solid-state, coprecipitation, microwave-assisted, microemulsion, and combustion methods (Sue et al., 2006). However, most of these methods are laborious, highly energy-consuming, and not environmental-friendly due to the involvement of high temperature calcination, long reaction time, and the requirement of toxic-stabilizing agents. Thus, facile, time-saving, green, and effective approaches are desirable.

The utilization of high-intensity ultrasound in the synthesis of nanomaterials including metal, metal oxide, metal chalcogenides, and carbides has been widely studied for many years. It has been considered as an alternative to satisfy the above-mentioned requirements (Bang and Suslick, 2010; Gedanken, 2004). Also, the ultrasound-assisted method could generate novel materials with unique properties. The primary physical phenomena associated with ultrasound that are responsible for materials synthesis are cavitation and nebulization, corresponding to the sonochemical and ultrasonic spray pyrolysis (USP)-assisted methods. This chapter focuses on the ultrasound-assisted synthesis of metal oxide nanoparticles with different functions.

4.2 Sonochemical Synthesis of Metal Oxide Nanoparticles

The high-intensity ultrasonic irradiation provides a unique interaction between energy and matter. The direct utilization of ultrasound in chemistry, namely sonochemistry, has been widely studied for many years. In the very early stage, efforts were focused on accelerating the organic reactions (Weissler et al., 1965). Since 1980s, the advent of inexpensive and reliable generators of high-intensity ultrasound makes it a routine method for the sonochemical synthesis of nanosized inorganic materials, especially metal oxides (Suslick, 1989).

Ultrasound wave is composed of compression and expansion cycles. The wavelength ranging from 10 to 0.1 cm is much longer than molecular dimensions (Suslick, 1990). Thus, the sonochemistry process does not involve a direct interaction between the sound field and molecular species. The chemical effects of ultrasound come from the acoustic cavitation: the formation, growth, and implosive collapse of bubbles within a liquid (Bang and Suslick, 2010). The formation of cavities in liquids is a nucleated process. Generally before applying ultrasonic irradiation, a gas should be introduced into the liquid to act as the nuclei for cavitation. When high-intensity ultrasound is applied to the liquids, the alternating expansive waves create bubbles and accumulate ultrasonic energy. Although the compressive acoustic waves undergo violent collapse, the concentrated energy is released within a short time (with a heating and cooling rate of $>10^{10}$ K/s). The collapsing bubbles generate localized hot spots. These localized hot spots possess a high temperature of nearly 5000°C and a high pressure of nearly 1000 bar. These extreme conditions created during acoustic cavitation are responsible for the sonochemical reactions (Ray et al., 2008b; Wang et al., 2002; Zolfaghari et al., 2007).

4.2.1 Methodology

A typical sonochemical synthesis procedure has been shown in Fig. 4.1. An inorganic precursor is dissolved in a solvent to form a gel. The concentration of precursors and pH of the reaction mixture can be adjusted to produce different products. Surfactants may also be added into the system to assist the formation of porous structure. Then, the mixed solution or gel is sonicated for a certain period of time. The as-prepared materials are collected, washed, and dried.

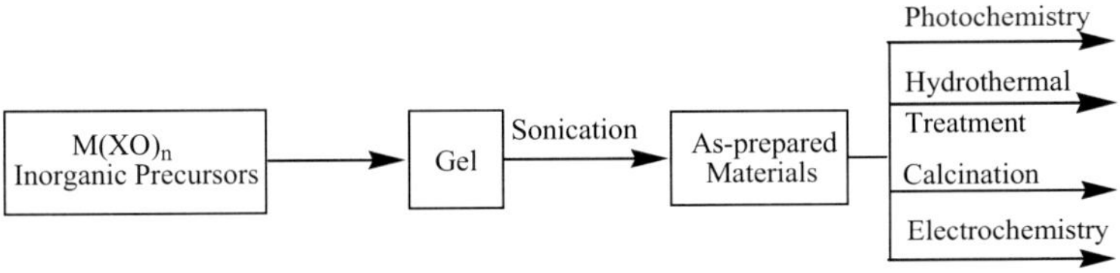

Figure 4.1 Typical procedure for the sonochemical synthesis.

The sonochemically as-prepared materials can be further treated by some conventional synthetic routes, including calcination, hydrothermal, photochemical, and electrochemical treatment. Final products with enhanced crystallinity, desirable composition, and interesting morphologies can be obtained.

4.2.2 Mechanism

4.2.2.1 Reaction regions

Sonochemical reactions may take place in three regions: (i) the inside of the collapsing bubbles where very high temperatures ($T > 5000°C$) are produced, causing a series of gas phase reactions; (ii) the interface between the bubble and liquid where the temperature is lower than that of inside of bubbles (about 1900°C), but still high enough to trigger the reactions; and (iii) the bulk solution at ambient temperature where reactions takes place between solute molecules with hydroxyl radicals or hydrogen atoms escaped from the interfacial region (Dhas and Gedanken, 1997; Ray et al., 2008b). For the volatile precursors, the gas phase reactions are predominant, while the non-volatile precursors prefer the reaction in the interface regions.

4.2.2.2 Reaction mechanism

Two different reaction mechanisms have been reported: sonochemical oxidation and sonochemical hydrolysis.

Sonochemical oxidation: Upon ultrasonic irradiation, H• and •OH radicals are generated from the sonolysis of water molecules. The produced H• and •OH radicals quickly recombine and the generated oxidants can initiate the oxidation of metal precursors. The sonochemical synthesis of Fe_3O_4, Co_3O_4, and MnO_2 nanoparticles follows this mechanism. The possible reaction steps for Fe_3O_4 can be summarized as follows (Kumar et al., 2000; Zolfaghari et al., 2007):

$$H_2O))) \rightarrow H^{\cdot} + {}^{\cdot}OH$$

$$H^{\cdot} + H^{\cdot} \rightarrow H_2$$

$${}^{\cdot}OH + {}^{\cdot}OH \rightarrow H_2O_2$$

$$Fe\,(II) + H_2O_2 \rightarrow Fe\,(III) + 2OH^-$$

Sonochemical hydrolysis: A sonochemical hydrolysis mechanism has been proposed to account for the reactions in which metal species are not oxidized. For example, the sonochemical synthesis of CuO and ZnO from Cu^{2+} and Zn^{2+} can be expressed as (Jung et al., 2007; Kumar et al., 2000):

$$Cu^{2+}\,(aq) + H_2O\,(l) \rightarrow CuO\,(s) + 2H^+\,(aq)$$

$$Zn^{2+}\,(aq) + H_2O\,(l) \rightarrow ZnO\,(s) + 2H^+\,(aq)$$

4.2.3 Advantages

4.2.3.1 High yield and fast reaction

As described above, sonochemical techniques could generate high temperature (>5000 K) and pressure (>1000 atm) with very fast cooling rate (>10^{10} K/s). It has been well documented that the local high temperatures could greatly accelerate the reaction rate in the preparation of metal oxides (Bang and Suslick, 2010). Rate enhancements of more than 10-fold are commonly observed. The conditions of high pressure and fast cooling are also favorable for driving many reactions to completion. Thus, yields are often greatly improved. Compared with the conventional methods such as sol–gel, coprecipitation, hydrothermal, and microemulsion, the sonochemical approach can generate large amounts of products within a much shorter period of time.

4.2.3.2 Small size and great monodispersity

During the sonochemical process, the extreme conditions allow the formation of a large amount of product nuclei. Meanwhile, the growth of the nuclei is suppressed. The resulting products exhibit ultrafine size. These extreme conditions could also help to avoid the agglomeration of the newly formed nanoparticles. Thus, the sonochemical method could produce highly dispersed nanoparticles without using any stabilizing agents.

4.2.3.3 Large surface area

Ultrafine nanoparticles produced by sonochemical method exhibit large surface area due to small size effect. The method has also been extended to the synthesis of porous materials (Kumar et al., 2000). The sol–gel reaction time needed to fabricate porous structure is significantly shortened to a few hours under ultrasonic irradiation, while the conventional methods require several days. It has been reported that sonochemical products also show better stability.

4.2.3.4 Pure phase

The other consequence of accelerated reaction is the formation of few by-products. Rapid cavitational cooling rate can be viewed as a quenching process. It could hinder the crystallization of the products and thus pure amorphous nanoparticles are obtained, especially in the case dealing with volatile precursors. Under these extreme conditions, the atomic level mixing of the constituent species is achieved in the amorphous phase so that the crystalline phase can be obtained by annealing at relatively low temperatures (Gedanken, 2004). If the precursor is non-volatile, crystallized particles could be directly obtained even without any calcination treatment.

4.2.3.5 Facile and green

Conventional approaches to prepare nanoparticles or porous materials involve many stabilizing and pore-creating surfactants, which must be removed by high temperature treatment. These methods are tedious and highly energy-consuming methods and they may also generate pollutants from the oxidation/decomposition of the surfactants. Sonochemical technique is a promising alternative. Normally, highly monodispersed nanoparticles could be obtained without any stabilizing agents or additives by using the sonochemical approach. Low temperature calcination (even no calcination) could be used to prepare crystalline particles.

Apart from the above advantages, sonochemical method is an effective tool to insert a nanosized material into another porous material. The mechanism is that the nanoparticles are deposited as a smooth layer on the inner mesoporous walls. The composite materials show better performance in catalysis, magnetic, and optical properties. Sonochemical method has been widely considered as a facile, environmentally friendly, cost-effective, and efficient approach to prepare ultrafine or porous metal oxides and composites.

4.2.4 Synthesis of Metal Oxides with Different Functions

4.2.4.1 Photocatalysts

Semiconductor-mediated photocatalysis is an attractive environmental remediation technology to treat pollutants both in gas and liquid phases because it makes use of the clean, safe, and renewable sunlight to tackle environmental problems (Li et al., 2011). Since 1972 when Fujishima and Honda discovered photocatalytic water splitting on a TiO_2 electrode, photocatalysis using semiconductors (e.g., TiO_2, Cu_2O, Bi_2O_3) has been widely studied (Fujishima and Honda, 1972).

Photocatalysis proceeds under the actions of both light and catalyst. When a semiconductor absorbs photons with energy higher than its band gap energy, electrons in the valence band would be excited to the conduction band, at the same time generating holes (h^+) in the valence band. The charge carriers could migrate to the surface of the semiconductor and then react with other species adsorbed on the surface. Photo-induced electrons are able to react with O_2 and H_2O to give $O_2^{-\cdot}$ and $H_2O^{\cdot}$, while the holes can react with H_2O or OH^- to give $^{\cdot}OH$. These radicals are highly reactive and participate in oxidation and reduction reactions, causing degradation of organic compounds (Hu et al., 2010).

According to the mechanism, the photocatalytic activity of a semiconductor is closely dependent on the surface properties of the semiconductor. The effects of size, morphology, and surface area on the photocatalytic activity have been widely studied (Chen and Mao, 2007; Hoffmann et al., 1995). As sonochemical method could produce ultrafine nanoparticles with large surface, sonochemistry-assisted formation of photocatalysts with these desirable properties has attracted great attention.

4.2.4.1.1 *Nanocrystalline structures*

Photocatalysts usually possess higher photocatalytic activity if the size of the semiconductor is in the nanoscale. It is due to its higher surface area to volume ratio that allows more adsorption of other species and faster charge transfer rates between surfaces. However, too small a particle size may result in increased rate of electron-hole recombination and more surface defects, which in turn lead to reduced photocatalytic activity. Hence, to prepare a photocatalyst with optimal particle size, using a proper synthetic method is critical.

Sonochemistry was applied to prepare TiO_2 nanoparticles and the resultant particle sizes were compared without applying ultrasonic radiation. Results showed that ultrasonic treatment reduced the particle size and enhanced the photocatalytic activity of

TiO_2 nanoparticles, no matter the solvent is pure water or ethanol/water-mixed solutions (Yu et al., 2001).

Zhang et al. (2006) obtained Bi_2O_3 nanocrystals by using a sonochemical method. Owing to the high energy provided by the ultrasound radiation, Bi_2O_3 with a high degree of crystallinity could be formed. The product exhibited better photocatalytic activity in the degradation of methyl orange than microsized Bi_2O_3 under visible light irradiation.

4.2.4.1.2 *Morphologies*

Morphology plays an important role in the physical properties of photocatalysts, as well as their photocatalytic activities. In recent years, there are studies on the morphology control of photocatalysts, aiming to increase their surface area to volume ratio compared with their bulk counterparts.

TiO_2, as the most promising photocatalyst, has been widely studied. In addition to nanoparticles, TiO_2 with different morphologies such as nanorods, nanotubes, and nanowires has been reported in the literature (Chen and Mao, 2007). It has been found that highly ordered TiO_2 nanotube arrays show higher performance than their bulk counterparts in several aspects (Ghicov and Schmuki, 2009). Recently, Wang and coworkers employed ultrasonic treatment to obtain a TiO_2 nanoring/nanotube-combined structure (Wang et al., 2011). TiO_2 nanotubes prepared by anodization were first treated by ultrasonic radiation, giving textured Ti surface with nanoring structure. After that, second anodization was carried out to obtain the combined structure. The combined structure enhanced the charge separation and transportation of photoinduced electrons.

4.2.4.1.3 *Porous structures*

Photocatalysts with porous structures exhibit higher activities owing to their increased surface area to volume ratio. It is noted that photocatalytic reactions occur on the surface of porous photocatalysts with large surface area to volume ratio and thus are able to adsorb more reaction molecules and so the reactions may occur more rapidly.

Meso-porous and macro-porous TiO_2 are synthesized by ultrasound-induced solvothermal method, and under sonication with different ultrasound intensities, TiO_2 nanoparticles with different mesoporosities have been obtained (Liu et al., 2008b). The high energy of the ultrasound can influence the formation of mesoporous structures. At room temperature, no porous structure could be observed in the sample prepared under the intensity of 100 W/cm^2, whereas samples with porous structure were obtained under the intensity of 250 W/cm^2. The porous TiO_2 exhibited higher photocatalytic activities than those having non-porous structures.

We have developed a direct sonochemical approach to prepare mesoporous TiO_2 with a bicrystalline framework (Yu et al., 2002). Under high-intensity ultrasound irradiation applied for 3 h, the products showed higher photocatalytic activities than a commercial TiO_2 photocatalyst, Degussa P25.

Xu et al. (2009) studied the formation mechanism of porous Cu_2O nanospheres by looking at their TEM images at different ultrasonic irradiation times. They suggested that ultrasonic cavitation was essential in the formation of pores in the interior of the nanospheres. They also compared the photocatalytic activities between porous Cu_2O nanospheres with their solid ones. It was found that porous Cu_2O nanospheres perform better in the photodegradation.

4.2.4.2 Electrode materials

Highly efficient electrical energy storage systems could transform renewable energy sources close to theoretical argument in our energy economy (Pumera, 2011). As the electrode materials are basic elements for energy storage device, the development of efficient electrode materials with high energy and power densities has attracted intensive research interest.

SnO is a promising candidate for anodes in lithium batteries because of their large reversible capacity compared with carbonaceous materials. In SnO-based anode, lithium can be inserted electrochemically and reversibly into tin to form Li-Sn compound, corresponding to a theoretical capacity of 790 mAh/g. However, the conventional approaches to prepare SnO are highly energy-consuming. Aurbach et al. (2002) developed the sonication-assisted technique to prepare amorphous SnO at room temperature. At relatively low calcination temperature (200°C), the amorphous structure could be transformed to a nanocrystalline phase. The performance of nanocrystalline SnO is superior over the amorphous phase. The reversible capacity of nanocrystalline SnO reached the theoretical value, corresponding to the formation of $Li_{17}Sn_4$ compound. Majumdar et al. (2008) investigated the effect of ultrasonic power on the phase of SnO formed. Pure nanocrystalline phase was obtained at high ultrasonic power. With the decrease of ultrasonic power, the mixed phase was observed. Kim et al. (2008) reported the ultrasound-assisted synthesis of nano-sized $LiCoO_2$, which is another excellent cathode material for lithium-rechargeable batteries. The synthesis was performed at 80°C without applying any high temperature treatment. According to their proposed formation mechanism, the phase and morphology of the products strongly depended on the sonication time and temperature. This method was also extended for the preparation of $LiMn_2O_4$ and $LiNi_{0.5}Mn_{0.5}O_2$ nanoparticles (Park et al., 2009; Zhang et al., 2007).

The lead-acid battery is a reliable, safe, and affordable power source. PbO has been widely used as a positive electrode in the lead-acid cells. The performance of the lead-acid cell is highly dependent on the discharge capacity and cycle life of the used PbO electrode. Porous and uniform nanostructured PbO (20–40 nm) was produced by applying constant frequency ultrasonic waves and by using polyvinyl pyrrolidone (PVP) as a structure-directing agent (Karami et al., 2008). The synthesized lead oxide, as anode and cathode of lead-acid batteries, showed very excellent discharge capacity (230 mAh/g) and cycle life.

MnO_2 is also a promising candidate for the electrochemical capacitors. Due to its abundance, low cost, and environmental friendliness, MnO_2 has been considered as a replacement for the conventionally used, expensive, and toxic RuO_2 materials. The smaller size and larger surface area facilitate the active surface area for faradic redox reaction and the charge transfer mobility. As shown in Fig. 4.2, a sonochemical technique was applied to prepare MnO_2 nanoparticles with flower-like and nanowire morphologies, which could be obtained by adjusting the pH (Nam et al., 2010). The prepared materials exhibited larger surface area (larger than 250 m^2/g). MnO_2 nanowires showed the maximum specific capacitance of 300 F/g at a scan rate of 5 mV/s and the minimum charge transfer resistance of 0.4 Ω. It is mentioned that the long chains of the nanowires could provide pathways and facilitate diffusion of protons in the electrolyte, thus resulting in an enhanced accessibility of ionic species to the active materials. Zolfaghari et al. (2007) made γ- and layered structured MnO_2 and they found out that layered structure MnO_2 showed better electrochemical performance. The maximum specific capacitance is 344 F/g at a scan rate of 5 mV/s. Mn_2O_3 ultrafine nanoparticles with the size of 10–15 nm have been prepared by a combined sonochemical and solvothermal synthesis process (Nathan et al., 2008).

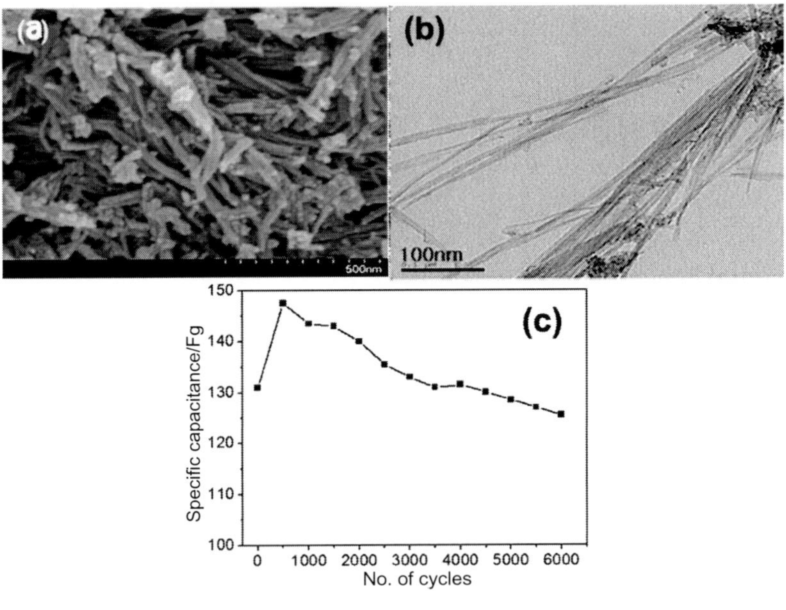

Figure 4.2 SEM (a), TEM (b) images, and cycling data for specific capacitance (c) of MnO_2 nanowires (Nam et al., 2010).

4.2.4.3 Gas sensors

The increasing attention on industrial health and safety and environmental monitoring creates a strong demand for the development of gas sensor devices. One of the promising gas sensors is the metal oxide semiconductor material. Now metal oxide semiconductor gas sensors have been widely used to measure and monitor trace amounts of environmentally important gases (such as CO, CO_2, alcohol, NO_2, SO_2, and so on) because of their high sensitivity, quick response time, excellent selectivity, and low cost (Fine et al., 2010; Mishra et al., 2004).

The gas sensing mechanism relies on several gas/semiconductor surface interactions, including adsorption-desorption process, reduction/oxidation reactions, and complex surface chemical reactions among the different adsorbed chemical species. These reversible interactions cause significant change in the conductivity, which can be readily detected (Fine et al., 2010; Mishra et al., 2004). Performance of the semiconductor oxide gas sensors is highly dependent on the surface area and particle size. It has been widely reported that the large surface area and small particle size are beneficial for the sensitivity. As an efficient approach to prepare ultrafine nanomaterials with larger surface area, sonochemical method has been used to synthesize metal oxide gas sensors.

The most widely used metal oxide sensor material is SnO_2. SnO_2 quantum dots synthesized by the sonochemical method show three times more sensitive response to CO than that of the conventionally synthesized product by the sol–gel method (Sedghi et al., 2010). The dramatic increase in the sensitivity observed is attributed to the grain model effect. The sensor also shows high selectivity to CO in the presence of methane below 375°C. Zhu et al. (2010) have used a biological template (cotton) to fabricate SnO_2 nanotubular materials by a sonochemical method. The resulting SnO_2 nanotubular materials show good selectivity and response to acetone. The recovery time to 20 ppm of acetone gas is only 9 s. They ascribed the high response and reversibility to nanotubular morphology with the larger surface area (Zhu et al., 2009). They have also prepared SnO_2 replica by using butterfly wings as templates. The sensitivity of SnO_2 replica to ethanol reaches as low as 50 ppm at 300°C, accompanied by a rapid response and recovery (around 8–15 s).

One-dimensional ZnO nanostructure is another excellent material for gas sensor. Oh et al. (2009) developed a simple sonochemical route for the synthesis of vertically aligned ZnO nanorods on various substrates, including Zn sheets, Si wafers, transparent glass, and flexible polymeric materials. As shown in Fig. 4.3, the prepared ZnO nanorod was highly sensitive to NO_2 gas. The detection limit was as low as 10 ppb and its response and recovery time were short. Chen et al. (2006) have obtained flower-like ZnO nanorods with diameter less than 15 nm by a sonochemical method. The prepared ZnO nanorods exhibited excellent ethanol-sensing properties. At the working temperature of 300°C, the sensitivity to ethanol vapor was 176.8–100 ppm. They also reported flower-like ZnO nanotubes as a highly sensitive sensor to ethanol (Chen et al., 2008).

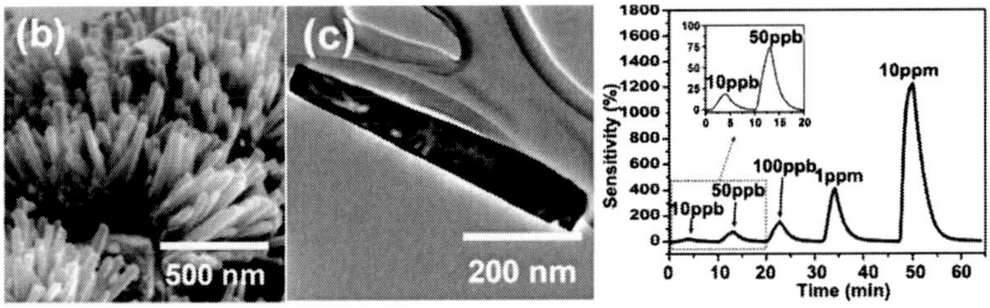

Figure 4.3 SEM (a), TEM (b) images, and isothermal response curve of ZnO nanorod arrays sensor toward NO_2 (Oh et al., 2009).

NiO nanoparticles exhibit good sensing properties to a wide range of chemical compounds. In order to further improve the performance, NiO nanoparticles have been prepared by the sonochemical hydrolysis of nickel acetate (Aslani et al., 2011). Compared with the conventional approaches, no calcination and additives are required in the sonochemical method. At low working temperature, the prepared NiO nanoparticles show a good sensitivity to nitrogen oxide (2ppm), while at temperature higher than 350°C, the materials are more sensitive to carbon monoxide.

Nanocrystalline Fe_2O_3 nanoparticles synthesized at 70°C using a sonication-assisted precipitation technique have also been studied as a useful gas-sensing material (Ray et al., 2008a). The prepared Fe_2O_3 nanoparticles show maximum response to 1000 ppm *n*-butane at about 300°C, and the response and recovery time were less than 12 s and 120 s, respectively.

In another study, Chang et al. (2011) synthesized orthorhombic $WO_3{\cdot}H_2O$ nanosheets in high yield by using a sonochemical method. They proposed that the sheet-like morphology is based on the acoustic cavitation effect as well as the crystalline structure of orthorhombic tungsten trioxide monohydrate. The tungsten trioxide monohydrate nanosheets exhibited ideal room temperature gas-sensing performances to various flammable organic vapors and harmful gases.

4.2.4.4 Magnetic materials

Metal oxide magnetic nanoparticles have attracted great interest because of their potential applications in magnetic liquids, magnetic refrigeration, and biotechnology. The magnetic property significantly depends on the particle size, surface area, and crystal phase. The conventional approaches to prepare magnetic nanoparticles require high annealing temperature. It is very difficult to control the particle size or to yield products of pure phases and monodispersity. Sonochemical methods offer some hope for addressing these difficulties.

Fe-based oxides are attractive magnetic materials, including amorphous, α-Fe_2O_3 and γ-Fe_2O_3 and Fe_3O_4. Amorphous Fe_2O_3 is a good candidate for gas sorption, electrode material, and suitable precursor for α-Fe_2O_3 and γ-Fe_2O_3 magnetic nanoparticles. The unique properties involving rapid cooling rates, atomic level mixing, and the nonequilibrium nature of the reactions render sonochemistry the most suitable method for the preparation of amorphous phases. Cao et al. (1997) reported the sonochemical synthesis of amorphous Fe_2O_3 nanoparticles by using $Fe(CO)_5$ as precursors. They found that the particle size was adjustable by changing the $Fe(CO)_5$ concentration. Fe_2O_3 nanoparticles were found to possess superparamagnetic character. Amorphous Fe_2O_3 nanoparticles were synthesized by the sonolysis of other iron precursors including $Fe(acac)_3$ and Iron(III) ethoxide (Pinkas et al., 2008).

As shown in Fig. 4.4, crystalline hollow α-Fe_2O_3 was fabricated by the ultrasonic decomposition of $Fe(CO)_5$ (Bang and Suslick, 2007). Carbon nanoparticles were used as templates. The proposed mechanism for the formation of the hollow α-Fe_2O_3 was that amorphous Fe_2O_3 produced from the sonochemical decomposition of $Fe(CO)_5$ forms shells around the pre-existing carbon nanoparticles. Upon exposure to oxygen and high temperature, the carbon particles were ignited. The released heat crystallized the iron oxide shells, converting them to α-Fe_2O_3 with hollow cores. The produced hollow α-Fe_2O_3 nanoparticles showed good ferromagnetism down to 5 K.

γ-Fe_2O_3 is widely used to construct magneto-optical devices because of its small coercivity. In the conventional approach to prepare γ-Fe_2O_3, the pH value, method of mixing, temperature, the nature, and concentration of the anions should be carefully controlled. Lei et al. (2007) used a facile sonochemical method to prepare colloidal γ-Fe_2O_3 nanoparticles at room temperature in the absence of any additional nucleation and surfactants. Post-hydrothermal treatment further improved the crystallinity of γ-Fe_2O_3. In another study, sonochemical method was used as an efficient and facile route for the quantitative coating of γ-Fe_2O_3 nanoparticles with octadecyltrihydrosilane (Shafi et al., 2002). Compared with the uncoated ones, the coated nanoparticles showed enhanced magnetization because of improved crystallinity.

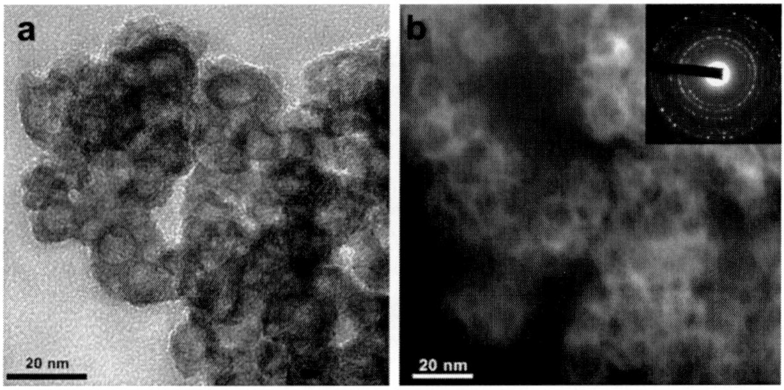

Figure 4.4 TEM images of hollow α-Fe_2O_3 nanoparticles (Bang and Suslick, 2007).

As the saturation of magnetization of Fe_3O_4 is larger than Fe_2O_3, much effort has been focused on the preparation of Fe_3O_4. Abu Mukh-Qasem and Gedanken (2005) reported the synthesis of Fe_3O_4 nanoparticles by sonolysing an aqueous solution of $Fe(CO)_5$ in the presence of sodium dodecyl sulfate. Wang et al. (2010) applied sonochemical technology to assist the synthesis of Fe_3O_4 nanoparticles by using a reverse coprecipitation method. Fe_3O_4 nanorods have been prepared by the sonochemical oxidation of the aqueous solution of iron (II) acetate in the presence of β-cyclodextrin as a template (Kumar et al., 2001).

The sonochemical method has also been widely used for the preparation of non-Fe based magnetic materials, including NiO, CoO, Co_3O_4, and Mn_3O_4 (Lei et al., 2007; Park et al., 2009; Zhang et al., 2007). The prepared magnetic nanoparticles have smaller particle size and larger surface area. Apparently, the sonochemical process facilitated the experimental procedure and enhanced the magnetic properties.

4.2.4.5 Other functional materials

Sonochemical synthesis followed by supercritical drying has been used to prepare MgO nanoparticles. The resulting nanoparticles exhibit very large surface area with the largest one as high as 1236 m^2/g. The fundamental effect of ultrasound on the specific surface of the MgO precipitates has been observed (Stengl et al., 2003). HgO nanocrystallites of different shapes have been prepared by the sonochemical method (Askarinejad and Morsali, 2009).

Sonochemical method is a powerful tool to insert an active nanoparticle to another porous material to prepare composites. Gedanken et al. (2000) reported the fabrication of a mesoporous iron oxide-titanium oxide composite. Other composites such as Fe_3O_4/SiO_2, Al_2O_3/ZnO, TiO_2/CdS, Eu_2O_3/TiO_2, Eu_2O_3/ZrO_2, and ZnO/CdS have also been reported (Bhattacharyya and Gedanken, 2008; Gao et al., 2005; Morel et al., 2008; Perkas et al., 2003; Pol et al., 2002; Wang et al., 2009).

4.2.5 Ultrasonic Spray Pyrolysis Synthesis of Metal Oxide Nanoparticles

USP is an interesting application of ultrasound in the synthesis of inorganic oxide materials (Patil, 1999). Unlike sonochemistry, the ultrasound employed in USP is not to introduce a series of chemical reactions, but to create the phase isolation of one microdroplet reactor from the other. Figure 4.5 shows a typical synthesis process (Bang and Suslick, 2010). The precursor solution is vaporized to form many droplets by ultrasonic nebulization. The size of the droplets can be controlled by changing the reaction parameters. Each droplet may be considered as an individual microreactor (submicrometer diameter). The aerosol is driven by a carrier gas to pass two furnaces. A low temperature (150–300°C) furnace is to remove the solvent and decompose the precursors. Then, the second high temperature (500–1000°C) is to oxidize the precursor and densify the solid products.

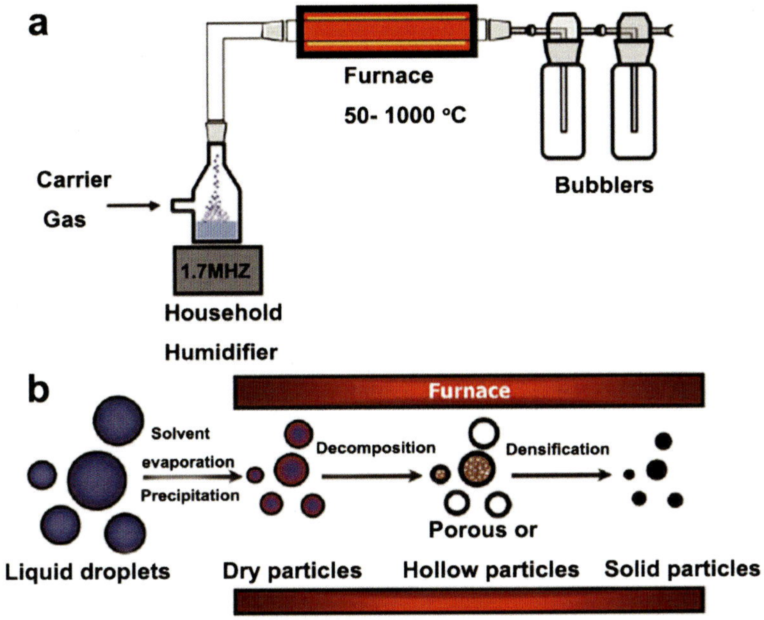

Figure 4.5 Schematic illustration of (a) typical USP apparatus and (b) a simplified USP process (Bang and Suslick, 2010).

USP exhibits great advantages to control the size and monodispersity of products, to prepare porous materials with large surface, and to obtain high product purity. USP also shows some unique advantages:

4.2.5.1 Easy thin film fabrication

Thin films are attractive materials in various applications including electrodes, gas sensors, optical devices, and photocatalysts. The conventional methods to fabricate films involve at least two steps: the synthesis of solid material and the fabrication of film. Special equipment is required to carry out processes, such as spinning coating and dip coating. USP is considered as a simple method to fabricate films. As shown in Fig. 4.5, if the furnace is coupled with a spray nozzle, the product could be easily sprayed onto a substrate to form a film. Thickness of the film can be controlled by changing the spray time. The products obtained are crystalline and no further calcination process is needed.

4.2.5.2 Potential for mass production

The equipment involved in USP is simple and the yield can be as high as tons every day. Thus, this method shows great potential for mass production in the industries.

4.2.6 Comparison between Sonochemistry and USP

Table 4.1 lists the major differences between sonochemistry and USP (Bang and Suslick, 2010).

Table 4.1 Comparison of the sonochemistry and ultrasonic spray pyrolysis (Bang and Suslick, 2010).

	Sonochemical method	**Ultrasonic spray pyrolysis**
Reaction site	Gas bubbles	Liquid microdroplets
Conditions	-5000 K, 1000 bar	500–1300 K, 1 bar
Reactants	Volatiles (primarily)	Nonvolatile solutes
Cooling rates	$>10^{10}$ K/s	10^4 K/s
Templating	Easy	Easy
Heating zone	Single extreme hot zone	Multiple hot zones possible
Scalability	Scalable: kg/day	Easily scalable: ton/day
Particle size control	Nano-and submicron-sized particles	Typically submicron-size, but nanoparticles possible
Composition control	Easy	Easy
Anisotropic shapes	Yes	No
Hollow structures	Yes	Yes
Core/shell structures	Yes	Yes

4.2.7 Synthesis of Metal Oxides with Specific Functions

4.2.7.1 Electrode materials

The family of lithium manganese transition metal oxides, LiM_xMn_yO (M=Fe, Cr, Co, Ni, and so on), are attractive cathode materials for lithium-ion batteries. These materials are conventionally synthesized by solid state reactions and sol–gel methods, which involve a long calcination time. The products are generally agglomerated and non-stoichiometric powders. USP method has been adopted to solve the undesirable problems.

Taniguchi's group did a lot of exploration in this field. As the simplest member of the above-mentioned family, $LiMn_2O_4$ was prepared by using USP method (Matsuda and Taniguchi, 2004; Taniguchi et al., 2002a). The effects of crystallite size, specific surface area, and morphology on the electrochemical properties were investigated. $LiMn_2O_4$ suffers from a serious capacity fading problem during charge/discharge cycle. The decomposition of the electrolyte at high-voltage region and the dissolution of Mn ions into the electrolyte are responsible for the problem. To improve the durability of $LiMn_2O_4$, many researchers have investigated the use of transition metals to produce partially substituted lithium manganese oxides. A comparison study revealed that the electrochemical performance of the substituted materials, $LiM_{1/6}Mn_{11/6}O_4$ (M = Co, Al, Ni), was as good as $LiMn_2O_4$, while the good cycle life was improved. The most stable material, $LiCo_{1/6}Mn_{11/6}O_4$, showed only 2% loss for the capacity after 30 cycles compared with 15% loss for $LiMn_2O_4$ (Taniguchi et al., 2002b). To further improve the durability, Cr-substituted lithium manganese oxide ($LiCr_xMn_{2-x}O_4$) nanoparticles were successfully synthesized by USP (Taniguchi, 2005). It was found that the capacity retentions improved as x increased in $LiCr_xMn_{2-x}O_4$. The capacity retention was 98% after 100 cycles for the nanostructured $LiCr_{0.2}Mn_{1.8}O_4$ particles.

As an effective tool to obtain multicomponent product, the USP method has been extended to the synthesis of more complex Li-based materials. $Li[Ni_{1/3+x}Co_{1/3}Mn_{1/3-2x}Mo_x]O_2$ cathode materials were developed to investigate the effect of Mo-doping on the electrochemical and durability performance (Park et al., 2005). The amount of Mo was adjusted in the $0 \leq x \leq 0.05$ region. As the content of molybdenum increased, the discharge capacity increased, whereas the cycling life decreased. Considering both electrochemical and durability performance, $Li[Ni_{1/3+0.01}Co_{1/3}Mn_{1/3-0.02}Mo_{0.01}]O_2$ was the best material with the discharge capacity of 175 mA h/g and unchanged capacity retention after 50 cycles.

Other Li-based cathode materials including $LiFePO_4$, $LiFe_{0.9}Mg_{0.1}PO_4$, and anode materials, including Co_3O_4, CuO, and NiO have also been prepared by using USP method (Oh et al., 2007; Teng et al., 2007; Yang et al., 2006).

4.2.7.2 Gas sensors

Nanostructured SnO_2-based gas sensors prepared by USP technique have been widely investigated. Patil et al. (2009) built a simple sono-tek spray nozzle system. After pyrolysis, SnO_2 nanoparticles were directly sprayed on the glass substrate to form SnO_2 thin film. The thickness could be easily controlled by changing the spraying time. Figure 4.6 shows that the produced SnO_2 film exhibited higher selectivity toward H_2 than other gases, including NH_3, CO_2, LPG, and C_2H_5OH. The response and recovery time were 2 s and 12 s, respectively. They optimized their system and prepared SnO_2 nanoparticles with diameter smaller than 20 nm. These ultrafine SnO_2 nanoparticles are also highly sensitive and selective to H_2. Sacrificial templates have also been used to assist the formation of SnO_2 porous materials. Hieda et al. (2008) introduced polymethylmethacrylate (PMMA) microspheres as templates in the USP method. The prepared SnO_2 porous materials show larger surface area, excellent thermal stability, and high sensitivity to H_2.

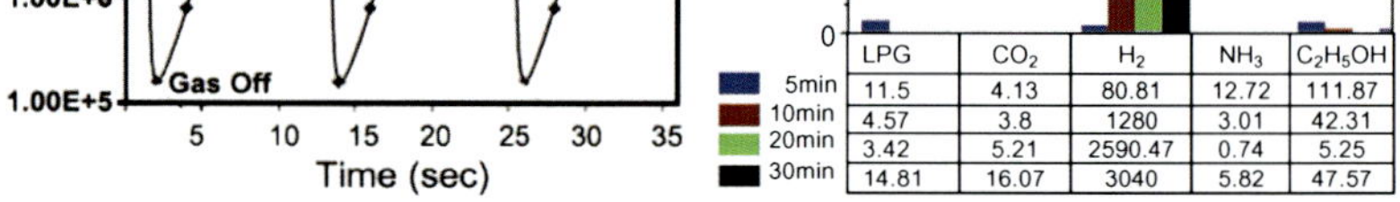

Figure 4.6 (a) Response and recovery of the sensor toward NO_2. (b) Selectivity of sensor to various gases (Patil et al., 2009).

USP deposition technique has also been extended to the preparation of various gas sensor materials, including nanocrystalline ZnO thin films with high sensitivity for LPG (Patil et al., 2010), nanocrystalline Fe_2O_3 films for CH_4 sensing (Wang et al., 2000), HfO_2 coatings as humidity and propane sensors (Avila-García and García-

Hipólito, 2008), and In_2O_3 porous particles for NO_2 sensing (Hyodo et al., 2010).

In general, USP favors two kinds of production: porous hierarchical microspheres composed of many small nanoparticles and films. These two products are beneficial for gas sensing because of the larger surface area.

4.2.7.3 Luminescent films

The advanced luminescent films have attracted considerable research interest owing to their promising applications in display devices, high-density optical storage, electroluminescent display devices, and so on. USP has been proven to be a powerful method to fabricate the films composed of metal oxide semiconductors. As well as a simple and inexpensive method for large-area deposition, USP could improve stoichiometry and reduce intrinsic defects that, in turn, improve luminescence performance.

Among the most widely studied metal oxide systems, ZnO has attracted special attention because of its unique optical and luminescence properties. ZnO films deposited on MgO substrates by using USP method exhibit ultraviolet (UV) emission peak (3.37 eV) from 18 K to 300 K with a reduced value of the peak energy [Choi and Ma, 2006]. To enhance the luminescence performance, selective elements were doped onto ZnO. It was found that the green emissions for Ni doped-ZnO films were remarkably increased (up to 10 times) (Abed et al., 2011). ZnO:Ag films show strong room temperature ultraviolet emission. The optimal atomic ratio of Ag : Zn is found to be 3% (Liu et al., 2008a).

Multi-rare earth element doped metal oxide films are very useful in advanced illumination technologies. However, conventional approaches have difficulty in precisely controlling the stoichiometry of the products. USP-assisted synthesis of $Al_2O_3:Ce^{3+}:Tb^{3+}:Mn^{2+}$ films has been recently reported (Martinez et al., 2010). Through precise control on the elements concentrations, a strong white light emission for the $Al_2O_3:Ce^{3+}(1.3\ at.\%):Tb^{3+}(0.2\ at.\%):Mn^{2+}(0.3\ at.\%)$ film under UV excitation is observed. A blue-yellow photoluminescence is observed in the $HfO_2:Ce^{3+}:Dy^{3+}$ films deposited by USP (Martinez et al., 2011).

Apart from the above photoluminescence system, the electroluminescent films have also been fabricated by USP methods.

In a study by Du et al. (2006), ZnO homojunction light-emitting films were grown on a single-crystal GaAs substrate by USP. The electroluminescence emission associated with defects was observed under forward current injection at room temperature. By using the similar procedure, a multilayered n-ZnMgO/ZnO/p-ZnMgO heterojunction device was fabricated (Bian et al., 2006). A strong visible electroluminescence with an emission peak at 450 nm was observed at room temperature. They proposed that the origin of electroluminescence emission was a radiative recombination through deep-level defects in the ZnO active layer. A fast-switching electrochromic Li^+-doped NiO films is also fabricated by USP method (Tenent et al., 2010). The analysis of potential step response data shows that the obtained lithium-doped nickel oxide film completes 90% of its total coloration change in 29 s, which is much faster than the conventionally obtained film.

4.3 Conclusions and Perspectives

In the sonochemical synthesis of metal oxide nanomaterials, transient localized hot spots are generated from the acoustic cavitation of bubbles in a reaction medium. The high temperature and pressure conditions cause the decomposition of the precursors to form metal oxides. This method is considered as a facile, green, and efficient approach to prepare ultrafine or porous metal oxides and composites. Many sonochemical systems are now commercially available to prepare materials on the laboratory scale. However, large-scale equipment remains uncommon. Further development of more efficient apparatus is needed for the industrial production.

The USP-assisted approach has been used to prepare metal oxide materials, such as electrodes, gas sensors, and luminescent films. Ultrasound provides energy to nebulizer precursor solution to produce micron-sized droplets, which serve as micro-reactors. This method shows unique advantages for the preparation of multicomponent materials and thin films. However, the final products are usually micron-sized spherical particles made up of a large amount of primary nanoparticles. The relatively large size may limit the applications of the products, particularly in biotechnology. Further progress must be made to develop an efficient method to generate nanosized droplets for the size control of final products.

Significant progress is expected to be made on the development of hyphenated methods, such as ultrasound/microwave, ultrasound/photochemistry, and ultrasound/electrochemistry, for materials fabrication. These interesting combinations of technologies may provide extremely efficient and green alternatives to the traditional approaches.

References

Abed, S., Aida, M. S., Bouchouit, K., Arbaoui, A., Iliopoulos, K., and Sahraoui, B. (2011). Non-linear optical and electrical properties of ZnO doped Ni thin films obtained using spray ultrasonic technique. *Opt. Mater.*, **33**, pp. 968–972.

Abu Mukh-Qasem, R., and Gedanken, A. (2005). Sonochemical synthesis of stable hydrosol of Fe_3O_4 nanoparticles. *J. Colloid Interface Sci.*, **284**, pp. 489–494.

Askarinejad, A., and Morsali, A. (2009). Synthesis and characterization of mercury oxide unusual nanostructures by ultrasonic method. *Chem. Eng. J.*, **153**, pp. 183–186.

Aslani, A. A. A., Oroojpour, V., and Fallahi, M. (2011). Sonochemical synthesis, size controlling and gas sensing properties of NiO nanoparticles. *Appl. Surf. Sci.*, **257**, pp. 4056–4061.

Aurbach, D., Nimberger, A., Markovsky, B., Levi, E., Sominski, E., and Gedanken, A. (2002). Nanoparticles of SnO produced by sonochemistry as anode materials for rechargeable lithium batteries. *Chem. Mater.*, **14**, pp. 4155–4163.

Avila-García, A., and García-Hipólito, M. (2008). Characterization of gas sensing HfO_2 coatings synthesized by spray pyrolysis technique. *Sensor Actuat. B-Chem*, **133**, pp. 302–307.

Bang, J. H., and Suslick, K. S. (2007). Sonochemical synthesis of nanosized hollow hematite. *J. Am. Chem. Soc.*, **129**, pp. 2242–2243.

Bang, J. H., and Suslick, K. S. (2010). Applications of ultrasound to the synthesis of nanostructured materials. *Adv. Mater.*, **22**, pp. 1039–1059.

Bhattacharyya, S., and Gedanken, A. (2008). Interplay of porosity in gamma-Al_2O_3-doped ZnO nanocomposites: A comparative study of sonochemical and microwave reaction routes. *J. Phys. Chem. C*, **112**, pp. 13156–13162.

Bian, J. M., Liu, W. F., Liang, H. W., Hu, L. Z., Sun, J. C., Luo, Y. M., and Du, G. T. (2006). Room temperature electroluminescence from the n-ZnMgO/

ZnO/p-ZnMgO heterojunction device grown by ultrasonic spray pyrolysis. *Chem. Phys. Lett.*, **430**, pp. 183–187.

Cao, X., Koltypin, Y., Prozorov, R., and Kataby, G. (1997). Preparation of amorphous Fe_2O_3 powder with different particle sizes. *J. Mater. Chem.*, **7**, pp. 2447–2451.

Chang, X. T., Sun, S. B., and Yin, Y. S. (2011). Green synthesis of tungsten trioxide monohydrate nanosheets as gas sensor. *Mater. Chem. Phys.*, **126**, pp. 717–721.

Chen, X., and Mao, S. S. (2007). Titanium dioxide nanomaterials: Synthesis, properties, modifications, and applications. *Chem. Rev.*, **107**, pp. 2891–2959.

Chen, Y. J., Zhu, C. L., and Xiao, G. (2006). Reduced-temperature ethanol sensing characteristics of flower-like ZnO nanorods synthesized by a sonochemical method. *Nanotechnology*, **17**, pp. 4537–4541.

Chen, Y. J., Zhu, C. L., and Xiao, G. (2008). Ethanol sensing characteristics of ambient temperature sonochemically synthesized ZnO nanotubes. *Sensor Actuat B-Chem.*, **129**, pp. 639–642.

Choi, M. H., and Ma, T. Y. (2006). Influence of substrate temperature on ultraviolet emission of ZnO films prepared by ultrasonic spray pyrolysis. *J. Mater. Sci.*, **41**, pp. 431–435.

Dhas, N. A., and Gedanken, A. (1997). Sonochemical synthesis of molybdenum oxide- and molybdenum carbide-silica nanocomposites. *Chem. Mater.*, **9**, pp. 3144–3154.

Du, G. T., Liu, W. F., Bian, J. M., Hu, L. Z., Liang, H. W., Wang, X. S., Liu, A. M., and Yang, T. P. (2006). Room temperature defect related electroluminescence from ZnO homojunctions grown by ultrasonic spray pyrolysis. *Appl. Phys. Lett.*, **89**, pp. 052113–052113.

Fine, G. F., Cavanagh, L. M., Afonja, A., and Binions, R. (2010). Metal oxide semi-conductor gas sensors in environmental monitoring. *Sensors*, **10**, pp. 5469–5502.

Fujishima, A., and Honda, K. (1972). Electrochemical photocatalysis of water at a semiconductor electrode. *Nature*, **238**, pp. 37–38.

Gao, T., Li, Q. H., and Wang, T. H. (2005). Sonochemical synthesis, optical properties, and electrical properties of core/shell-type ZnO nanorod/CdS nanoparticle composites. *Chem. Mater.*, **17**, pp. 887–892.

Gedanken, A. (2004). Using sonochemistry for the fabrication of nanomaterials. *Ultrason. Sonochem.*, **11**, pp. 47–55.

Gedanken, A., Reisfeld, R., Sominski, E., Palchik, O., Koltypin, Y., Panczer, G., Gaft, M., and Minti, H. (2000). Sonochemical preparation and characterization of europium oxide doped in and coated on ZrO_2 and yttrium-stabilized zirconium (YSZ). *J. Phys. Chem. B*, **104**, pp. 7057–7065.

Ghicov, A., and Schmuki, P. (2009). Self-ordering electrochemistry: a review on growth and functionality of TiO(2) nanotubes and other self-aligned MO(x) structures. *Chem. Commun.*, pp. 2791–2808.

Hieda, K., Hyodo, T., Shimizu, Y., and Egashira, M. (2008). Preparation of porous tin dioxide powder by ultrasonic spray pyrolysis and their application to sensor materials. *Sensor Actuat. B-Chem*, **133**, pp. 144–150.

Hoffmann, M. R., Martin, S. T., Choi, W. Y., and Bahnemann, D. W. (1995). Environmental applications of semiconductor photocatalysis. *Chem. Rev.*, **95**, pp. 69–96.

Hu, X. L., Li, G. S., and Yu, J. C. (2010). Design, fabrication, and modification of nanostructured semiconductor materials for environmental and energy applications. *Langmuir*, **26**, pp. 3031–3039.

Hyodo, T., Inoue, H., Motomura, H., Matsuo, K., Hashishin, T., Tamaki, J., Shimizu, Y., and Egashira, M. (2010). NO2 sensing properties of macroporous In_2O_3-based powders fabricated by utilizing ultrasonic spray pyrolysis employing polymethylmethacrylate microspheres as a template. *Sensor Actuat. B-Chem*, **151**, pp. 265–273.

Jung, S. H., Oh, E., Lee, K. H., Park, W., and Jeong, S. H. (2007). A sonochemical method for fabricating aligned ZnO nanorods. *Adv. Mater.*, **19**, pp. 749–753.

Karami, H., Karimi, M. A., Haghdar, S., Sadeghi, A., Mir-Ghasemi, R., and Mahdi-Khani, S. (2008). Synthesis of lead oxide nanoparticles by sonochemical method and its application as cathode and anode of lead-acid batteries. *Mater. Chem. Phys.*, **108**, pp. 337–344.

Kim, K.-H., and Kim, K.-B. (2008). Ultrasound assisted synthesis of nano-sized lithium cobalt oxide. *Ultrason. Sonochem.*, **15**, pp. 1019–1025.

Kumar, R. V., Diamant, Y., and Gedanken, A. (2000). Sonochemical synthesis and characterization of nanometer-size transition metal oxides from metal acetates. *Chem. Mater.*, **12**, pp. 2301–2305.

Kumar, R. V., Koltypin, Y., Xu, X. N., Yeshurun, Y., Gedanken, A., and Felner, I. (2001). Fabrication of magnetite nanorods by ultrasound irradiation. *J. Appl. Phys.*, **89**, pp. 6324–6328.

Lei, S. J., Tang, K. B., Fang, Z., and Sheng, J. (2007). One-step synthesis of colloidal Mn(3)O(4) and gamma-Fe(2)O(3) nanoparticles at room temperature. *J. Nanopart. Res.*, **9**, pp. 833–840.

Li, C. H., Wang, F., and Yu, J. C. (2011). Semiconductor/biomolecular composites for solar energy applications. *Energy Environ. Sci.*, **4**, pp. 100–113.

Liu, K. P., Yang, B. F., Yan, H. W., Fu, Z. P., Wen, M. W., Chen, Y. J., and Zuo, J. (2008a). Strong room-temperature ultraviolet emission from nanocrystalline ZnO and ZnO: Ag films grown by ultrasonic spray pyrolysis. *Appl. Surf. Sci.*, **255**, pp. 2052–2056.

Liu, Y., Li, Y., Wang, Y. T., Xie, L., Zheng, J., and Li, X. G. (2008b). Sonochemical synthesis and photocatalytic activity of meso- and macro-porous TiO_2 for oxidation of toluene. *J. Hazard. Mater.*, **150**, pp. 153–157.

Majumdar, S., Chakraborty, S., Devi, P. S., and Sen, A. (2008). Room temperature synthesis of nanocrystalline SnO through sonochemical route. *Mater. Lett.*, **62**, pp. 1249–1251.

Martinez, M. R., Alvarez, E., Speghini, A., Falcony, C., and Caldino, U. (2010). White light generation in Al(2)O(3):Ce(3+):Tb(3+):Mn(2+) films deposited by ultrasonic spray pyrolysis. *Thin Solid Films*, **518**, pp. 5724–5730.

Martinez, M. R., Lira, A. C., Speghini, A., Falcony, C., and Caldino, U. (2011). Blue-yellow photoluminescence from Ce(3+) -> Dy(3+) energy transfer in HfO(2):Ce(3+):Dy(3+) films deposited by ultrasonic spray pyrolysis. *J. Alloys Compd.*, **509**, pp. 3160–3165.

Matsuda, K., and Taniguchi, I. (2004). Relationship between the electrochemical and particle properties of $LiMn_2O_4$ prepared by ultrasonic spray pyrolysis. *J. Power Sources*, **132**, pp. 156–160.

Mishra, S., Ghanshyam, C., Ram, N., Bajpai, R. P., and Bedi, R. K. (2004). Detection mechanism of metal oxide gas sensor under UV radiation. *Sensor Actuat B-Chem.*, **97**, pp. 387–390.

Morel, A. L., Nikitenko, S. I., Gionnet, K., Wattiaux, A., Lai-Kee-Him, J., Labrugere, C., Chevalier, B., Deleris, G., Petibois, C., Brisson, A., and Simonoff, M. (2008). Sonochemical approach to the synthesis of Fe_3O_4@SiO_2 core-shell nanoparticles with tunable properties. *ACS Nano*, **2**, pp. 847–856.

Nam, H.-S., Yoon, J.-K., Ko, J. M., and Kim, J.-D. (2010). Electrochemical capacitors of flower-like and nanowire structured MnO_2 by a sonochemical method. *Mater. Chem. Phys.*, **123**, pp. 331–336.

Nathan, T., Cloke, M., and Prabaharan, S. R. S. (2008). Electrode Properties of Mn_2O_3 nanospheres synthesized by combined sonochemical/ solvothermal method for use in electrochemical capacitors. *J. Nanomater.*, **2008**, pp. 1–8.

Oh, E., Choi, H. Y., Jung, S. H., Cho, S., Kim, J. C., Lee, K. H., Kang, S. W., Kim, J., Yun, J. Y., and Jeong, S. H. (2009). High-performance NO(2) gas sensor based on ZnO nanorod grown by ultrasonic irradiation. *Sensor Actuat B-Chem.*, **141**, pp. 239–243.

Oh, S. W., Bang, H. J., Bae, Y. C., and Sun, Y.-K. (2007). Effect of calcination temperature on morphology, crystallinity and electrochemical properties of nano-crystalline metal oxides (Co_3O_4, CuO, and NiO) prepared via ultrasonic spray pyrolysis. *J. Power Sources*, **173**, pp. 502–509.

Park, J. P., Kim, S. K., Park, J.-Y., Hwang, C. H., Choi, M.-h., Kim, J. E., Ok, K. M., Kwak, H.-Y., and Shim, I.-W. (2009). Syntheses of Mn_3O_4 and $LiMn_2O_4$ nanoparticles by a simple sonochemical method. *Mater. Lett.*, **63**, pp. 2201–2204.

Park, S.-H., Oh, S. W., and Sun, Y.-K. (2005). Synthesis and structural characterization of layered Li[Ni1/3+xCo1/3Mn1/3−2xMox]O_2 cathode materials by ultrasonic spray pyrolysis. *J. Power Sources*, **146**, pp. 622–625.

Patil, L. A., Bari, A. R., Shinde, M. D., and Deo, V. (2010). Ultrasonically prepared nanocrystalline ZnO thin films for highly sensitive LPG sensing. *Sensor Actuat. B-Chem*, **149**, pp. 79–86.

Patil, L. A., Shinde, M. D., Bari, A. R., and Deo, V. V. (2009). Highly sensitive and quickly responding ultrasonically sprayed nanostructured SnO_2 thin films for hydrogen gas sensing. *Sensor Actuat. B-Chem*, **143**, pp. 270–277.

Patil, P. S. (1999). Versatility of chemical spray pyrolysis technique. *Mater. Chem. Phys.*, **59**, pp. 185–198.

Perkas, N., Palchik, O., Brukental, I., Nowik, I., Gofer, Y., Koltypin, Y., and Gedanken, A. (2003). A mesoporous iron-titanium oxide composite prepared sonochemically. *J. Phys. Chem. B*, **107**, pp. 8772–8778.

Pinkas, J., Reichlova, V., Zboril, R., Moravec, Z., Bezdicka, P., and Matejkova, J. (2008). Sonochemical synthesis of amorphous nanoscopic iron(III) oxide from Fe(acac)(3). *Ultrason. Sonochem.*, **15**, pp. 257–264.

Pol, V. G., Reisfeld, R., and Gedanken, A. (2002). Sonochemical synthesis and optical properties of europium oxide nanolayer coated on titania. *Chem. Mater.*, **14**, pp. 3920–3924.

Pumera, M. (2011). Graphene-based nanomaterials for energy storage. *Energy Environ. Sci.*, **4**, pp. 668–674.

Ray, I., Chakraborty, S., Chowdhury, A., Majumdar, S., Prakash, A., Pyare, R., and Sen, A. (2008b). Room temperature synthesis of γ-Fe_2O_3 by sonochemical route and its response towards butane. *Sensor Actuat. B-Chem*, **130**, pp. 882–888.

Sedghi, S. M., Mortazavi, Y., and Khodadadi, A. (2010). Low temperature CO and CH(4) dual selective gas sensor using SnO(2) quantum dots prepared by sonochemical method. *Sensor Actuat B-Chem.*, **145**, pp. 7–12.

Shafi, K. V. P. M., Ulman, A., Dyal, A., Yan, X. Z., Yang, N. L., Estournes, C., Fournes, L., Wattiaux, A., White, H., and Rafailovich, M. (2002). Magnetic enhancement of gamma-Fe_2O_3 nanoparticles by sonochemical coating. *Chem. Mater.*, **14**, pp. 1778–1787.

Stengl, V., Bakardjieva, S., Marikova, M., Bezdicka, P., and Subrt, J. (2003). Magnesium oxide nanoparticles prepared by ultrasound enhanced hydrolysis of Mg-alkoxides. *Mater. Lett.*, **57**, pp. 3998–4003.

Sue, K., Suzuki, M., Arai, K., Ohashi, T., Ura, H., Matsui, K., Hakuta, Y., Hayashi, H., Watanabe, M., and Hiaki, T. (2006). Size-controlled synthesis of metal oxide nanoparticles with a flow-through supercritical water method. *Green Chemistry*, **8**, pp. 634–638.

Suslick, K. S. (1989). The chemical effects of ultrasound. *Scientific American*, **260**, pp. 80–86.

Suslick, K. S. (1990). Sonochemistry. *Science*, **247**, pp. 1439–1445.

Taniguchi, I. (2005). Physical and electrochemical properties of spherical nanostructured LiCrxMn2-xO4 particles synthesized by ultrasonic spray pyrolysis. *Ind. Eng. Chem. Res.*, **44**, pp. 6560–6565.

Taniguchi, I., Lim, C. K., Song, D., and Wakihara, M. (2002a). Particle morphology and electrochemical performances of spinel $LiMn_2O_4$ powders synthesized using ultrasonic spray pyrolysis method. *Solid State Ionics*, **146**, pp. 239–247.

Taniguchi, I., Song, D., and Wakihara, M. (2002b). Electrochemical properties of LiM1/6Mn11/6O4 (M = Mn, Co, Al and Ni) as cathode materials for Li-ion batteries prepared by ultrasonic spray pyrolysis method. *J. Power Sources*, **109**, pp. 333–339.

Tenent, R. C., Gillaspie, D. T., Miedaner, A., Parilla, P. A., Curtis, C. J., and Dillon, A. C. (2010). Fast-switching electrochromic Li(+)-Doped NiO films by ultrasonic spray deposition. *J. Electrochem. Soc.*, **157**, pp. H318–H322.

Teng, T., Yang, M., Wu, S., and Chiang, Y. (2007). Electrochemical properties of LiFe0.9Mg0.1PO4/carbon cathode materials prepared by ultrasonic spray pyrolysis. *Solid State Commun.*, **142**, pp. 389–392.

Wang, F., Liu, Y., Dong, W., Shen, M. R., and Kang, Z. H. (2011). Tuning TiO(2 photoelectrochemical properties by nanoring/nanotube combined structure. *J. Phys. Chem. C*, **115**, pp. 14635–14640.

Wang, H., Zhu, J.-J., Zhu, J.-M., Liao, X.-H., Xu, S., Ding, T., and Chen, H.-Y. (2002). Preparation of nanocrystalline ceria particles by sonochemical and microwave assisted heating methods. *Phys. Chem. Chem. Phys.*, **4**, pp. 3794–3799.

Wang, N., Zhu, L., Wang, D., Wang, M., Lin, Z., and Tang, H. (2010). Sono-assisted preparation of highly-efficient peroxidase-like Fe_3O_4 magnetic nanoparticles for catalytic removal of organic pollutants with H_2O_2. *Ultrason. Sonochem.*, **17**, pp. 526–533.

Wang, S. Y., Wang, W., Wang, W. Z., Jiao, Z., Liu, J. H., and Qian, Y. T. (2000). Characterization and gas-sensing properties of nanocrystalline iron(III) oxide films prepared by ultrasonic spray pyrolysis on silicon. *Sensor Actuat B-Chem.*, **69**, pp. 22–27.

Wang, Z. Y., Yu, Z. P., Liu, B., Ye, B., Qian, G. D., and Fan, X. P. (2009). Sonochemical synthesis of core/shell structured CdS/TiO(2) nanocrystals composites. *J. Wuhan Univ. Technol.*, **24**, pp. 698–701.

Weissler, A., Pecht, I., and Anbar, M. (1965). Ultrasound chemical effects on pure organic liquids. *Science*, **150**, pp. 1288–1289.

Xu, L., Jiang, L. P., and Zhu, J. J. (2009). Sonochemical synthesis and photocatalysis of porous Cu(2)O nanospheres with controllable structures. *Nanotechnology*, **20**, pp. 045605–045605.

Yang, M.-R., Teng, T.-H., and Wu, S.-H. (2006). $LiFePO_4$/carbon cathode materials prepared by ultrasonic spray pyrolysis. *J. Power Sources*, **159**, pp. 307–311.

Yu, J. C., Yu, J. G., Ho, W. K., and Zhang, L. Z. (2001). Preparation of highly photocatalytic active nano-sized TiO_2 particles via ultrasonic irradiation. *Chem. Commun.*, pp. 1942–1943.

Yu, J. C., Zhang, L. Z., and Yu, J. G. (2002). Direct sonochemical preparation and characterization of highly active mesoporous TiO_2 with a bicrystalline framework. *Chem. Mater.*, **14**, pp. 4647–4653.

Zhang, B., Chen, G., Xu, P., and Lv, Z. (2007). Effect of ultrasonic irradiation on the structure and electrochemical properties of cathode material $LiNi_{0.5}Mn_{0.5}O_2$ for lithium batteries. *Solid State Ionics*, **178**, pp. 1230–1234.

Zhang, L. S., Wang, W. Z., Yang, J. O., Chen, Z. G., Zhang, W. Q., Zhou, L., and Liu, S. W. (2006). Sonochemical synthesis of nanocrystallite Bi_2O_3 as a visible-light-driven photocatalyst. *Appl. Catal. A-Gen.*, **308**, pp. 105–110.

Zhou, L., Xu, J., Li, X., and Wang, F. (2006). Metal oxide nanoparticles from inorganic sources via a simple and general method. *Mater. Chem. Phys.*, **97**, pp. 137–142.

Zhu, S. M., Zhang, D., Chen, Z. X., Gu, J. J., Li, W. F., Jiang, H. B., and Zhou, G. (2009). A simple and effective approach towards biomimetic replication of photonic structures from butterfly wings. *Nanotechnology*, **20**, pp. 315303–315303.

Zhu, S. M., Zhang, D., Gu, J. J., Xu, J. Q., Dong, J. P., and Li, J. L. (2010). Biotemplate fabrication of SnO(2) nanotubular materials by a sonochemical method for gas sensors. *J. Nanopart. Res.*, **12**, pp. 1389–1400.

Zolfaghari, A., Ataherian, F., Ghaemi, M., and Gholami, A. (2007). Capacitive behavior of nanostructured MnO_2 prepared by sonochemistry method. *Electrochim. Acta*, **52**, pp. 2806–2814.

Chapter 5

Synthesis of Nanomaterials Using Hydrodynamic Cavitation

D. V. Pinjari, P. R. Gogate, K. Prasad, and A. B. Pandit
Chemical Engineering Department, Institute of Chemical Technology, Mumbai, Maharashtra, India
ab.pandit@ictmumbai.edu.in

This chapter aims to provide overview of synthesis of nanomaterials by using hydrodynamic cavitation technique as platform technology. Hydrodynamic cavitation has shown its potential in reducing the millimeter or micrometer-sized particles to nanoparticles. In hydrodynamic cavitation setup, size reduction can be controlled by the parameters such as pump discharge pressure, flow rate of the liquid, design of orifice plate, that is, free flow area for the particles to travel through the constriction, concentration of the solution, and use of additives, which essentially decide the cavity dynamics. Hydrodynamic cavitation technique can also be used for hard, brittle, rubbery, and also highly crystalline materials.

Cavitation: A Novel Energy-Efficient Technique for the Generation of Nanomaterials
Edited by Sivakumar Manickam and Muthupandian Ashokkumar

ISBN 978-981-4411-54-7 (Hardcover), 978-981-4411-55-4 (eBook)
www.panstanford.com

5.1 Introduction

Nanotechnology is a transformational technology that utilizes inherently unique properties of matter in reduced dimensions. Applications of nanotechnology range from material science, medicine, energy, environment, communications, and electronics among others. Some of the applications in the context of health and environment vis-à-vis risk scenario have been now discussed in some details. Nanoengineered platforms possess unprecedented potential for early detection, diagnosis, and site-specific and time-controlled cure of diseases (Vaseashta, 2009). Multifunctionality provides a key advantage to nanoengineered platforms, where targeting, imaging, site and time-specific therapeutics and drug delivery, and numerous other functionalities can be integrated to allow for site-specific and targeted molecular imaging, therapy, and diagnostics probes, demonstrating tremendous potential to enhance the sensitivity, specificity, and signaling capabilities of various biomarkers in human diseases. Furthermore, nano-biotechnology is likely to help identify cell-source to be divided by specific cellular phenotypes, thus improving strategies for in vivo detection and monitoring of the cellular implants, techniques to enhance integration of the transplant within the host tissue, and deliver genes to cells. Nanomaterials have also been used for bactericidal, antibiotics, and drug-resistant strains.

Nanomaterials have been studied in the context of environmental applications to solve problems such as pollution and issues that support sustainability. Nanomaterials have been used as active, anti-reflective, and spectrum-shifting layers to produce high-efficiency solar cells inexpensively. Work has also been reported for the use of nanomaterials to reduce energy load by using lightweight materials for vehicles; materials and geometries that contribute to more effective temperature control; technologies that improve manufacturing process efficiency; materials that increase the efficiency of electrical components and transmission lines; materials that contribute to new and improved fuel cells (FCs)—for hydrogen economy; and biodegradable materials for reduced and benign landfill volume. Nanoscale materials are used for cosmetics like sunscreens, house paints, clothing, and computers. Despite their abundant use and potential, there exists a societal concern that nanomaterials pose health hazards, thus necessitating risk assessment and guidelines for using engineered nanomaterials. The challenge of nanotechnology is to deploy the benefits with minimal risks.

The methods for synthesizing nanoparticles can be divided into two distinct categories, viz physical methods and chemical methods.

5.2 Physical Methods

The physical methods of nanoparticle synthesis, as the name suggests, are methods that use majorly mechanical energy for synthesizing the nanoparticles, although some methods in this category do utilize a combination of mechanical and chemical routes for the synthesis. The various physical nanoparticle synthesis techniques are solid-state technique, vapor phase synthesis, and laser ablation technique.

5.2.1 Solid-State Synthesis

Solid-state synthesis generally involves a heat treatment step (in order to achieve the desired crystal structure), which is followed by media milling. In this method, it is difficult to generally obtain particle sizes less than 100 μm. However, if the raw material used is having particle size around 200 μm, then it is possible to synthesize particles having size 30 nm and below with this method. In conventional high-energy ball milling, the creation and self-organization of dislocations to high-angle grain boundaries within the powder particles during the milling process leads to a reduction in the grain size by a factor of about 10^4.

The solid-state synthesis is essentially mechano-chemical in nature and involves using dry milling in order to initiate chemical reaction in a salt matrix, which is then removed by washing. The particles obtained have fine crystalline structure and ultrafine dispersion (Payne et al., 2006). In addition to refining the grain size, the milling process introduces ultrafine dispersion, which improves the high temperature creep resistance of the materials (Luton et al., 1989).

5.2.2 Vapor Phase Synthesis

Gas condensation as a technique for producing nanoparticles refers to the formation of nanoparticles in the gas phase, that is, condensing atoms and molecules in the vapor phase. There are various subcategories in case of vapor phase synthesis.

Inert gas condensation (IGC) system (Fig. 5.1) consists of (a) a vacuum chamber with a vacuum system, (b) a heating power supply with a temperature control system, (c) an atmospheric control system to obtain the product powder with a desired composition, and (d) a cooling system with a liquid nitrogen piped cold trap. The vacuum chamber includes a tungsten boat for melting the bulk manganese, a vacuum pumping system, a vacuum gauge, and a cold trap for powder condensation.

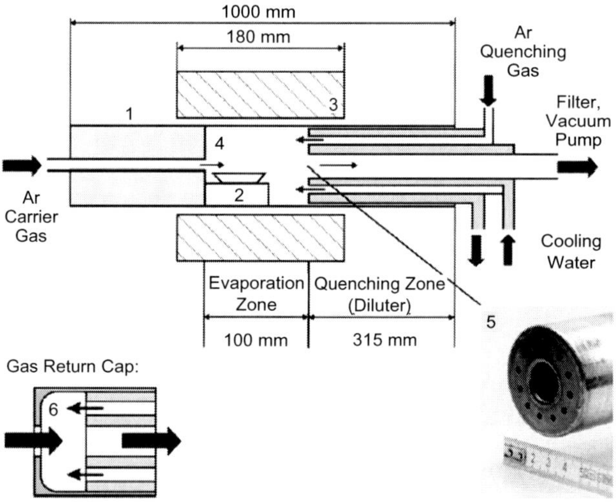

Figure 5.1 Schematic representation of setup based on Inert gas condensation. Reprinted from Wegner et al., Copyright (2002), with permission from Elsevier.

Plasma-based synthesis (PBS) of nanoparticles makes use of ionized gas (plasma) as the source of heating (Fig. 5.2). A piece of metal is mounted on a water-cooled copper hearth and heated by a plasma jet flame. The gas atmosphere is helium (a few hundred torr), mixed with about 15% hydrogen. The evolving smoke flows into a cold cone onto which the ultrafine particles are deposited. They are collected in an ampoule, which is sealed-off without exposing the collected sample to air.

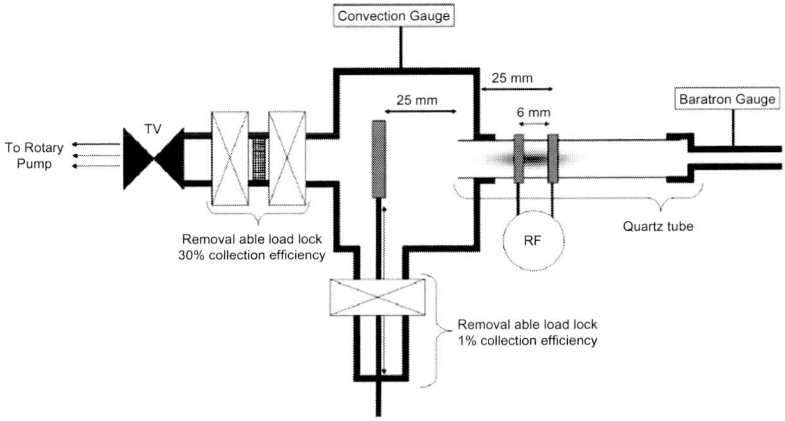

Figure 5.2 Schematic representation of system for plasma based synthesis of nanoparticles. Reprinted from Kendrick et al., Copyright (2014), with permission from Elsevier.

Flame-based synthesis of nanoparticles makes use of hydrocarbon-oxygen flame to pyrolyze the chemical precursor species to produce nanoparticles (Fig. 5.3). Particle sizes less than 50 nm can be obtained by this process (Jang and Kim, 2003).

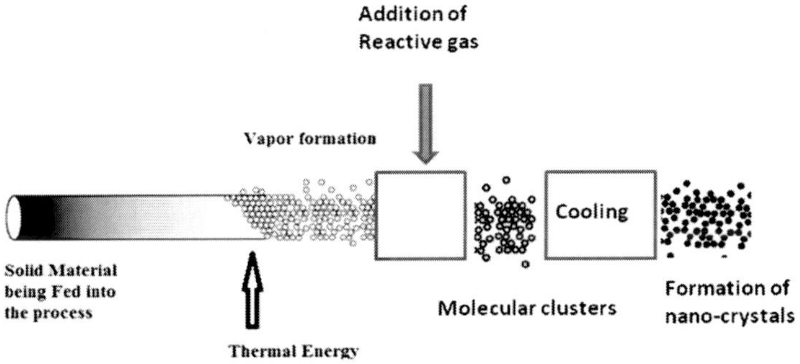

Figure 5.3 Schematic representation of process for flame-based synthesis of nanoparticles.

Spray pyrolysis (Fig. 5.4) combines both methods of gas-phase processing and solution precipitation.

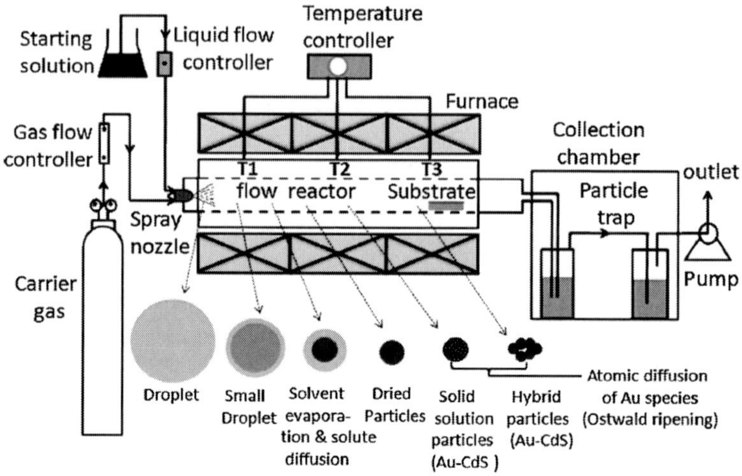

Figure 5.4 Schematic representation of continuous spray pyrolysis setup. Reprinted from Kumar et al., Copyright (2014), with permission from Elsevier.

Precursor solutions of metal nitrates, metal chlorides, and metal acetates are atomized into fine droplets and sprayed into the thermal zone. Inside the heating zone, the solvent evaporates and reactions occur within each particle to form a product particle. Spherical, dense particles in the 100–1000 nm range can easily be formed in large volume by this method. The principal advantage of the spray pyrolysis method is the ability to form multicomponent nanoparticles, as solutions of different metal salts can be mixed and aerosolized into the reaction zone. Also this method is the only method via which encapsulated nanoparticles may be obtained (Nagaraja et al., 2008).

5.2.3 Laser Ablation

This method uses pulsed laser beams to attain the vaporization of a plume of metallic material that is condensed resulting in the formation of nanoparticles. This process can produce only a small amount of nanoparticles at a time but can easily vaporize materials that are not easily vaporized. The infrared CO_2 laser is utilized for this purpose. Pulsed laser shortens reaction time and allows preparation of even smaller particles.

5.3 Chemical Methods

The physical methods of nanoparticle synthesis have a few inherent disadvantages, which limits their usage. They cannot prepare nanoparticles having controlled size distribution and also the sizes that may be synthesized are often at the 100 nm and above level, which is not good enough for the newer demands of the nanoparticle industry.

5.3.1 Sol–Gel Processing

The sol–gel process is a versatile solution process for making ceramic and glass materials. The sol–gel process, as the name implies, involves the evolution of inorganic networks through the formation of a colloidal suspension (sol) and gelation of the sol to form a network in a continuous liquid phase (gel). A sol is a dispersion of the solid particles, with the diameter of 1–1000 nm (Hamer et al., 2005) in a liquid in which only the Brownian motions suspend the particles. Sol–gel synthesis has been used widely for synthesizing nanoparticles of TiO_2 (Khali and Zakir, 2001; Choi *et al.*, 2006; Hosseinia et al., 2008), WO_3 (Wang et al., 2009), SiO_2 hybrids (Grasset et al., 2006; Li et al., 2009), and other metallic oxides. The sol–gel process is of interest in preparing these materials due to its mild conditions in terms of low temperature and pressure. This process provides a convenient route to combine inorganic and organic components as a homogeneous hybrid material (Periyat et al., 2008).

In these hybrid materials, organic and inorganic components can be chemically bonded or just physically mixed. A significant feature to enhance compatibility in hybrid materials is the formation of covalent bonding between organic–inorganic polymers and inorganic components (Iangroudi et al., 1997; Hsiue et al., 2000). In addition, the use of organically-modified precursors provides unique opportunities to tailor the physical and chemical properties of the hybrid materials and has the potential to incorporate chemically tailored features.

5.3.2 Precipitation Method

In the precipitation method, an inorganic metal salt (e.g., chloride, nitrate, acetate, or oxychloride) is dissolved in water. Metal cations in water exist in the form of metal hydrate species, such as $Al(H_2O)^{3+}$ and $Fe(H_2O)_6{}^{3+}$. These species are hydrolyzed by adding a base solution, such as NaOH or NH_4OH (Rao and Cheetam, 2006). The hydrolyzed species condense with each other to form either a metal hydroxide or hydrous metal oxide precipitate on increasing the concentration of OH^- ions in the solution. The precipitate is then washed, filtered, and dried. The dried powder is subsequently calcined to obtain the final crystalline metal oxide phase. The major advantage of this process is that it is relatively economical and is used to synthesize a wide range of single and multicomponent oxide nanopowders. One of the major drawbacks of the process, as described above, is the inability to control the size of particles and their subsequent aggregation.

5.3.3 Water-in-Oil Micro-Emulsions (Reverse Micelle) Method

Uniform and size-controlled nanoparticles of metal, semiconductor, and metal oxides can be produced by the water-in-oil (w/o) microemulsion (also called reverse micelle) method. In a w/o microemulsion, nanosized water droplets, stabilized by a surfactant, are dispersed in an oil phase (Fig. 5.5). Nanosized water droplets act as a microreactor, wherein particle formation occurs and helps in controlling the size of nanoparticles. A unique feature of the reverse micelle process is that the particles are generally nanosized and monodispersed. This is because the surfactant molecules that stabilize the water droplets also adsorb on the surface of the nanoparticles, once the particle size approaches that of the water droplet.

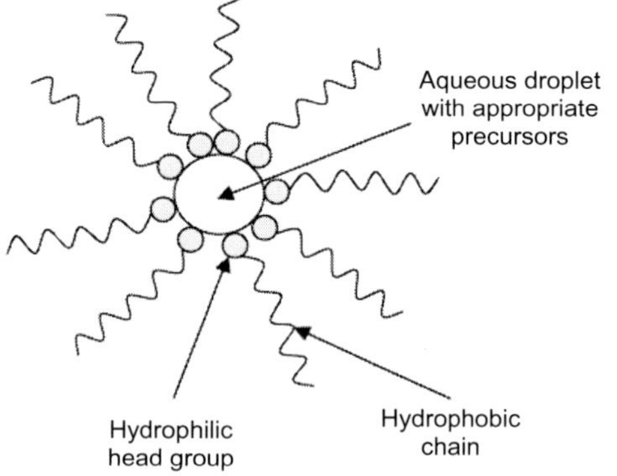

Figure 5.5 w/o Microemulsion (Pinjari, 2012).

A technique, which in some ways combines the aspects of different solution precipitation techniques, is the hydrodynamic cavitation process. Nanocrystalline oxide ceramic particles in the range of 100 nm to a few micrometers have been produced by hydrodynamic cavitation in a microfluidizer. The method of producing oxide nanoparticles by the hydrodynamic cavitation process begins with the coprecipitation of the metal oxide components. The precipitated slurry stream is then drawn into a device in which it is immediately elevated to high pressures within a small volume. The precipitated gel experiences ultrashear forces and cavitational heating. These two aspects lead to the formation of nanophase particles and high-phase purity in complex metal oxides (Moser, 1995).

The aim of the current chapter is to make available a balanced, concise yet comprehensive perspective of manufacturing of nanomaterials with the use of hydrodynamic cavitation technology. In hydrodynamic cavitation, nanoparticles are generated through creation and release of gas and/or bubbles in solution. By rapidly pressurizing in a supercritical drying chamber and exposing to cavitational disturbance and high temperature heating, the solution is mixed. The imploded/oscillating cavitation bubbles are responsible for nucleation, growth, and also terminating the growth of the nanoparticles. Particle size can be controlled by adjusting the pressure and the solution retention time in the cavitation chamber. The extreme transient conditions generated in the vicinity and within the collapsing cavitational bubbles result in the size reduction of the material to the nanoscale. Nanoparticle synthesis can be achieved using various techniques such as sonochemical processing, cavitation processing, and high-energy ball milling. In sonochemistry, an acoustic cavitation process can generate a transient localized hot zone with extremely high temperature gradient and pressure (Suslick et al., 1996). Such sudden changes in temperature and pressure assist the destruction of the sonochemical precursor (e.g., organometallic solution) and result in the formation of nanoparticles. The technique in principle can be used to produce a large volume of material for industrial applications, but no such industrial scale reports are available in the open literature due to the problems associated with scale-up and efficient operation of large-scale sonochemical reactors. Use of the acoustic cavitation for the formation of nanoparticles has been reported (Suslick et al., 1996). The sonication of $Fe(CO)_5$ either as a neat liquid or in a decalin solution resulted in the formation of 10–20 nm size amorphous iron particles. Similar results have been reported (Gedanken, 2004) for the synthesis of nanoparticles of many other inorganic materials using acoustic cavitation. Mesoporous silica with Au nanoparticles inside its pores was synthesized using ultrasonic irradiation at room temperature (Chen et al., 2001). A new method for synthesizing carbon nanomaterials by an electric plasma discharge in an ultrasonic cavitation field in benzene liquid produced amorphous carbon nanoparticles, which are less than 30 nm in size (Sergienko et al., 2006). Gold nanoparticles were synthesized using tetrachloroauric acid and wide range of alcohols (methanol to hexanol) using relatively high-frequency (200 kHz) ultrasound (Mori and Kitamoto, 2005). Mo and Co oxides were precipitated under ultrasonication treatment from $Mo(CO)_6$ and $Co(CO)_3NO$ dissolved in decalin. Introduction of wide pore Al-MCM-41 materials with an average pore diameter of 8.3 nm and a surface area of 840 m^2/g increased the Mo oxide precipitation rate by an order of magnitude. The observed results have been attributed to the ultrasonically induced chemical interaction between metal carbonyl (oxide) and the surface silica atomic layer yielding surface silicates (Landau et al., 2001).

5.4 Basic Aspects of Hydrodynamic Cavitation

5.4.1 Background of Generation of Hydrodynamic Cavitation

During the past few decades, a number of new technologies have been developed, each of which offers a hope of providing chemists with innovative routes of product synthesis and chemical engineers with sophisticated methods of introducing energy to bring about the desired chemical change. Cavitation is one such emerging technology with immense potential in a variety of fields such as physical and chemical processing, textile industry, food industry, industrial wastewater treatment, petroleum industry, and so on. Cavitation is the phenomenon of sequential formation, growth, and collapse of millions of microscopic vapor bubbles (voids) in the liquid. The collapse or implosion of these cavities creates high localized temperatures roughly of about 10000 K and a pressure of about 1000 atm and results into short-lived, localized hot-spot in cold liquid. Thus, cavitation serves as a means of concentrating the diffused energy in the fluid. The extreme transient conditions thus generated in the vicinity and within the collapsing cavitational bubbles open many unique target points to direct the process, and thus it becomes a new and novel way to drive physical, chemical and biological transformations.

Generally, cavitation is classified into four types on the basis of the mode of generation of the cavitating conditions:

(1) Acoustic cavitation: Cavitation is induced by passing high frequency (16 kHz–1 MHz) sound waves, that is, ultrasound.
(2) Hydrodynamic cavitation: Cavitation is produced by pressure variations obtained by changing the geometry of the system.
(3) Optic cavitation: It is produced by passing the photons of high-intensity (laser) light through the liquid.
(4) Particle cavitation: It is produced by bombardments of any other type of elementary particles such as proton in the liquid.

On the basis of the mechanism of generation of cavities, it can be said that hydrodynamic and acoustic cavitation are the results of tensions prevailing in the liquid while optic and particle cavitation is a consequence of local deposition of energy (Lauterborn, 1979) as shown in the schematic representation (Fig. 5.6). Out of the four types mentioned above, only acoustic and hydrodynamic cavitation methods are potentially important for industrial purposes.

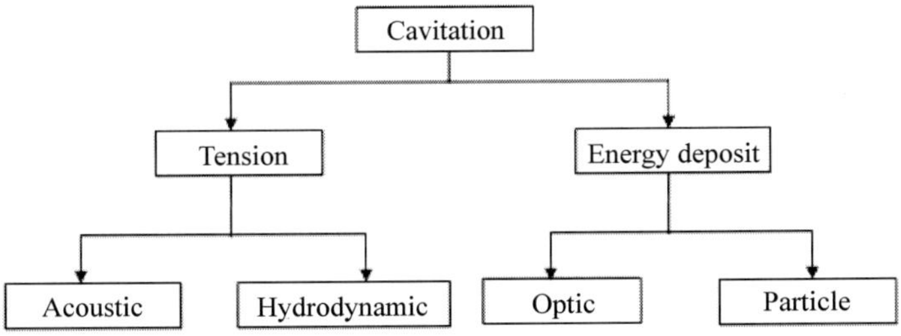

Figure 5.6 Classification scheme of different types of cavitation (Pinjari, 2012).

After giving a brief background about the cavitation phenomena in general, we will now concentrate on hydrodynamic cavitation considering the focus of the present chapter. Hydrodynamic cavitation is generated by passing the liquid through a constriction such as throttled valve, orifice plate, venturi, and so on. When the liquid passes through the constriction, the kinetic energy of the liquid increases at the expense of the pressure, and if the throttling is sufficient, the pressure around the point of vena contracta will fall below the threshold pressure for cavitation (usually the vapor pressure of the liquid at the operating temperature) and cavities are generated. Subsequently, as the liquid jet expands, the pressure recovers and this results in the collapse of the cavities. Figure 5.7(a) shows the mechanism of the generation of hydrodynamic cavitation schematically, whereas Fig. 5.7(b) shows the pressure variations expected downstream of the constriction. A very high intensity of fluid turbulence is generated downstream of the constriction and its intensity depends on the magnitude of the pressure drop generated in the system, which again depends on the geometry of the constriction/pipe and the flow conditions of the liquid. The intensity of the turbulence has a profound effect on the cavitation intensity (Moholkar and Pandit, 1997). A dimensionless number known as cavitation number (C_v) is used to relate the flow conditions with the cavitation intensity as shown in the following equation (5.1):

$$C_v = \frac{P_2 - P_v}{\frac{1}{2}\rho v_o^2} \tag{5.1}$$

where P_2 is the fully recovered downstream pressure, P_v is the vapor pressure of the liquid, and v_o is the liquid velocity at the throat of cavitating device that can be estimated from the knowledge of the flow rate through the main line and flow area available through the cavitating device. The maximum cavitation number at which the inception of cavitation occurs is known as the cavitation inception number C_{vi}. Ideally speaking, the cavitation starts when C_{vi} is 1.0 or lower. However, the inception of the cavitation can occur over a range of C_v from 1.0 to 2.5 (Harrison and Pandit, 1992). This has been attributed to the presence of the dissolved gases in the flowing liquid and the fluctuating pressure component in the fluid due to turbulence. Although cavitation can be achieved even at higher cavitation numbers, for maximum benefit from the reactor, the flow conditions and the geometry should be adjusted in such a way that the cavitation number lies in the range of 0.1–1, as very low operating cavitation numbers can lead to super-cavitation resulting in vapor locking and no cavitational collapse.

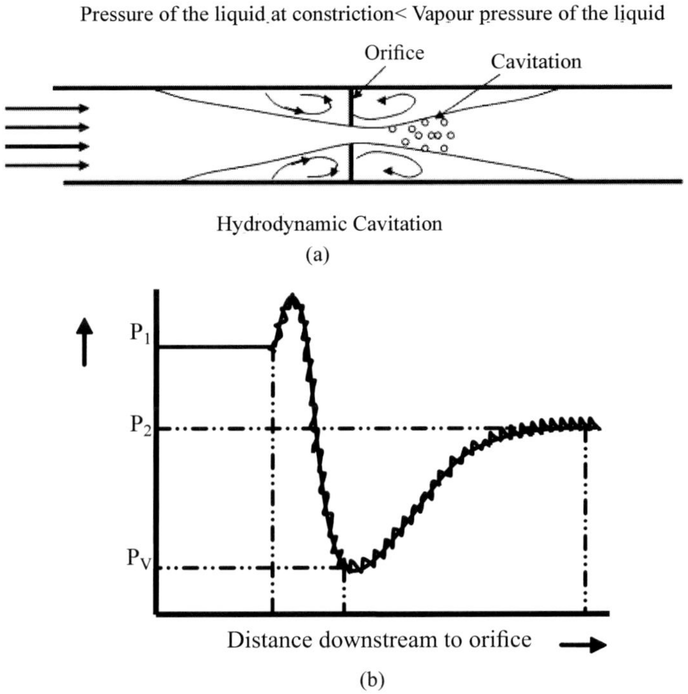

Figure 5.7 (a) Generation of hydrodynamic cavitation, (b) pressure variations in the hydrodynamic cavitation reactor (Pinjari, 2012).

Hydrodynamic cavitation results in number of effects (both chemical and physical) in the system. The chemical effects involve the generation of local hot spots and free radicals that control the intensification observed in the case of chemical processing applications. With reference to the nanomaterial synthesis application, the physical effects of cavitation phenomena are more important and hence have been discussed in more details here. The known major physical effects of hydrodynamic cavitation that are dominantly observed in the heterogeneous systems are as follows:

- Generation of shock waves
- Liquid microjet formation
- Creation of interfacial turbulence

In heterogeneous systems, the asymmetrically collapsing cavity results in the formation of microjet at the interface due to the asymmetric rush of the liquid. In solid–liquid systems, this microjet of high velocity (>100 m/s) disturbs the boundary layer at the solid surface, leading to the breakdown of the liquid film responsible for the resistance to the mass transfer. If the associated energy of the microjet is sufficiently high, it leads to the fragmentation of the solid particle or surface erosion. The oscillating cavity, whose size is typically of the same order of magnitude as the film thickness (based on film theory), results in the generation of interfacial turbulence, when it approaches the interface. In liquid– liquid systems, the action of cavitation (oscillation or cavitation) is by way of liquid micro jets leading to the breakage of droplets generating large interfacial areas and by generation of interfacial turbulence at the liquid–liquid interface. The role of cavitation in the enhancement of the heat transfer can also be explained along the above lines. The heat transfer rate almost doubles in the presence of cavitation in a conduit (Kataoka, 1985; Kimoto and Sumita, 1986).

5.4.2 Mechanism for Nanomaterial Synthesis

The exact mechanism for the formation of nanoparticles during the cavitation phenomenon can be explained on the basis of Hot-Spot theory. The theory postulated that very high temperatures (5000–14000 K) are obtained upon the collapse of the bubbles. As this collapse occurs in few microseconds (final adiabatic phase), very high cooling rates (in excess of 10^{11} K/s), can be obtained locally around the vicinity of the cavity collapse. These high cooling rates hinder the organization and crystallization of the products. For this reason, in all the cases dealing with volatile precursors, where gas phase reactions are predominant, amorphous nanoparticles have been obtained (Gedanken, 2004). Although the explanation for the creation of amorphous products is well understood, the reason for the formation of nanostructured products under the cavitational effects is not yet clear. One possible explanation is that the fast kinetics does not permit the growth of the nuclei, and in each collapsing bubble, a few nucleation centers are formed whose growth is limited by the short cavity collapse time. On the contrary, if the precursor is a non-volatile compound, the reaction would occur in a

200 nm spherical shell surrounding the collapsing bubble (Suslick et al., 1991). The temperature in this liquid shell is lower than that of inside the collapsing bubble, but higher than the temperature of the bulk liquid. The temperature in the ring region has been estimated as being around 1900°C (Suslick et al., 1996). The formation of nanoamorphous or nanocrystalline particles depends on the existing fluid temperature in the spherical fluid shell region where the reaction takes place. In any case, in almost all the sonochemical reactions leading to inorganic products, nanomaterials have been obtained, though there is a variation in the size, shape, structure, and nature of the solid product such as amorphous or crystalline (Gedanken, 2004).

In hydrodynamic cavitation assisted sol–gel operations, nanoparticles are generated through the creation and release of gas bubbles inside the sol–gel solution (Sunstrom et al., 1996). By rapidly pressurizing the solution in a supercritical drying chamber and exposing it to the cavitational disturbance and localized heating, the sol–gel solution is rapidly mixed. The erupting hydrodynamically generated cavitating bubbles are responsible for the nucleation, the growth of the nanoparticles, and also for their quenching to the bulk operating temperature. Particle size can be controlled by adjusting the pressure and the solution retention time in the cavitation chamber. A similar application has been developed on the basis of the use of hydrodynamic cavitation, generated mechanically using a high-pressure fluid system in which the precipitant stream experiences a large pressure drop (from 1400 atm to 1 atm) across an interaction chamber. The work has been effectively explored for the application of this technique for synthesis of several nanostructured metal oxides (Sunstrom et al., 1996).

In the hydrodynamic or acoustic cavitation setup, there are two possible reasons for the observed size reduction. One of the possibilities is that, when a cavity collapse takes place, the shock wave generated travels through the liquid media generating local pressure gradient and fluid shear causing attrition of the solid particles and the reduction in the particle size. Other possibility is that when the cavity collapses asymmetrically on the surface of the solid surface, it produces a high-velocity liquid jet pointing toward the particle surface, which results into an action similar to the liquid jet cutting. It is not known conclusively which of the mechanism could be the dominating mechanism in the synthesis of nanomaterials though some information can be obtained based on the magnitudes of energy dissipation in the hydrodynamic cavitation reactor.

5.4.3 Types of Hydrodynamic Cavitation Reactors

We will now discuss the different types of reactor configurations that have been used for intensification of different chemical and physical processing applications. It is always important to choose an optimum design configuration of the hydrodynamic cavitation reactor so as to maximize the cavitational effects with a cost-effective operation. In this section, we will in general depict the available reactor configurations, whereas some of the specific configurations that have been used for nanomaterial synthesis will be discussed in detail in the Case Studies section later.

5.4.3.1 High-pressure homogenizer

A commonly used device based on hydrodynamic cavitation phenomena is the high-pressure homogenizer, which is, in essence, a high-pressure positive displacement pump with a throttling device that operates according to the principle of high-pressure relief valve. In this, liquid is pumped at very high pressure through a constriction that converts its pressure energy into kinetic energy. At extremely high kinetic energy levels, the pressure reduces below the liquid vapor pressure generating cavitation. A high pressure homogenizer is designed to operate at pressure ranging from 50 atm to as high as 300 atm. Typically, a high-pressure homogenizer reactor consists of a feed tank and two throttling valves, designated as first stage and second stage homogenization, to control the operating pressure in the hydrodynamic cavitation reactor. There is a critical discharge pressure at which cavitation inception occurs, and significant cavitational yields are obtained beyond this discharge pressure. Apart from cavitation, high-pressure homogenization also utilizes the intense shear produced due to fluid flow and impingement of high-velocity liquid on solid wall, for the desired physical or chemical transformations.

5.4.3.2 High-speed homogenizer

Cavitation can also be generated in rotating equipments. When the tip speed of the rotating device (impeller) reaches a critical speed, the local pressure near the periphery of the impeller drops and approaches the vapor pressure of the liquid. This results in the generation of vaporous cavities. Subsequently, as the liquid moves away from the impeller to the boundary of the tank, the liquid pressure recovers at the expense of the velocity head. This causes the cavities that have traveled with the bulk liquid to collapse. Again, similar to the high-pressure homogenizer, there exists a critical rotational speed for the inception of cavitation. It should be noted that the energy consumption in these types of reactors is much higher, and flexibility over the design parameters is lesser than the reactors based on the use of low-pressure cavitation devices such as an orifice or a venturi. These devices operate at rotating speeds of 4000 rpm to as high as 20000 rpm. A high-speed homogenizer usually consists of a stator-rotor assembly preferably made of stainless steel, with flexibility in the design of stators and rotor. A plate with holes attached to the stator can be provided for inserting baffles so as to avoid vortex formation and surface aeration in the bulk liquid. The speed of rotation of the homogenizer can be varied by changing the applied voltage.

5.4.3.3 Low-pressure hydrodynamic cavitation reactor

In this type of reactor, the flow through the main line passes through a constriction where the local velocities suddenly increase due to a reduction in the flow area resulting in lower pressures that may even go below the vapor pressure of the liquid medium generating the cavities. The constriction can be a venturi, a single-hole orifice, or multiple holes on an orifice plate. Multiple hole orifice plates having different combinations of number, diameter, and shape of holes can be constructed. Such an arrangement helps to achieve different intensities of cavitation, and also the number of cavitational events generated in the reactor can be manipulated to control the intensity. Thus, these reactors present tremendous flexibility in terms of the operating (control of inlet pressure, inlet flow rate, temperature) and geometric conditions (different arrangements of holes on the orifice plates). Depending on the type of application and requirements, geometry and operating conditions can be selected in the hydrodynamic cavitation reactor to maximize the cavitational effects at minimum possible energy consumption. A new conceptual design of a novel hydrodynamic cavitation reactor that uses a converging–diverging nozzle for creating pressure variation in the flow necessary for driving bubble motion instead of the orifice plates as discussed earlier has also been reported (Sampathkumar and Moholkar, 2007). The cavitation bubbles or nuclei are introduced in the water flow externally, upstream of the nozzle, using a sparger. Different gases can be used for the introduction of the bubbles. Also, the size of the gas distributor (usually a glass frit), flow rate of gas, and the pressure of gas in the reservoir (or source) from which gas is withdrawn can be suitably controlled to control the initial size of the cavitational nuclei and the gaseous composition of the cavity, which significantly affects the resultant cavitational intensity. The aim should be to generate the smallest size of nuclei as possible to maximize the intensity and hence the net cavitational effects.

From the above discussion about various hydrodynamic cavitation reactors, it can be easily concluded that the orifice plate setup offers maximum flexibility and can also be operated at relatively larger scales of operation. It should be also noted that the scale-up of such reactors is relatively easier, as the efficiency of the pump increases with an increase in size (flow rate and discharge rate), which will necessarily result into higher energy efficiencies. After looking at the basic aspects of the hydrodynamic cavitation reactors and their different configurations, we now look into different case studies in which hydrodynamic cavitation reactors have been effectively used for the synthesis of nanomaterials.

5.5 Synthesis of Cellulose Nanofibrils

5.5.1 Introduction

Cellulose is an integral part of all plant life. It is the primary structural material of the cell walls of higher plants and the world's most abundant organic biopolymer with an estimated capacity of more than 1×10^{11} tons being produced each year (Brown, 1996). Cellulose is the main constituent in woody plants and the most renewable bioresource. Cellulose is a colloidal substance, a long chain polymer polysaccharide carbohydrate, and is a water-insoluble natural polymer. The strong hydrogen bonding network within the molecules plays an important role in deciding the properties of cellulose. Electrical properties can be seen in dry fibers as they are said to be good electrical insulators. Cellulose is extensively used as filler or blends with other polymers in plastics, fibers, wood and paper products, and coatings. Its structure consists of aggregated particles with extensive pores and provides a varied range of applications such as as a dispersing agent, thickener, emulsifier, anti-caking agent, gelling agent, and stabilizer (Kennedy et al., 1985; Tonnesen and Karlsen, 2002). Typically, cellulose polymers aggregate to form substructures, microfibrils, which in turn aggregate to form cellulosic fibers. Using new effective methods, these fibrils can be disintegrated from the fibers to form uniform nanosized material. These nanofibrils have extremely good structured strength and potential in many nanotechnology applications, such as strength reinforcement in nanocomposites (Berglund, 2005).

The properties of cellulose microfibrils and nanofibrils differ greatly from the properties of wood pulp fibers mainly because of the large specific surface area and high aspect ratio of the fibrils. In aqueous suspensions, the fibrils form a highly entangled network that behaves as a pseudoplastic gel (Herrick et al., 1983). The rheological properties enable the use of the fibril materials as thickeners or stabilizers in suspensions or emulsions (Turbak et al., 1983; Andresen and Stenius, 2007). The Young's modulus of cellulose fibrils has been reported to be 138 GPa and extremely good strength and thermal stability of the fibrils make them suitable for use as reinforcement in nanocomposites (Berglund, 2005; Hubbe et al., 2008) and paper products (Turbak et al., 1983). Hydroxyl groups on cellulose fibril surfaces offer many possibilities for chemical modification, such as hydrophobization (Andresen et al., 2006; Lu et al., 2008), and these new functionalities can also be effectively utilized (Andresen et al., 2006). Homogeneous thin films of cellulose fibrils can be prepared upon drying and these films can be utilized, for example, in paper coatings. Films from antimicrobial-grafted microfibrils have shown antibacterial behavior, and hence these can be used for wound healing or air filtration applications (Andresen et al., 2007).

Cellulose nanofibrils can be effectively synthesized using hydrodynamic cavitation reactors. We now discuss a case study (Pinjari and Pandit, 2010) in which aqueous suspensions of cellulose (natural cellulose) with higher size (micrometer) were cavitationally processed to achieve particle size reduction (cavitational milling). Due to cavitation, the particle size of aqueous suspensions of cellulose particles (natural cellulose) decreased from micrometer to nanometer size.

5.5.2 Experimental Details

Cavitation is generated with the help of hydrodynamic cavitation setup, which has been shown in Fig. 5.8. The details of setup and the processing parameters used in the reported work are as follows:

(1) Operating capacity: 50×10^{-3} m^3/batch
(2) Circulating pump: centrifugal pump coupled with 7.5 hp (5.0 KW) electric motor
(3) Diameter of pipe: 3.5×10^{-2} m
(4) Diameter of orifice hole: 1×10^{-3} m
(5) No. of holes on the orifice plate: 33
(6) Pump discharge pressure: 7.8 atm

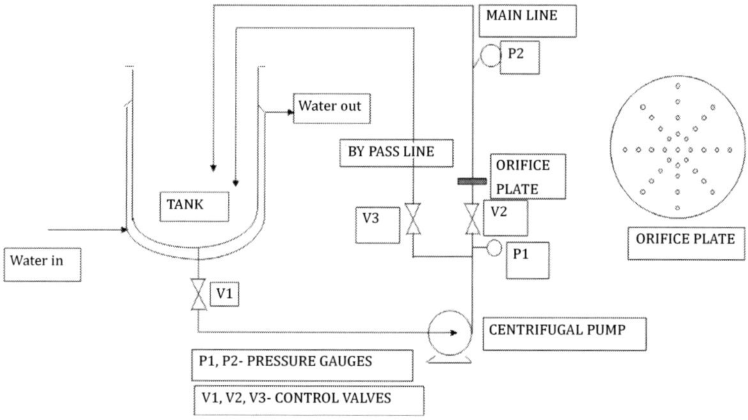

Figure 5.8 Schematic representation of hydrodynamic cavitation reactor setup (HC1). Reprinted from Pinjari and Pandit, Copyright (2010), with permission from Elsevier.

The reactor configuration was based on an arrangement to bypass the flow to control the inlet pressure (upstream) and the liquid flow rate through the main line, which holds the multiple-hole orifice as a cavitating device. The outlet pressure (downstream) is always kept at atmospheric pressure. The arrangement of the holes on the cavitation plate is also shown in Fig. 5.8. The inlet pressure (upstream) was maintained at 7.8 atm to get the liquid flow rate through the orifice plate as 274.7×10^{-6} m^3/s. The temperature of the process was maintained at room temperature (35°C) by circulating water in outer jacket of the tank. The average velocity at the orifice was 10.81 m/s giving the cavitation number as 2.61.

The aqueous solution of 1% (0.5 kg solids in 50 × 10^{-3} m^3) (by weight/volume) of natural cellulose was used as a starting liquid suspension. Initial mean particle size was found to be 63 μm. The solution was recirculated for 6 h through cavitation zone and sample was taken out for measuring its characteristic parameters using particle size measurement (PSM), X-ray diffraction (XRD), differential scanning calorimetry (DSC), and scanning electron microscopy (SEM).

5.6 Results and Discussion

5.6.1 Particle-Size Analysis

PSM was carried out for the analysis of the suspension sample before and after the treatment. It was found that particle size reduced from initial 63 μm to 1.3 μm in case of hydrodynamically processed sample in 6 h after which it remained constant. The reason for the size reduction is the asymmetric cavity collapse on the cellulose particle surface producing a high-velocity liquid jet pointed toward the material (particle) surface resulting in a form of liquid jet cutting. Another possible reason can be that when the cavity collapse occurs, the resulting shock wave propagates through the fluid media generating local pressure gradient and fluid shear causing attrition of the solid materials (particles) and the material (particle) size reduction takes place. Figure 5.9 shows the images of the initial and treated sample.

5.6.2 X-Ray Diffraction Method

Crystallinity of the samples was determined by an XRD pattern analysis method (Prasad et al., 2010). It has been found that as the exposure time to cavitation increased, crystallinity of the cellulose decreased. Initially, natural cellulose sample was having crystallinity of 86.56% (Table 5.1) which was observed to decrease when exposed to cavitational treatment.

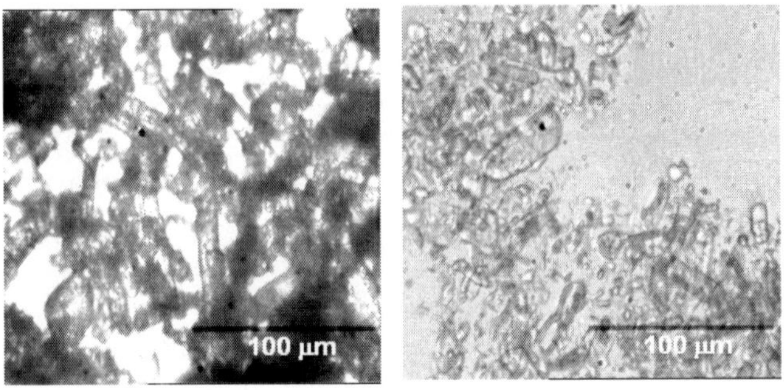

Figure 5.9 Photographic images of (a) initial sample and (b) hydrodynamically processed sample. Reprinted from Pinjari and Pandit, Copyright (2010), with permission from Elsevier.

Table 5.1 2θ angle, d-spacing, intensity, and crystallinity for the processed samples (Pinjari and Pandit, 2010)

Sample	2θ angle	d-spacing	Intensity (cps)	Crystallinity (Xc)
Initial natural sample	22.60	3.93	3855	86.56
Hydrodynamically processed sample	22.60	3.93	3545	60.65

The observed results can be possibly attributed to the inherent structure of cellulose. Cellulose structure consists of crystalline and amorphous regions. Crystallinity is a function of thermal shock. Energy dissipated during the cavitation directly affects the crystalline nature because of the generation of hot spots. Because of hot spots developed during the cavitation (temperature in the range of 5000–10000 K), the crystalline region of the cellulosic material is converted to amorphous region, resulting into lowering of the crystallinity.

5.6.3 Differential Scanning Calorimetry (Melting Temperature, T_m)

It has been reported that due to the cavitation treatment, melting temperature (T_m) of the processed cellulose was reduced. The melting point of the initial raw sample was 101.8°C (Fig. 5.10) while it was found that the melting point of hydrodynamically processed sample was 81.6°C (Fig. 5.11). The reason for this reduction in the melting point is attributed to the lowering of the extent of crystallinity of the cellulose. Highly crystalline cellulosic material has an ability to withstand higher temperatures than the amorphous cellulose. As the crystallinity decreases, the biological decomposition rate is also expected to increase, which ultimately helps in faster environmental degradation of cellulose after its use (Yokota et al., 1999). Thus, the samples processed using hydrodynamic cavitation offer more advantages than the natural cellulose samples.

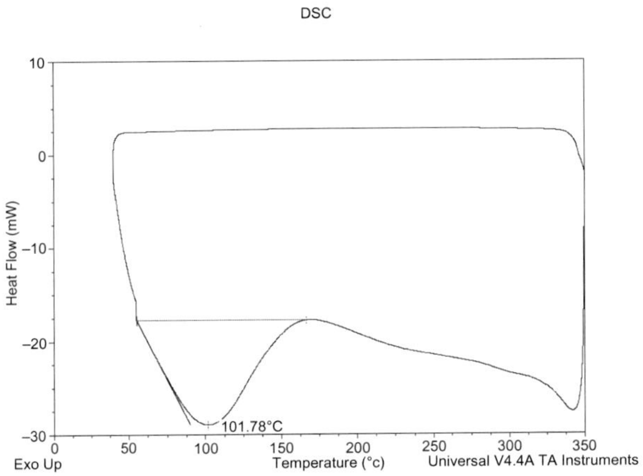

Figure 5.10 DSC (T_m) of initial sample. Reprinted from Pinjari and Pandit, Copyright (2010), with permission from Elsevier.

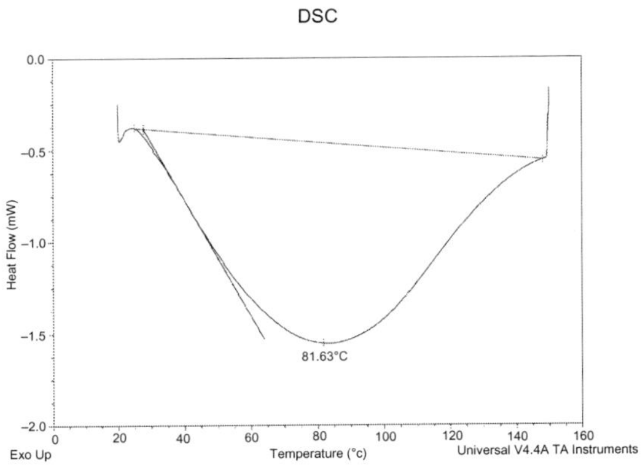

Figure 5.11 DSC (T_m) of hydrodynamically processed sample. Reprinted from Pinjari and Pandit, Copyright (2010), with permission from Elsevier.

5.6.4 Scanning Electron Microscopy

SEM images for the initial sample and the hydrodynamically processed sample are shown in Fig. 5.12. Magnifications of SEM images for initial and hydrodynamically processed sample were 100× and 1000×, respectively. The magnification of these samples has been differently chosen so as to obtain clear SEM images that show the fiber–fiber alignment and its shape and size within the cellulose particles. It can be seen from the images that before cavitational processing, the material consists of larger particles (approximately 200 μm in size, in image of Fig. 5.12a) and its average particle size estimated by PSM was found to be 63 μm. SEM image of the hydrodynamically processed sample (Fig. 5.12b) shows the alignment of the fiber and their arrangement, seemingly in a randomly distributed manner with abrupt ruptures (separated clumps).

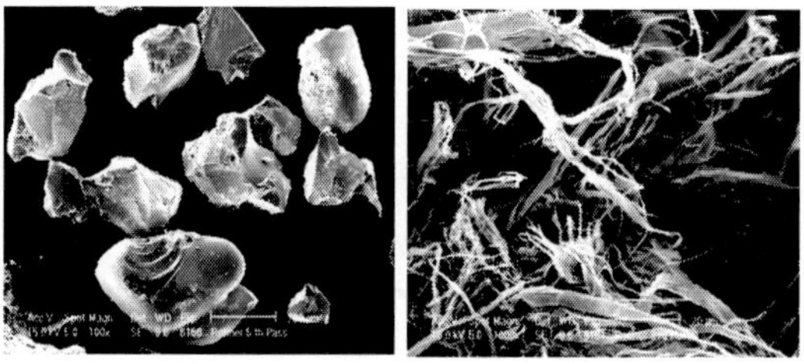

(a) (b)

Figure 5.12 SEM images of (a) initial sample and (b) hydrodynamically processed sample. Reprinted from Pinjari and Pandit, Copyright (2010), with permission from Elsevier.

The SEM image of hydrodynamically processed sample at 10000× shows some fibers with one of its dimension at nanolevel and also having declumped appearance (Fig. 5.13b).

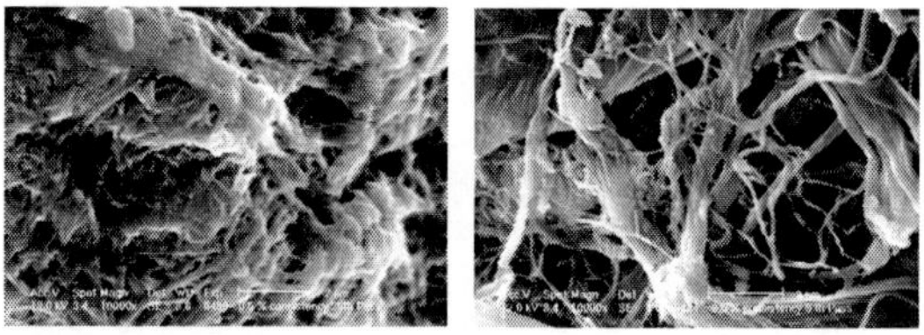

(a) (b)

Figure 5.13 SEM images of (a) initial sample at 10000× and (b) hydrodynamically processed sample at 10000×. Reprinted from Pinjari and Pandit, Copyright (2010), with permission from Elsevier.

5.6.5 Energy Balance Calculations

The energy requirement is calculated by two methods: (1) actual energy dissipated per unit volume and (2) total electric energy utilized to dissipate the actual energy, which are reported in Appendix A.

5.6.6 Conclusions

It has been established that hydrodynamic cavitation is effective in reducing the size of the natural cellulose particles efficiently along with declumping of the compact particle structure to more open fibrous structure. In hydrodynamic cavitation setup, size reduction can be controlled by the parameters such as pump discharge pressure, flow rate of the liquid, design of orifice plate, that is, free flow area for the particles to travel through the constriction, and concentration of the solution, which essentially decide the cavity dynamics. Finally, the hydrodynamic cavitation technique appears to be a very efficient tool for reducing the size from millimeter to nanometer and can also be used for hard, brittle, rubber,y and also highly crystalline materials.

5.7 Synthesis of Styrene–Butadiene Rubber Nanoemulsions

5.7.1 Introduction

Nanosuspensions have emerged as a promising strategy for an efficient delivery of hydrophobic drugs because of their versatile features such as very small particle size. Techniques such as media milling and high-pressure homogenization have been used commercially for producing nano-suspensions (Patravale et al., 2004). The unique features of nano-suspensions have enabled their use in various dosage forms, including specialized delivery systems such as mucoadhesive hydrogels. Rapid strides have been made in the delivery of nano-suspensions by peroral, ocular, and pulmonary routes and also efforts are being directed to extend their applications in site-specific drug delivery (Patravale et al., 2004). One such nanosuspension that has been synthesized using hydrodynamic cavitation is of styrene–butadiene rubber (SBR).

Styrene-butadiene or SBR is a synthetic rubber copolymer consisting of styrene and butadiene. It has good abrasion resistance and good aging stability when protected by additives, and is widely used in automobile industry applications, where it is blended with natural rubber. Another major use of styrene butadiene nanoemulsion is also in the applications in which controlled release of the constituents is required. Styrene butadiene copolymer can be used for the preparation of nanocapsules on the basis of the method of emulsion-coacervation (Lutter et al., 2008). The incorporation of SBR emulsion improves the chloride penetration resistance along with the general ionic permeability of Portland cement, while increasing its ionic transport resistance and decreasing the electric capacitance (Yang et al., 2009). SBR can also be used as a toughening agent for plastics such as polystyrene (Sardelis et al., 1987).

In a recent study performed by our group (Patil and Pandit, 2007), attempts have been made to reduce the size of the rubber latex particles (SBR) present in the form of aqueous suspension with 275 μm particle initial size to the nano-scale by utilizing a hydrodynamic cavitation process.

5.7.2 Experimental Details

The equipment used for hydrodynamic cavitation studies was essentially based on the use of multiple-hole orifice plate, as the cavitation chamber and the experimental setup were similar to that described earlier. The operating pressure at the inlet of the cavitation chamber was 4.2 atm and the liquid flow-rate through the orifice plate was 3.33×10^{-4} m^3/s. The average fluid (suspension) velocities at the orifice were 12.84 m/s giving the cavitation number, as defined in Eq. (5.1) as 1.87. The estimated energy dissipation in the cavitation zone was 7.10×10^2 kW/m^3 based on the volume of the cavitation zone downstream of the orifice plate. The method of estimation of the energy dissipation has been discussed in Section 5.7.4. The aqueous suspension of 6% (3 kg solids in 0.05 m^3) by weight rubber latex suspension was used as a starting liquid. Initial mean particle size of the latex in the suspension was 275 μm. The suspension was re-circulated for 3 h (equivalent to 32 passes) and sample was taken for the PSM.

A second hydrodynamic cavitation experimental setup (HC 2 setup) has also been used in the investigation and the configuration details are as follows:

(1) Operating capacity: 50×10^{-3} m^3/batch
(2) Circulating pump: plunger pump coupled with 1 hp electric motor
(3) Diameter of pipe: 3.175×10^{-2} m
(4) Diameter of orifice hole: 0.6×10^{-3} m
(5) No. of holes on the orifice plate: 1
(6) Operating pressure: 11 atm

A schematic representation of this setup has been given in Fig. 5.14. Alternative hydrodynamic cavitation setup was designed to achieve higher orifice velocities and lower cavitation number to improve the cavitation intensity. The liquid flow rate was maintained at 34.8×10^{-3} m^3/h using a metering pump giving an orifice velocity of 34 m/s. The cavitation number at these operating conditions was estimated to be 0.18. The energy dissipation level was estimated to be 16.6 kW/m^3 (estimated by the method explained in the Section 5.7.4). For the HC 2 system, the already processed 6% aqueous suspension from HC 1 was used as a starting material. It had a mean particle size of 129 nm. The samples were taken out after each circulation for analysis.

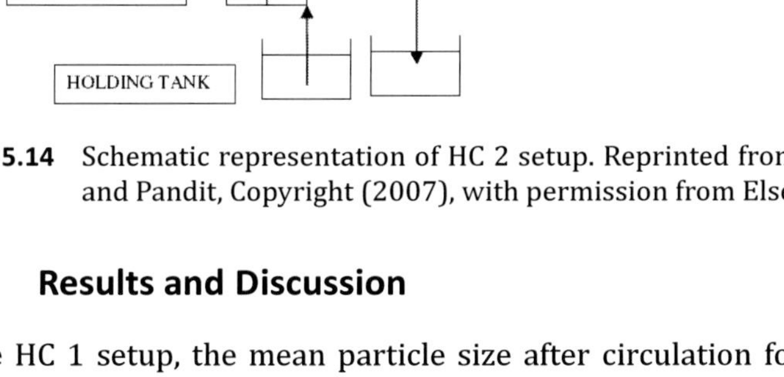

Figure 5.14 Schematic representation of HC 2 setup. Reprinted from Patil and Pandit, Copyright (2007), with permission from Elsevier.

5.7.3 Results and Discussion

In the HC 1 setup, the mean particle size after circulation for 3 h (72 passes) was 129 nm with less than 2% variation. This was the least size of rubber particles that could be obtained for the energy dissipation level in the HC 1 setup. The important consideration in the HC 1 experiment was the lowest variation (2%) with an increased scale of operation (50×10^{-3} m^3 of suspension). The rate of production (processing) for HC 1 setup works out to be 1 kg of the solids per hour, which is substantially higher than the acoustic cavitation method as well as conventional methods such as ball mill or gas-phase synthesis. However, the desired size of 40 nm could not be achieved in HC 1 setup, and hence, the processed sample from HC-1 was subjected to repetitive hydrodynamic cavitation conditions in the HC 2 setup.

In the HC 2 setup, working with 50×10^{-3} m^3 of the suspension volume, the mean particle size obtained after first circulation (one pass through the cavitating conditions) was 80 nm with 8–10% variation in the size starting with 129 nm initial mean size and 2% variation. This 80 nm suspension was used for the second circulation and mean particle size was reduced to 70 nm, again with 8–10% variation in the size distribution. The third circulation resulted in a mean size of 35 nm with less than 2% variation. For the fourth circulation, this 35 nm suspension was used as the initial liquid and a mean particle size of 20 nm again with less than 2% variation was obtained. The fourth liquid circulation was carried out to check the minimum size (equilibrium) of the particles obtained in HC 2 setup under the given set of operating conditions. Fifth circulation (with a starting size of 20 nm) of the solution gave the particle size of 50 nm. The reason behind this increased size can be explained on the basis of the cavitational effect. When the high-velocity intraparticle collision takes place, if the collision is at a direct angle, particle collision can occur at very high velocities, which induces effective melting at the point of collision. The PSM in such case gives higher size value. In Fig. 5.15, this effect can be seen clearly. In the present situation, rubber latex of 20 nm size may be the limiting size (equilibrium with reference to a local energy dissipation rate achieved in the setup). The sample (having mean particle size of 35 nm, third pass) prepared using HC 2 was again analyzed for the mean particle size 2 months after the preparation. The samples were stored at room temperature in airtight plastic cans. The size of the particles was observed to be the same as that obtained immediately at the time of preparation.

In the case of hydrodynamic cavitation, the particle size variation seems to be lower. This indicates a very high degree of mixing and the uniformity within the cavitation zone. Also, energy distribution in the cavitation zone appears to be more uniform in the case of the hydrodynamic cavitation. The reason for this observation may be better spatial distribution of the collapsing or oscillating cavities in the cavitating medium in the former. Hydrodynamic cavitation has already been proved to be very efficient for the generation of the cavities (Gogate et al., 2001). A proper distribution of the orifices on the plate is responsible for improved spatial distribution of the collapsing cavities. It can be seen that by reducing the number of holes to 1 (HC 2 setup), the variation from the mean increases again to 8–10%, possibly attributed due to the localization of the cavitation effect due to a single hole orifice plate in the HC 2 setup. This initial variation can be reduced by subjecting the suspension to multiple passes through this single hole.

5.7.3.1 Limiting/equilibrium particle size

The lowest particle size observed in each of the setups and dependency on the operating conditions can be explained on the basis of the dynamic behavior of the cavity as indicated by numerical simulations (Gastgar, 2004). Although the energy dissipated per unit volume for HC 1 setup (7.1×10^2 kW/m^3) was one order of magnitude larger than that of HC 2 setup (16.6 kW/m^3), the final size obtained in the latter was smaller. With higher orifice velocities in the HC 2 setup, the size reduction is possibly taking place more by attrition in the zone downstream of the orifice due to fluid shear, which appears to be a more efficient way of producing smaller particles for a solid material such as rubber having some elasticity. As the simulated collapse pressure pulses are also lower in the case of the HC 2 setup, the above explanation is possibly correct. Sample calculations showing this effect are given in Section 5.7.4 and one can see from these calculations that rather than the mean (averaged) energy dissipation values, peak (maximum) energy dissipation values decide this limiting reduced particle diameter. This indicates that the average energy dissipation is not the only parameter that decides the final size, but also the distribution and the form of the energy dissipation are equally responsible. Numerical simulation studies (Gastgar, 2004) indicate that except in HC 2 setup, in all the cases, cavities are transient (single oscillations), and in HC 2 setup, cavities are in stable mode (multiple oscillations). Transient cavitation gives violent collapse with very high magnitude of temperature and pressure. After collapse of the cavities, shock waves are generated and the size reduction takes place depending upon the strength of the shockwave. In stable cavitation, cavity oscillates many times before collapse and gives low collapse pressures and temperatures compared with transient cavitation. The latter, though associated with slow collapse pressures, generate very large fluid shear gradients around the fast oscillating cavities giving rise to mechanical effects such as size reduction, as the bubble wall velocities are in the range of sound velocities (1500–2000 m/s) in the liquid in an alternating directions (toward the center of the cavity and radially outward), rather than unidirectional velocities in the case of collapsing cavities.

5.7.3.2 Efficacy of energy utilization

For the comparison of the performance of the two types of cavitating systems over the selected operating ranges, the energy dissipated per kilogram of the solids for each case has been calculated. The energy taken into consideration is the total energy supplied to the system. On the basis of the solid content of the suspension, energy dissipated per kilogram of the solid size reduction has been calculated (J/kg). Sample calculation is again shown in the Section 5.7.4. The energy dissipated per kg of the solids processed (Table 5.2) clearly indicates that hydrodynamic cavitation is an energy-efficient process. For the hydrodynamic cavitation setup, these values vary from 2.026×10^4 to 6.316×10^5 J/kg. The fluid velocities in the case of the hydrodynamic cavitation setup range from 10 to 30 m/s, and thus, it suggests that rather than impact grinding (cavity collapse pressure, releasing shock wave), shear grinding or turbulent shear as a result of the stable oscillating cavity appears to be the controlling mechanism of size reduction in this case.

Energy required to create new surface area is calculated for each of the experiments and the results are reported in Table 5.2. For the same system (rubber latex suspension), the energy required to create new surface are observed to be a function of the type of the equipment used, operating parameters, and the solid concentration. There is a definite role of the type of grinding mechanism in deciding the relation between the energy supply and the increase in the surface energy (area $\times$ interfacial tension) of the system.

Work index (WI) is calculated on the basis of the total energy supplied to the system for the reduction in size and increase in the particle surface area. From the knowledge of the WI for the various equipments (for the same material), it is possible to find out the right kind of the equipment with optimized operating parameters for a specific grinding operation. It also suggests the possible mechanism responsible for the observed size reduction based on the comparison of the WI values (Perry, 2008). The mode or the form and the intensity of energy dissipation are more important than the total quantum of energy dissipation to obtain particles of a specific size.

The cost of size reduction includes the operating cost (electricity consumption). The amount of electricity required to run the equipment was calculated knowing the rated power of the equipment to reduce the material from an initial size to the final size. The presented data from Table 5.2 show the operating cost per kilogram of the processed solids of given initial sizes to the final sizes. Sample calculation is given in Section 5.7.4. From Table 5.2, it can be seen that the hydrodynamic cavitation setup appears significantly cost-effective in reducing the size of the elastic material such as rubber to the nanoscale.

5.7.4 Parameter Calculation

To calculate the various parameters associated with size reduction, some assumptions were made. Firstly, size was assumed to be the mean size and variation was taken as zero (i.e., all particles were of same size and no variation at all). The particles were assumed to be perfectly spherical; the SEM image (Fig. 5.15) clearly indicates the sphericity. Sample calculations for cases HC 1 and HC 2 have been reported in Appendix B.

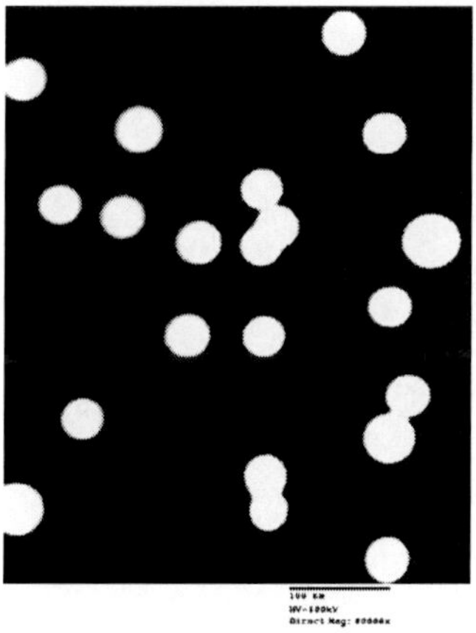

Figure 5.15 SEM image of the nanoemulsion. Reprinted from Patil and Pandit, Copyright (2007), with permission from Elsevier.

Table 5.2 Energy, WI, and operating cost calculation (Patil and Pandit, 2007)

Case no.	Condition and time of operation	Particle size (nm) Initial size	Final size	Variation in final size	Energy dissipated/volume (W/m^3)	Energy dissipated/kg solids (J/kg)	Energy required to create new surface area (N/m)	WI calculated index	Operating cost/kg solids (Rs)	Energy dissipated/particle	Reduction ratio
1	HC1 setup, 4.2 atm pressure drop, 3 h, 50 L 6% suspension	2.75×10^5	129	<2%	7.10×10^5	6.316×10^5	13.6	203	22.2	6.88×10^{-3}	2131.8
2	HC2 setup, 11 atm pressure drop, 1st circulation, 34.8 L 6% suspension	129	80	8–10%	1.66	2.026×10^4	0.712	47.17	1.42	2.28×10^{-14}	1.61
3	HC2 setup, 11 atm pressure drop, 2nd circulation, 34.8 L 6% suspension	80	70	8–10%	1.66×10^4	2.026×10^4	1.89	145	1.42	5.44×10^{-15}	1.14

Case no.	Condition and time of operation	Particle size (nm)		Variation in final size	Energy dissipated/ volume (W/m^3)	Energy dissipated/ kg solids (J/kg)	Energy required to create new surface area (N/m)	WI calculated index	Operating cost/kg solids (Rs)	Energy dissipated/ particle	Reduction ratio
		Initial size	Final size								
4	HC2 setup, 11 atm pressure drop, 3rd circulation, 34.8 L 6% suspension	70	35	<2%	1.66×10^4	2.026×10^4	0.237	22.63	1.42	3.64×10^{-15}	2
5	HC2 setup, 11 atm pressure drop, 4th circulation, 34.8 L 6% suspension	35	20	<2%	1.66×10^4	2.026×10^4	0.158	20.53	1.42	4.55×10^{-16}	1.75

5.7.5 Conclusions

(1) The hydrodynamic cavitation has proved to be very effective in reducing the size of the elastic materials such as rubber efficiently. HC2 is more efficient than all the other equipments tested in this work. The time of operation varies depending on the final required size and the extent of permissible variation.

(2) The mechanism of size reduction appears to be shear and attrition, caused by the cavitation phenomenon. As compared with the transient cavitation (single high-magnitude pressure pulse), stable cavitation (multiple oscillations and high fluid shear) shows higher efficacy for size reduction, as in stable cavitation, fluid shear direction changes along with the oscillations of the cavity (toward the center of the cavity during contraction and in outward direction during the expansion of the cavity).

(3) The stability of the suspension prepared by HC2 setup was very good even after 2 months.

(4) The systems used in this work can also be used for the size reduction of hard and brittle materials but may show a different mechanism of the size reduction and hence different equipment (may be transient cavitation) may show higher energy efficiency.

5.8 Synthesis of Nano Calcium Carbonate

5.8.1 Introduction

Calcium carbonate synthesis is one typical example of the carbonation process, in which CO_2 gas is bubbled through $Ca(OH)_2$ slurry, which is also known as reactive crystallization process. The process is limited by mass transfer (Liu et al., 2006) confirmed by the investigation of the mass transfer effect for the reactive crystallization of calcite crystals in the presence of sodium tripolyphosphate in a multiphase system. The effect of ultrasound on such crystallization processes has been extensively studied (He et al., 2005). Nanosized $CaCO_3$ particles were synthesized using a sonochemical carbonation process in which CO_2 gas was passed through a hole of an ultrasonic horn to achieve effective micromixing during the synthesis (Sonawane et al., 2008).

Nano calcium carbonate is widely used in the plastic industry as filler and in the nanostructured form as a reinforcing agent to create high-grade plastic products. It also has applications as a rheology modifier for PVC blends improving the mechanical properties of the PVC material (Chen et al., 2006). As a reinforcing agent for plastics, it adds toughness to the plastic and improves the plastic's bending strength and modulus, heat distortion temperature, and dimensional stability. Nano calcium carbonate also finds wide applications in the rubber industry (Chen et al., 2004; Yu et al., 2006; Gao et al., 2009). Using nano $CaCO_3$ in making rubber nanocomposites greatly increases the thermal stability of the rubber (Jin and Park, 2008). Nano calcium carbonate is also used in the printing ink industry and gives excellent dispersion, transparency, gloss, and ink absorption. When used as a pigment, it offers a delicate, uniform coloration with high whiteness, and good optical properties.

The paper industry is a major user of nano calcium carbonate. At present, nano calcium carbonate finds use in the manufacture of sanitary products. Using nano calcium carbonate as filler gives many advantages such as increase in opacity, brightness, whiteness, and color fastness of paper products with reduction of cost.

Nanostructured calcium carbonate can also be effectively synthesized using hydrodynamic cavitation (Sonawane et al., 2008) where the effect of various reaction parameters and reactor design on the product characteristics have also been reported. We now present some details from this work as a case study for better understanding of the efficacy of hydrodynamic cavitation reactors.

5.8.2 Experimental Details

The experimental assembly (Fig. 5.16a) consists of a closed loop reactor with a holding tank of 5 L volume, a centrifugal pump (2880 RPM, 0.5HP/0.37 KW) to recycle the $Ca(OH)_2$ slurry, and two flanges for holding the orifice plate. Downstream pipe (20 mm dia) was made up of a transparent acrylic material in order to observe the generation of cavities. Upstream pipe was made of (SS 316) stainless steel. Pressure gauges were provided to measure the inlet pressure (P1) and fully recovered pressure (P2). Different orifice plates varying in geometry were used in the study (Fig. 5.16b).

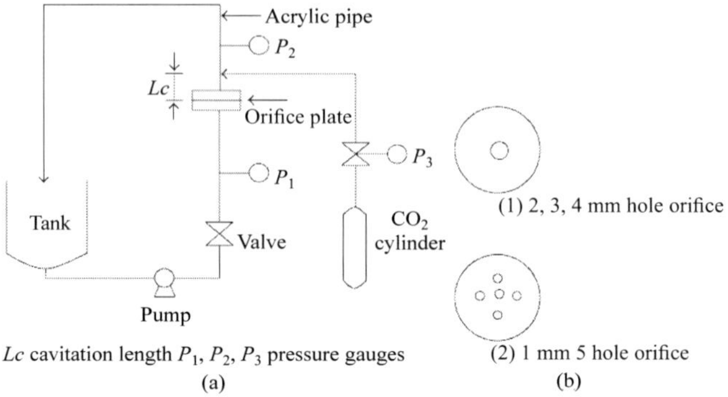

Figure 5.16 (a) Hydrodynamic cavitation setup for nano-$CaCO_3$ production. (b) Orifice configurations used for cavitation (Sonawane et al., 2010).

CO_2 (99.9% pure) gas has been introduced at different flow rates from a hole drilled at the downstream part of pipe. The recycling of slurry enabled the particles to be in suspension without settling at the bottom. The cavitation number (Cv) values have been calculated from P_2 (downstream pressure), ρ (density of water), V_0 (average velocity near orifice), and Pv (vapor pressure of water) (Mishra and Peles, 2006).

Structural properties of the $CaCO_3$ synthesized by the hydrodynamic cavitation process were evaluated using powder XRD (Philips, 1800). The Cu-Kα radiation (LFF tube 35 kV, 50mA) was selected for the analysis. Transmission electron microscopy (TEM) was performed on Technai G20-stwin working at 200 kV. The measurement of particle size distribution of the nanocrystals was done by dynamic light scattering technique (via Laser input energy of 632 nm).

5.8.3 Results and Discussion

Initially, experiments have been carried out to establish the patterns for the consumption of $Ca(OH)_2$ and the corresponding changes in the solution pH and conductivity. Preliminary experiments were carried out at 4% $Ca(OH)_2$ and 5 L/min flow rate of CO_2 and the obtained results have been shown in Fig. 5.17. It can be easily seen from the figure that three distinct regions corresponding to an induction period, nucleation, and precipitation are observed for all the parameter variations as shown in Fig. 5.17(a–c). The first region shows the time required for onset of nuclei formation and can be considered as an induction period. In the second region, growth around the crystal nuclei takes place, whereas in the third region precipitation process occurs.

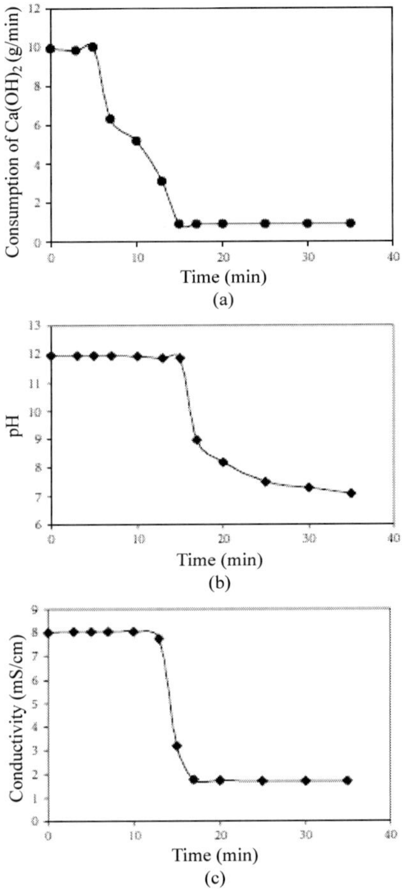

Figure 5.17 (a) Consumption of $Ca(OH)_2$ during the carbonation process without orifice (4% $Ca(OH)_2$ and 5 L/min CO_2 flow rate). (b) Change in pH during the carbonation process without orifice (4% $Ca(OH)_2$ and 5 L/min CO_2 flow rate). (c) Change in conductivity during the carbonation process without orifice [4% $Ca(OH)_2$ and 5 L/min CO_2 flow rate] (Sonawane et al., 2010).

The initial induction period (Fig. 5.17a) in which no change in [Ca^{2+}] was observed indicates that there was no spontaneous nucleation when the reaction was carried out without the orifice. The nucleation process appears to be slow and it takes relatively longer time (about 15 min) for the reaction to complete when the reaction was carried out in the pipe without orifice. It can also be noticed that the carbonation process was not 100% complete and there was some residual $Ca(OH)_2$ present at the end of the reaction. It was also observed that the particles synthesized without the orifice are coarse in nature and they tend to agglomerate during drying. It is observed that without the orifice, there is a delay in the pH drop (Fig. 5.17b) compared with the [Ca^{2+}] drop. This behavior is expected, as the carbon dioxide was fed with a constant flow rate with a high initial concentration of $Ca(OH)_2$, which led to no change in pH. These data are also supported by the conductivity measurement (Fig. 5.17c), which is an indirect measure of the total concentration of the charged species in the liquid phase. The first region of the conductivity plot gives information about the "induction time," in which the nuclei are formed by absorption of CO_2 onto the surface of $Ca(OH)_2$. The formation of $CaCO_3$ leads to a decrease in the concentration of dissolved ionic species and hence there is a drop in the conductivity. The relatively slow precipitation process in the absence of the orifice increases the size (101 nm) of the crystals.

5.8.3.1 Effect of orifice diameter and geometry of the orifice on the rate of consumption of calcium hydroxide

Different configurations of orifice plates (three with single hole of varying diameter as 2, 3, and 4 mm and one orifice plate having five holes of 1 mm diameter) were used for generating the cavitation events and the configurations have been represented in Fig. 5.18(b). Cavitation number (C_v) and β (ratio of orifice diameter to pipe diameter) for the above orifices have been given in Table 5.3.

Table 5.3 Hydrodynamic conditions of different orifice plate used for experiments (Sonawane et al., 2010)

S. No.	Diameter of orifice (mm)	Ratio of orifice diameter to pipe diameter	Cavitation number (C_v)
1	2	0.10	0.15
2	3	0.15	0.75
3	4	0.20	2.15
4	5 holes of 1 mm diameter	0.05	0.21

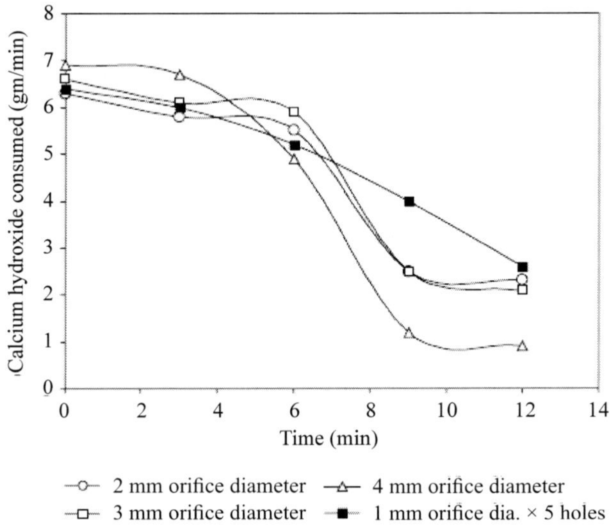

Figure 5.18 Effect of different orifice on the consumption of $Ca(OH)_2$ slurry with respect to time (showing constant rate period and falling rate periods) (4% $Ca(OH)_2$ and 5 L/min CO_2 flow rate) (Sonawane et al., 2010).

The consumption profiles of $Ca(OH)_2$ slurry with different orifice plates have been shown in Fig. 5.18. It can be seen that the rate of reaction is much faster in the presence of any orifice than the rate observed in the absence of an orifice (Fig. 5.18). During the precipitation process, the rate of nucleation step is considered as the rate-limiting step. It is observed that the nucleation is enhanced by the hydrodynamic cavitation and hence enhances the overall rate of precipitation.

Acoustic cavitation generated physical effects of microstreaming significantly influence the precipitation of calcium carbonate (Nishida, 2004). This investigation predicts the same effect on the calcium carbonate precipitation by hydrodynamic cavitation. Cavitation has a major effect on the primary nucleation of potassium sulfate (Lyczko et al., 2002). The induction period is drastically reduced in the presence of acoustic cavitation. Similar reduction in induction period and enhancement of primary nucleation (no external addition of seed particles) is observed in our case for calcium carbonate precipitation due to hydrodynamic cavitation. As in the case of without orifice, there is an induction period prior to the consumption of $Ca(OH)_2$. However, the length of induction period is shortest for the 4 mm diameter orifice compared with the rest. This indicates that the 4 mm orifice initiates the reaction due to a relatively stronger shear forces generated by the hydrodynamic cavitation process. The larger diameter orifice might have generated the stronger hydrodynamic forces or higher flow rate of circulation. On changing the geometry of orifice to five holes of 1 mm size (Fig. 5.18b) the reaction is initiated without any induction period. This arrangement also yielded the smallest particle size during the synthesis of calcium carbonate in the carbonation process.

The grain size (crystallite size) was calculated for all the samples by using Debye Scherrer formula. As shown in Table 5.4, the orifice with 5 holes of 1 mm size generated the smallest crystal size of 37 nm. The powder XRD is found to be preferably oriented along (1010) plane for samples including the $CaCO_3$ without the orifice. No effect is observed on phases due to change in orifice diameter and geometry of orifice. Single calcite phase is observed in all the samples.

Table 5.4 Effect of slurry concentration and CO_2 flow rates, orifice diameter on crystallite size (calculated from XRD), and particle size distribution (from TEM) (Sonawane et al., 2010)

$Ca(OH)_2$ slurry %	CO_2 (L/min)	Orifice diameter (mm)	Crystallite size (nm)	Particle size distribution (nm)
Effect of change in CO_2 flow rate				
4	3	4	74	65–92
4	5	4	54	62–53
4	7	4	47	35–55

$Ca(OH)_2$ slurry %	CO_2 (L/min)	Orifice diameter (mm)	Crystallite size (nm)	Particle size distribution (nm)
		Effect of change in $Ca(OH)_2$ concentration		
2	5	4	50	69–52
4	5	4	54	62–53
6	5	4	61	60–72
		Effect of change in orifice diameter		
4	5	1 × 5 holes	37	29–38
4	5	2	39	30–41
4	5	3	49	43–56
		Without orifice		
4	5	4	54	62–53
4	5	–	101	90–168

The confirmation of calcite phase observed in the XRD data is confirmed by the TEM images.

Figure 5.19 TEM image of nano-$CaCO_3$ synthesized by using hydrodynamic cavitation (4 mm orifice, 4% $Ca(OH)_2$ slurry, 5 L/min, CO_2 flow rate) (scale 500 nm) (Sonawane et al., 2010).

As shown in Fig. 5.19, the TEM image shows the cubic shape of calcite particles synthesized using 4 mm orifice diameter at 4% slurry concentration and 5 L/min CO_2 flow rate. The rate of consumption of calcium hydroxide slurry during $CaCO_3$ synthesis using hydrodynamic cavitation is significantly dependent on the geometry of the orifice and its cavitation number. The use of orifice geometry significantly reduces the grain size, hence increasing the surface area of the synthesized calcite and making it more suitable for the surface coatings applications.

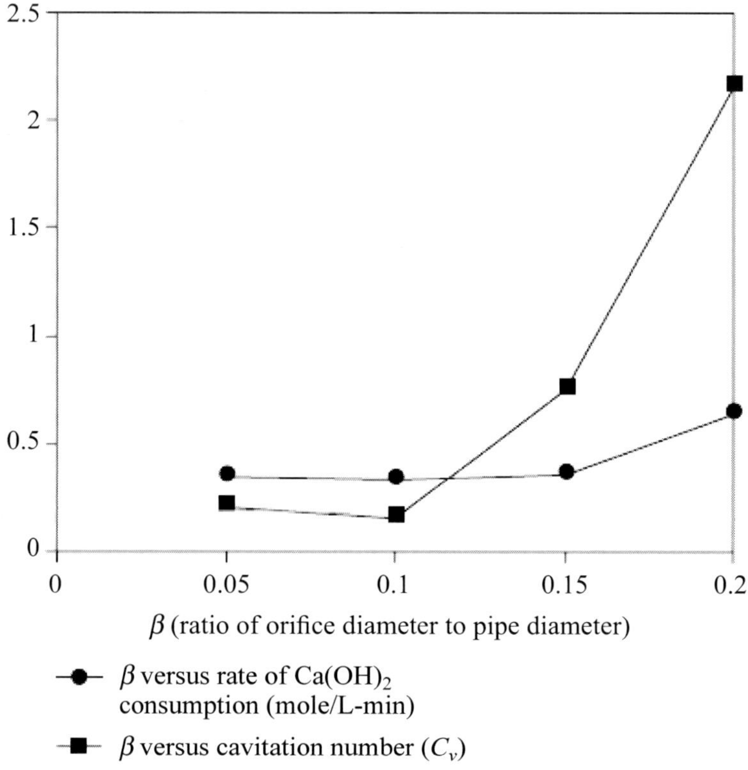

Figure 5.20 ffect of β factor onto the rate of calcium hydroxide consumption and cavitation number (C_v) (Sonawane et al., 2010).

Figure 5.20 shows the effect of the ratio of orifice diameter to pipe diameter (β factor) on the rate of consumption of $Ca(OH)_2$ slurry and the cavitation number. It is found that the rate of consumption of $Ca(OH)_2$ increases with an increase in the ratio of orifice diameter to plate diameter (β). $Ca(OH)_2$ consumption rate increases exponentially with an increase in the β factor, which indicates that there is a significant effect of the geometry of orifice plate. We speculate that the β factor has significant influence on the generation of hydrodynamic cavitation and the overall cavitational intensity. Generally for effective cavitation, it is recommended that the cavitation number should be less than 1.0 and it is important to know that severity of cavitation increases with decreasing the cavitation number (Gogate et al., 2001), which is the case for all orifices.

5.8.3.2 Effect of calcium hydroxide slurry concentration on the rate of reaction

At higher $Ca(OH)_2$ concentration (6%), the drop in the conductivity takes longer time in comparison to 2% $Ca(OH)_2$ slurry concentration, which indicates that there is longer induction time for higher concentrations of $Ca(OH)_2$. At 2% of slurry concentration, the drop in conductivity occurs very fast (within 5 min). It is also found that the available CO_2 gas concentration is more for the 2% slurry and hence induces the massive nuclei formation. The effect of calcium hydroxide slurry concentration on the pH values (indicator of the completion of reaction) was also studied. The reaction was carried out at constant flow rate of CO_2 (5 L/min) using 4 mm diameter of orifice. It is seen that the completion of reaction takes longer time (20 min) with 6% slurry concentration, and on the contrary, 2% slurry takes less than 10 min for completion. The constant period zone and falling rate zone are observed in the case of 4% and 6% slurry, while 2% slurry shows only falling rate period. As shown in Table 5.4, with an increase in the concentration of calcium hydroxide, there is an increase in the crystallite size from 50 nm for 2% concentration to 61 nm for 6% concentration. This observation suggests that increasing the $Ca(OH)_2$ has no favorable effect on the crystallite size. The XRD patterns for calcite synthesized at different $Ca(OH)_2$ concentration indicate that the calcite powder is crystalline in nature and exhibits pure calcite phase (Montes-Hernandez et al., 2007).

5.8.4 Conclusions

The case study has demonstrated that nanocalcite can be synthesized by using a hydrodynamic cavitation reactor, without formation of vaterite phase. The average grain size of the calcite synthesized without cavitation was found to be 101 nm. It is seen that there is a wide distribution of particle size in the range 90–168 nm for operation in the absence of hydrodynamic cavitation. The geometry of orifice has a significant effect on the crystallite size. In the case of geometry with five holes of 1 mm size orifice, minimum size of crystallite was obtained as 37 nm.

5.9 Overall Conclusions

The synthesis of various nano-systems can be effectively achieved using hydrodynamic cavitation. It was found that the production of nanoscale materials was achieved efficiently for systems as varied as biopolymeric (cellulose), synthetic polymeric (SBR), and inorganic salt ($CaCO_3$). One major advantage of hydrodynamic cavitation is that cavitation is uniform throughout the reactor due to efficient mixing and hence there is no problem of directional sensitivity. Another advantage is that cavitation is at the shear layer of the bulk liquid and hence the material erosion problems are less severe. Hydrodynamic cavitation provides an avenue through which the outstanding advantages of cavitation technology may be introduced into an industrial environment. Hydrodynamic cavitation technology results in yield and high throughput, accelerated reaction rates, thus improving processing capability and giving superior product quality. The relatively higher capacity, smaller footprint of the system, and simplicity of design reduces facilities' costs. Application of hydrodynamic cavitation technology reduces power consumption, waste, and provides access to green/clean technology investment capital.

Appendix A

1. Actual Energy Dissipated per Unit Volume

(7.8 atm pressure, 6 h, flow rate = 274.7×10^{-6} m^3/s)

- Mechanical energy dissipated by the pump
  = absolute inlet pressure × flow rate
  = $(101325 \times 7.8)\ N/m^2 \times (274.7 \times 10^{-6})\ m^3/s$
  = 217.10 W = 4.68×10^6 J
- Cavitational volume = Total processed volume through cavitation zone in 6 h = 50 L = 50×10^{-3} m^3
- **Energy dissipated per unit volume of liquid**
  = Energy associated with the liquid/cavitational volume
  = $(4.68 \times 10^6\ J)/(50 \times 10^{-3}\ m^3)$
  = $\mathbf{9.4 \times 10^4\ KJ/m^3}$

2. Total Electric Energy Utilized to Dissipate the Actual Energy

(5 KW power of electric pump, 6 h, 50 L)

- Electric energy input by pump in 6 h
  = 5 KW = 5000 J/s × 6 h × 3600 s
  = 108000 KJ
- Process volume = 50 L = 50×10^{-3} m^3
- **Energy utilized** = Electric energy input by pump in 6 h/ process volume
  = 108000000 J/50×10^{-3} m^3
  = **2.1×10^6 KJ/m^3**

Appendix B

1. Case 1

HC 1 setup, 4.2 atm pressure drop, 3 h

Total electrical energy consumed = 5.994×10^7 J (for 3 h)

Energy associated with the liquid = absolute inlet pressure × flow rate = 526890 N/m^2 × (0.333×10^{-3}) m^3/s = 175.45 W

Efficiency = (energy associated with the liquid/energy supplied to the pump) × 100 = (175/5550) × 100 = 3.16%

(The pressure recovery takes place within the length of eight times pipe diameter. The cavitational volume is considered as the volume of the liquid in pressure recovery zone, that is, eight times of the pipe diameter.) (Moholkar and Pandit 1997)

Cavitational volume = π/4 × (diameter of pipe)^2 × (length, that is, 8 × diameter of pipe) = $\pi/4 \times (34 \times 10^{-3})^2 \times (8 \times 34 \times 10^{-3}) = 2.47 \times 10^{-4}$ m^3

Energy dissipated per unit volume of liquid = $(175.45)/2.47 \times 10^{-4} = 7.1 \times 10^5$ W/m^3

Energy required for size reduction in terms of work index calculations:

$$\frac{P}{m} = 0.3162 \times W_i \times \left(\frac{1}{\sqrt{D_{pb}}} - \frac{1}{\sqrt{D_{pa}}} \right)$$

Where, P is the power required in kW; m is the solid flow rate in tons/h; D_{pa} and D_{pb} is the initial and final particle size in mm, W_i is the work index.

Here,

$P = 5.55$ kW

$m = 3 \times 10^{-3}$ ton/3 h = 10^{-3} ton/h

$D_{pb} = 129 \times 10^{-6}$ mm

$D_{pa} = 275 \times 10^{-3}$ mm

Therefore, W_i after calculation comes out to be 203.

Total energy dissipated = 175.45 W = 175.45 × 3 × 3600 = 1.89×10^6 J

Initial volume of the each particle (considering the spherical particle) (initial size = 275 μm) = (1/6) × π × $(275 \times 10^{-6})^3 = 1.089 \times 10^{-11}$ m^3

Density of the solids = 1000 kg/m^3

Initial mass of each particle = (1.089×10^{-11}) m^3 × 1000 kg/m^3 = 1.089×10^{-8} kg

Initial number of the particles = 3 kg/1.089×10^{-8} kg/particle = 2.75×10^8

Initial surface area of each particle = π × $(275 \times 10^{-6})^2 = 2.38 \times 10^{-7}$ m^2

Initial total surface area = 2.38×10^{-7} m^2 × 220373 = 65.45 m^2

Final volume of each particle (considering the spherical particle) (final size = 129 nm) = (1/6) × π × $(129 \times 10^{-9})^3 = 1.124 \times 10^{-21}$ m^3

Final mass of each particle = (1.124×10^{-12}) m^3 × 1000 kg/m^3 = 1.124×10^{-18} kg

Final number of the particles = 3 kg/1.124×10^{-18} kg/particle = 2.669×10^{18}

Final surface area of each particle = π × $(129 \times 10^{-9})^2 = 5.22 \times 10^{-14}$ m^2

Final total surface area = $(2.669 \times 10^{18}) \times (5.22 \times 10^{-14})$ = 139533 m^2

Increase in the surface area = 139533 – 65.45 = 139467.8 m^2

Total energy utilized for increasing the surface area = 1.89×10^6 J

So, energy utilized to create new surface area = 1.89×10^6 J/139467.8 m^2 = 13.55 N/m.

Energy utilized to create new surface area = 13.55 N/m

Cost of the electricity =22.2 Rs./h

Total hours operated = 3 h

Total electrical cost = Rs. 66.6

Solid concentration = 6% (3 kg solids).

So, the operating cost for getting 129 nm rubber particles from initial size of 275 μm in the form of suspension is 22.2 Rs./kg of the solids processed

Reduction ratio = initial particle size/final particle size = 2131.78

Cost/kg of the solids processes/unit size reduction = 22.2/2131.78 = 0.010 Rs./kg

2. Case 2

On similar lines the calculation of parameters for the HC 2 setup was done,

HC 2 (plunger pump setup), 11 atm pressure drop, 1st circulation

Total electrical energy consumed = 2664000 J

Energy associated with the liquid = 13.137 W

Efficiency = 1.78%

(The cavitational volume is considered as the volume of the liquid in pressure recovery zone, that is, eight times of the pipe diameter.) (Moholkar and Pandit 1997)

Cavitational volume = 2.01×10^{-4} m^3

Energy dissipated per unit volume of liquid = 6.54×10^4 W/m^3

Energy dissipated/kg of solids processed = 2.026×10^4 J/kg

Energy utilized to create new surface area = 0.712 N/m

Work index for the rubber latex particles came to be 47.17

The operating cost for getting 70 nm rubber particles from initial size of 129 nm in the form of suspension is 1.42 Rs./kg of the solids

Reduction ratio = initial particle size/final particle size = 1.84

Cost/kg of the solids processes/unit size reduction = Rs. 0.77

References

Andresen M. and Stenius P. (2007). Water-in-oil emulsions stabilized by hydrophobized microfibrillated cellulose, *Journal of Dispersion Science and Technology*, 28, pp. 837– 844.

Andresen M., Johansson L., Tanem B.S. and Stenius, P. (2006). Properties and characterization of hydrophobized microfibrillated cellulose, *Cellulose,* 13, pp. 665– 677.

Andresen, M., Stenstad, P., Moretro, T., Langsrud, S., Syverud, K., Johansson, L. and Stenius P. (2005). Nonleaching antimicrobial films prepared from surface-modified microfibrillated cellulose, *Biomacromolecules*, 8, pp. 2149–2155.

Berglund, L. (2005). Cellulose-based nanocomposites, In: Natural Fibers, Biopolymers, and Biocomposites, Mohanty, A., Misra, M., Drzal, L., Eds., CRC Press, Boca Raton, FL, pp 807–832.

Brown, R. M. (1996). The biosynthesis of cellulose, *Pure and Applied Chemistry*, 10, pp. 1345–1373.

Chaudhary, N. K., Gulia, R. and Gupta, S. (2007). Sol-gel technology for sensor applications, *Defence Science Journal*, 57, pp. 241–253.

Chen, N., Wan, C. Y., Zhang, Y. and Zhang, Y. X. (2004). Effect of nano-$CaCO_3$ on mechanical properties of PVC and PVC/Blendex blend, *Polymer*, 45, pp. 5985–5994.

Chen, W., Cai, W., Zhang, L., Wang, G. and Zhang, L. (2001). Sonochemical processes and formation of Gold nanoparticles within pores of mesoporous silica, *Journal of Colloid and Interface Science*, 238, pp. 291–295.

Chen, X., Li, C., Xu, S., Zhang, L., Shao, W. and Du, H. L. (2006). Interfacial adhesion and mechanical properties of PMMA-coated $CaCO_3$ nanoparticle reinforced PVC composites, *China Particuology*, 4, pp. 25–30.

Choi, H., Stathatos, E. and Dionysiou, D. D. (2006). Sol–gel preparation of mesoporous photocatalytic TiO_2 films and TiO_2/Al_2O_3 composite membranes for environmental applications, *Applied Catalysis B: Environmental,* 63, pp. 60–67.

Gao, Y., Liu, L. and Zhang, Z. (2009). Mechanical performance of nano-$CaCO_3$ filled polystyrene composites, *Acta Mechanica Solida Sinica*, 22, pp. 555–562.

Gastgar, S. N. (2004). M. Chem. Engg. Thesis, MUICT, Mumbai, India.

Gedanken, A. (2004). Using sonochemistry for the fabrication of nanomaterials, *Ultrasonics Sonochemistry*, 11, pp. 47–55.

Gogate, P. R. and Pandit A. B. (2005). A review and assessment of hydrodynamic cavitation as a technology for the future, *Ultrasonics Sonochemistry*, 12, pp. 21–27.

Gogate, P. R., Shirgaonkar, I. Z., Sivakumar, M., Senthilkumar, P., Vichare, N. P. and Pandit A. B. (2001). Cavitation reactors: Efficiency assessment using a model reaction, *AIChE Journal*, 47, pp. 2326–2338.

Grasset F., Marchand, R., Marie, A-M., Fauchadour, D. and Fajardie, F. (2006). Synthesis of CeO_2/SiO_2 core–shell nanoparticles by water-in-oil microemulsion: Preparation of functional thin film, *Journal of Colloid and Interface Science*, 299, pp. 726–732.

Green, D. W. and Perry, R. H. (2008). *Perry's Chemical Engineer's Handbook*, McGraw-Hill, pp. 8–11.

Hamer, M. A., Hendrickson, W. A. and Pryor, G. R. (2005). Organogel particles, US Patent 6858666.

Harrison, S. T. L. and Pandit, A. B. (1992). The disruption of microbial cells by hydrodynamic cavitation, Presented at 9th International Biotechnology Symposium, Washington, USA.

He, M., Forssberg, E., Wang, Y. and Han Y. (2005). Ultrasonication assisted synthesis of calcium carbonate nanoparticles, *Chemical Engineering Communications*, 192, pp. 1468–1481.

Herrick, F. W., Casebier, R. L., Hamilton, J. K. and Sandberg, K. R. (1983). Microfibrillated cellulose: Morphology and accessibility, *Journal of Applied Polymer Science: Applied Polymer Symposium*, 37, pp. 797–813.

Hosseinnia, A., Keyanpour-Rad, M., Kazemzad, M. and Pazouki, M. (2009). A novel approach for preparation of highly crystalline anatase TiO_2 nanopowder from the agglomerates, *Powder Technology*, 190, pp. 390–392.

Hsiue, G. H., Kuo, W. J., Huang, Y. P. and Jeng, R. J. (2000). Microstructural and morphological characteristics of PS–SiO_2 nanocomposites, *Polymer*, 41, pp. 2813–2825.

Hubbe, M. A., Rojas, O. J., Lucia, L. A. and Sain, M. (2008). Cellulosic nanocomposites: A review, *BioResources*, 3, pp. 929–980.

Jang, H-D. and Kim, S-K. (2003). Method for producing nanometer sized ultrafine titanium dioxide, U.S. Patent 6613301 B2.

Jin, F. L. and Park, S. J. (2008). Thermo-mechanical behaviors of butadiene rubber reinforced with nano-sized calcium carbonate, *Materials Science and Engineering: A*, 478, pp. 406–408.

Kataoka, K. (1985). Special flow techniques for heat transfer enhancement, *Kagaku Kagaku*, 49, pp. 269–275.

Kendrick, C., Klafehn, G., Guan, T., Anderson, T., Shen, H., Redwing, J. and Collins, R. (2014). Controlled growth of SiNPs by plasma synthesis. *Solar Energy Materials & Solar Cells*, 124, pp. 1–9.

Kennedy, J. F., Phillips, G. O., Wedlock, D. J. and Williams, P. A. (1985). Cellulose and its derivatives, chemistry, biochemistry and applications, Halsted Press: New York.

Khalil, K. M. S. and Zaki M. I. (2001). Preparation and characterization of sol–gel derived mesoporous titania spheroids, *Powder Technology*, 120, pp. 256–263.

Kimoto, H. and Sumita, Y. (1986). Heat transfer characteristics of a circular cylinder in a conduit under cavitation, *Nippon Kikai Gakkai Ronbunshu*, 52, pp. 2312.

Kumar, N., Komarala, V.K. and Dutta, V. (2014) In-situ synthesis of Au -CdS plasmonic photocatalyst by continuous spray pyrolysis and its visible light photocatalysis. *Chemical Engineering Journal*, 236, pp. 66–74.

Landau, M. V., Vradman, L., Herskowitz, M., Koltypin, Y. and Gedanken A. (2001). Ultrasonically controlled deposition-precipitation Co-Mo HDS catalysts deposited on wide-pore MCM material, *Journal of Catalysis*, 201, pp. 22–36.

Langroudi, A. E., Mai, C., Vigier, G. and Vassoile, R. (1997). Hydrophobic hybrid inorganic-organic thin film prepared by sol-gel process for glass protection and strengthening applications, *Journal of Applied Polymer Science*, 65, pp. 2387–2393.

Lauterborn, W. (1979). Cavitation and inhomogeneities in underwater acoustics, Proceedings of the First International Conference, Gottingen, Federal Republic. Germany, Springer-Verlag, pp. 332.

Li, J., Hao, Y., Li, H., Xia, M., Sun, X. and Wang, L. (2009). Direct synthesis of CeO_2/SiO_2 mesostructured composite materials via sol–gel process, *Microporous and Mesoporous Materials*, 120, pp. 421–425.

Lin, R., Zhang, J. and Bai, Y. (2006). Mass transfer of reactive crystallization in synthesizing calcite nanocrystal, *Chemical Engineering Science*, 61, pp. 7019–7028.

Lu, J., Askeland, P. and Drzal, L. T. (2008). Surface modification of microfibrillated cellulose for epoxy composite applications, *Polymer*, 49, pp. 1285–1296.

Luque de Castro, M. D. and Priego-Capote, F. (2007). Ultrasound assisted crystallization (sonocrystallization), *Ultrasonics Sonochemistry*, 14, pp. 717–724.

Luton, M. (1989). Method of extruding aluminium-base oxide dispersion strengthened, U.S. Patent 4818481.

Lutter, S., Koetz, J., Tiersch, B., Boschetti de Fierro, A. and Abetz, V. (2008). Formation of gold nanoparticles in triblock terpolymer-modified inverse Microemulsions, *Colloids and Surfaces A*, 329, pp. 160–176.

Lyczko, N., Espitalier, F., Louisnard, O. and Schwartzentruber, J. (2002). Effect of ultrasound on the induction time and the metastable zone widths of potassium sulphate, *Chemical Engineering Journal*, 86, pp. 233–241.

Mishra, C. and Peles, Y. (2006). An experimental investigation of hydrodynamic cavitation in micro-Venturis, *Physics of Fluids*, 18, pp. 103–109.

Moholkar, V. S. and Pandit, A. B. (1997). Bubble behavior in hydrodynamic cavitation, effect of turbulence, *AIChE Journal*, 43, pp. 1641–1648.

Moholkar, V. S., Senthilkumar, P. and Pandit, A. B. (1999). Hydrodynamic cavitation for sono-chemical effect, *Ultrasonics Sonochemistry*, 6, pp. 53–65.

Montes-Hernandes, G., Renard, F., Geoffroy, N., Charlet, L. and Pironon, J. (2007). Calcite precipitation from CO_2-H_2O-$Ca(OH)_2$ slurry under high pressure of CO_2, *Journal of Crystal Growth*, 308, pp. 228–236.

Mori, Y. and Kitamoto, N. (2005). Effect of kind of alcohol as additives on preparation of Gold Nanoparticles by sonochemistry, *Journal of Chemical Engineering of Japan*, 38, pp. 283–288.

Morison, R. and Hutchinson, C. A. (2009). Limiting of the Weissler reaction as model reaction for measuring the efficiency of hydrodynamic cavitation, *Ultrasonics Sonochemistry*, 16 pp. 176–183.

Moser, W. R. (1995). Process for the preparations of solid state materials, U.S. Patent 5,466,646.

Nagaraja, B. M., Haznan, A., Jung, K. D. and Yoo, K. S. (2008). Preparation of mesostructured barium sulphate with high surface area by dispersion method and its characterization, *Bulletin of the Korean Chemical Society*, 29, pp. 1007–1012.

Nishida, I. (2004). Precipitation of calcium carbonate by ultrasonic irradiation, *Ultrasonics Sonochemistry*, 11, pp. 423– 428.

Pandit, A. B. and Joshi, J. B. (1993). Hydrolysis of fatty oils: effect of cavitation, *Chemical Engineering Science*, 48, pp. 3440–3442.

Patravale, V. B., Date, A. A. and Kulkarni, R. M. (2004). Nanosuspensions: a promising drug delivery strategy, *Journal of Pharmacy and Pharmacology*, 56, pp. 827–840.

Payne, T., Dodd, A., Meiser, F., Postma, A., Cammarano, R., Caruso, F., Williams, J. and Mccormick, P. Nanoparticle composition and methods for synthesis thereof, WO/2006/069419.

Patil, M. N. and Pandit, A. B. (2007) Cavitation – A novel technique for making stable nano-suspensions, *Ultrason. Sonochem.*, 14, pp. 519-530

Pinjari, D. V. (2012). Cavitation based synthesis of nanoparticles, Ph.D. Thesis submitted to University of Mumbai.

Pinjari, D. V. and Pandit, A. B. (2010). Cavitation milling of natural cellulose, *Ultrason. Sonochem.*, 17, pp. 845–852.

Prasad, K., Pinjari, D. V., Pandit, A. B. and Mhaske, S. T. (2010). Phase transformation of nanostructured titanium dioxide from anatase-to-rutile via combined ultrasound assisted sol-gel technique, *Ultrasonics Sonochemistry*, 17, pp. 409–415.

Sampathkumar, K. and Moholkar, V. S. (2007). Conceptual design of a novel hydrodynamic cavitation reactor, *Chemical Engineering and Science*, 62, pp. 2698–2711.

Sardelis, K., Michels, H. J. and Allen, G. (1987). Toughened polystyrene containing block, graded block and randomized copolymers of butadiene-styrene, *Polymer*, 28, pp. 244–250.

Senthilkumar, P. and Pandit, A. B. (1997). Modelling hydrodynamic cavitation, *Chemical Engineering & Technology*, 22, pp. 1017–1027.

Senthilkumar, P., Sivakumar, M. and Pandit, A. B. (2000). Experimental quantification of chemical effects of hydrodynamic cavitation, *Chemical Engineering Science*, 55, pp. 1633–1639.

Sergiienko, R., Shibata, E., Suwa, H., Nakamura, T., Akase, Z., Murakami, Y. and Shindo, D. (2006). Synthesis of amorphous carbon nanoparticles and carbon encapsulated metal nanoparticles in liquid benzene by an electric plasma discharge in ultrasonic cavitation field, *Ultrasonics Sonochemistry*, 13, pp. 6–12.

Skandan, G. and Singhal, A. (2006). Nanomaterials Handbook, In: Gugotsi Y. (Ed.), Taylor Francis, NY.

Sonawane, S. H., Shirsath, S. R., Khanna, P. K., Pawar, S., Mahajan, C.M., Paithankar, V., Shinde, V., Kapadnis, C. V. (2008). An innovative method for effective micro-mixing of CO_2 gas during synthesis of nano-calcite crystal using sonochemical carbonization, *Chemical Engineering Journal*, 143, pp. 308–313

Sonawane, S. H., Gumfekar, S. P., Kate, K. H., Meshram, S. P., Kunte, K. J., Ramjee, L., Mahajan, C. M., Parande, M. G. and Ashokkumar, M. (2010). Hydrodynamic cavitation-assisted synthesis of nanocalcite, *International Journal of Chemical Engineering*, Article ID 242963.

Sunstrom, I. E., Moser, W. R. and Guerts, B. M. (1996). General route to nanocrystalline oxides by hydrodynamic cavitation, *Chemistry of Materials*, 8, pp. 2061–2067.

Suslick, K. S., Mdleleni, M. M. and Ries, J. T. (1997). Chemistry induced by hydrodynamic cavitation, *Journal of the American Chemical Society*, 119, pp. 9303–9304.

Suslick, K. S., Choe, S. B., Cichowlas, A.A. and Grinstaff, M. W. (1991). Sonochemical synthesis of amorphous iron, *Nature*, 353, pp. 414–416.

Suslick, K. S., Hyeon, T. W. and Fang, M. W. (1996). Nanostructured materials generated by high-intensity ultrasound: Sonochemical synthesis and catalytic studies, *Chemistry of Materials*, 8, pp. 2172–2179.

Suslick, K. S., Hyeon, T. W. and Fang, M. W. (1996). Nanostructured materials generated by high-intensity ultrasound: Sonochemical synthesis and catalytic studies, *Chemistry of Materials*, 8, pp. 2172–2179.

Tonnesen, H. H. and Karlsen, J. (2002). Alginate in drug delivery systems, *Drug Development & Industrial Pharmacy*, 28, pp. 621–630.

Tracton, A. A. (2006). Coatings Technology Handbook, 3rd ed, CRC Press.

Turbak, A. F., Snyder, F. W. and Sandberg, K. R. (1983). Microfibrillated cellulose, a new cellulose product: Properties, uses, and commercial potential, *Journal of Applied Polymer Science: Applied Polymer Symposium*, 37, pp. 815–827.

Vaseashta, A. (2009). *Nanomaterials: Applications, Risks, Ethics and Society*, In: Linkov, I. and Steevens, J. (Eds.), Nanomaterials: Risks and Benefits, Springer Science + Business Media, B.V.

Wang, W., Pang, Y. and Hodgson, S. N. B. (2009). XRD studies of thermally stable mesoporous tungsten oxide synthesised by a templated sol–gel process from tungstic acid precursor, *Microporous and Mesoporous Materials*, 121, pp. 121–128.

Wegner, K., Walker, B., Tsantilis, S. and Pratsinis, S.E. (2002). Design of metal nanoparticle synthesis by vapor flow condensation. *Chemical Engineering Science*, 57, pp. 1753–1762.

Yang, Z., Shi, X., Creighton, A. T., Peterson, M. M. (2009). Effect of styrene-butadiene rubber latex on the chloride permeability and microstructure of Portland cement mortar, *Construction and Building Materials*, 23, pp. 2283–2290.

Yokota, Y. and Marechal, H. (1999). Processability of biodegradable poly (butylene) succinate and its derivates. A case study, In: Biopolymer Conference, Würzburg (Germany).

Yu, H. J., Wang, L., Shi, Q., Jiang, G. H., Zhao, Z. R. and Dong, X. C. (2006). Study on nano-$CaCO_3$ modified epoxy powder coatings, *Progress in Organic Coatings*, 55, pp. 296–300.

Chapter 6

Sonoelectrochemical Synthesis of Nanomaterials

Jonathan P. Metters,[a] Craig E. Banks,[a] and Bruno G. Pollet[b]
[a]*Faculty of Science and Engineering, School of Chemistry and the Environment, Division of Chemistry and Environmental Science, Manchester Metropolitan University, Chester Street, Manchester M1 5GD, Lancs, UK*
[b]*South African Institute for Advanced Materials Chemistry (SAIAMC), University of the Western Cape, Robert Sobukwe Road, Private Bag X17, Bellville 7535, Cape Town, South Africa*
bgpollet@uwc.ac.za

This chapter discusses the concept of fabricating nanomaterials using power ultrasound coupled with electrochemistry that can give rise to some intriguing nanostructures, which typically are unable to be produced via conventional wet chemical fabrication approaches. We highlight the diverse range of nanomaterials that can be produced via this unique methodology and also focus on noble metals used as electrocatalysts within fuel cells.

Cavitation: A Novel Energy-Efficient Technique for the Generation of Nanomaterials
Edited by Sivakumar Manickam and Muthupandian Ashokkumar

ISBN 978-981-4411-54-7 (Hardcover), 978-981-4411-55-4 (eBook)
www.panstanford.com

6.1 Introduction

Acoustic cavitation is generated in liquids at high and moderate intensities of ultrasonic irradiation. Dissolved gas nuclei in a liquid are forced to grow by rectified diffusion and coalescence pathways in an acoustic field. Bubbles continue to grow until they reach a critical size, which depends on the ultrasonic frequency. The bubbles then grow to a maximum size and collapse within one acoustic cycle. The collapse of bubbles is very rapid, and due to the adiabatic nature of the collapse, localized hot spots are generated. Thus, cavitation can create extreme physical and chemical effects within the bubbles and at the bubble/solution interface despite the normal temperature conditions in the bulk solution. Hence, sonochemistry offers specific experimental conditions to perform synthetic chemical processes either within the bubble or at the bubble interface with extreme and non-equilibrium conditions, which could result in the formation of new materials with unique properties. Thus, the advent of promising sonochemical synthesis of nanomaterials meets the demand of nanotechnology as well as other physical sciences. A number of reviews and monographs devoted to various aspects of sonication have been reported in the last decade (Bang and Suslick, 2010). The main focus of this chapter is to review the sonochemical synthesis of robust noble monometallic and bimetallic nanomaterials with a particular attention to the resulting structures and properties of these nanomaterials. There is a clear and distinct requirement for nanoarchitectures at suitable sizes with functional building blocks within the fields of nano-engineering, bioelectronics, and electronics (Shipway et al., 2000; Panyam and Labhasetwar, 2003). The fabrication of nanomaterials is a highly competitive field in which a metal salt is reduced by the addition of a reducing agent, such as sodium citrate, borohydride, and variants of these (Turkevich et al., 1951; Handley, 1989; Ahmadi et al., 1996; Yee et al., 1999). Other methods to achieve this are available such as through the use of laser, electrodeposition, and gas phase synthesis with many reviews in the literature overviewing these techniques (Shipway et al., 2000).

One particularly exciting area that can facilitate the fabrication of nanoparticles is electrochemistry. Electrodeposition is a well-known technique where a target metal, say M^{n+}, is electrodeposited at an electrode surface through the application of a suitable negative potential, where the analyte is contained within a suitable electrolyte such that the following transition is favorably induced:

$$M^{n+}_{(aq)} + ne^{-} \rightarrow M_{(electrode)} \quad (6.1)$$

Typically, such an approach is the basis of anodic stripping voltammetry where, following an accumulation of the target analyte in the form of nucleated metal, the process is reversed to induce the stripping of the metal back to M^{n+}. This approach can reach analyte detection limits of 10^{-12} mol/dm^3 under favorable conditions (Wang, 2006; Banks and Compton, 2007). Unique surface structures can be readily observed and are then used in a plethora of applications such as electronic, optical, and sensing applications (Shipway et al., 2000), but can have limitations, as the material required to bring about the process in Eq. 6.1 needs to have favorable kinetics to induce metal electron transfer, but may not be ideally suited to the intended application. One approach to harness the benefits of electrochemistry is to use sonoelectrochemistry. The sonoelectrochemical method is the use of power ultrasound applied to various electrochemical processes leading to both chemical and physical effects, such as increase in mass transport, improved surface cleaning, increase in deposit hardness, deposit thickness, deposition rates and efficiencies, as well as improved deposit adhesion, and also promotes the formation of radicals via sonolysis (Pollet and Hihn, 2011).

6.2 Experimental Configuration

Figure 6.1 shows a typical sonoelectrochemical approach (Mancier et al., 2008; Pollet, 2010) wherein the use of power ultrasound is harnessed with electrochemistry and the ultrasonic tip acts as an electrode as well as the source of ultrasound. Note that this is quite unique, as usually the ultrasonic horn tip is placed opposite the working electrode (Banks and Compton, 2004), as is generally the case in sonoelectrochemistry, particularly sonoelectroanalysis or configured in other geometries (Compton et al., 1997).

Note that this experimental approach can be followed back to the work of Reisse et al. (1994), which also lead to the development of the sonotrode, used extensively in sono-electroanalysis, as pioneered by Compton (Eklund et al. 1995; Akkermans et al., 1998; Floate et al., 2002). In this widely used sonoelectrochemical configuration, the ultrasonic and electrochemical processes are pulsed.

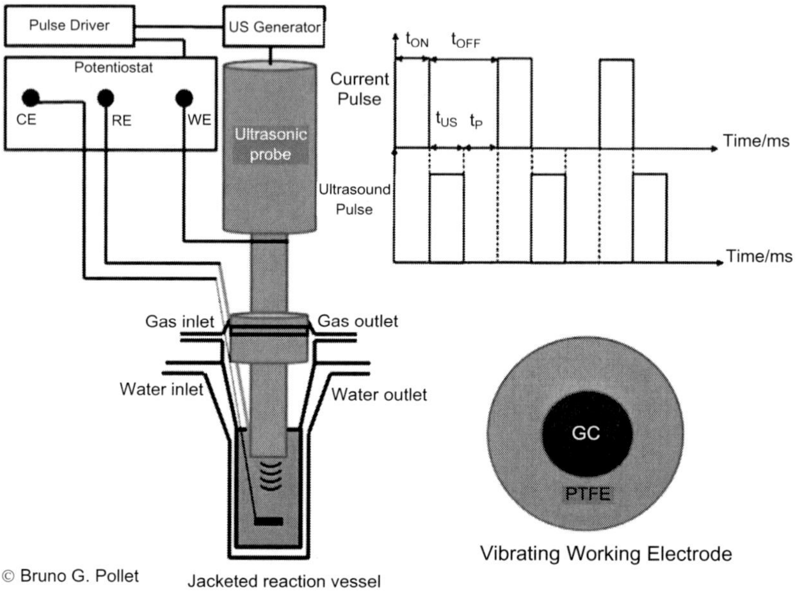

Figure 6.1 Pulsed sonoelectrochemical setup using the vibrating tip as the WE for the fabrication of nanomaterials. WE: working electrode; CE: counter electrode; RE: reference electrode

Figure 6.2 shows a typical regime where the potential is turned on (t_{on}) while the ultrasonic power is turned off (t_{off}). Following a short time, for example, hundreds of milliseconds (Mancier et al., 2008), the electrodeposition, a process as described by Eq. 6.1 for simple target analytes, occurs and metallic deposits are immobilised at the ultrasonic tip/electrode. Following this, the potential is turned off (t_{off}) and the power ultrasound sources (20 kHz) are turned on (t_{us}), typically anywhere between 100 and 400 ms (Mancier et al., 2008; Zin et al., 2009a). Such an approach is termed "out-of-phase ultrasound," as it is not being applied during the electrochemical process and serves to "renew" the electrode free from metallic deposits such that new deposits can be rapidly formed, from cavitational processes, that is, production and implosion of microbubbles that last for a few milliseconds as well as acoustic streaming arising from the acoustic field manifesting itself into the liquid medium (Banks and Compton, 2003). This also results in "refreshing" the diffusion layer of electroactive species as well as clearing any gas bubbles residing on the electrode surface that would inhibit nucleation in the repeat process. The clear advantages of this approach are that significant amounts of particles can be rapidly formed that can be collected from the solution and be applied into different technological areas. In addition, the tailoring of the current density along with the accumulation time (t_{on}) allows one to rapidly synthesize a variety of nanostructures.

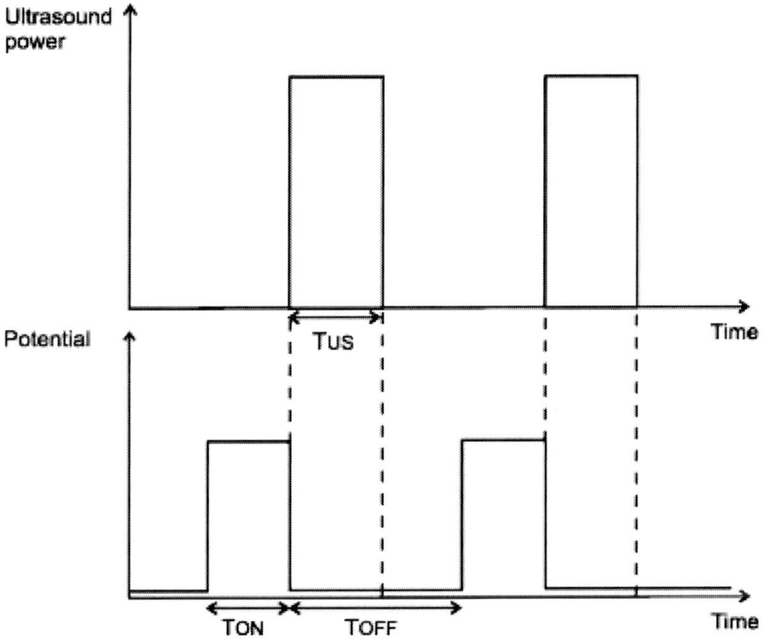

Figure 6.2 Out-of-phase electrochemical and ultrasound pulses. Reproduced from Mancier et al. (2008) with permission from Elsevier.

This chapter seeks to impart onto readers the true advantages of the sonoelectrochemical approach and highlight prominent literature examples while demonstrating how this can be applied for the generation of fuel cell electrocatalysts.

6.3 Metallic Systems

6.3.1 Copper and Platinum

Formation of copper oxide nanoparticles has been extensively explored, as it has been reported that the nanoparticle size and semiconductor properties can be controlled (Balamurugan et al., 2004; Mancier et al., 2008). Mancier et al. (2008) have reported elegantly the synthesis of 8 nm diameter Cu_2O nanoparticles. Interestingly, their work demonstrated that varying the t_{on} and t_{us} had no effect on the diameter of the copper nanoparticles, suggesting that once nucleation on active sites of the ultrasonic probe/electrode had been saturated, further deposition of copper upon already initially deposited copper results in an unfavorable nucleation and electron transfer kinetics.

Haas et al. (2008) have reported elegant copper dendrites from aqueous solutions of copper ions and polyvinyl alcohol. Figure 6.3 depicts the observed structure and distribution of particle sizes that were obtained. This work is of significant importance, as it sets-out to answer the question as to whether the dendrite structure obtained by sonoelectrochemistry is formed at one of the following three stages: (1) on the electrode (cathode) after the electrodeposition on the metallic copper; (2) in the solution after the removal of the copper from the electrode surface; and (3) on the TEM copper/carbon grid. The investigation also revealed that the sonoelectrochemical parameters and the nanoparticle concentration and size determine dendrite formation. Elegant images and careful experiments revealed that the dendrites are not formed on the electrode and probably not in the solution either, as proven by SEM images of the electrode following one pulse, but rather are formed at a later stage, that is, when metal-colloidal solutions are deposited on the copper/carbon grid under specific sonoelectrochemical parameters. The parameters upon which dendrite formation relied were determined to be as follows: the concentration of the copper nanoparticles and the size of the copper nanoparticles. These two parameters and the nature of the substrate influence the surface evaporation conditions. In addition, the experimental results are consistent with predictions made previously by Rabani et al. (Rabani et al., 2003; Sztrum et al.,

2005; Sztrum and Rabani, 2006; Yosef and Rabani, 2006; Haas et al., 2008).

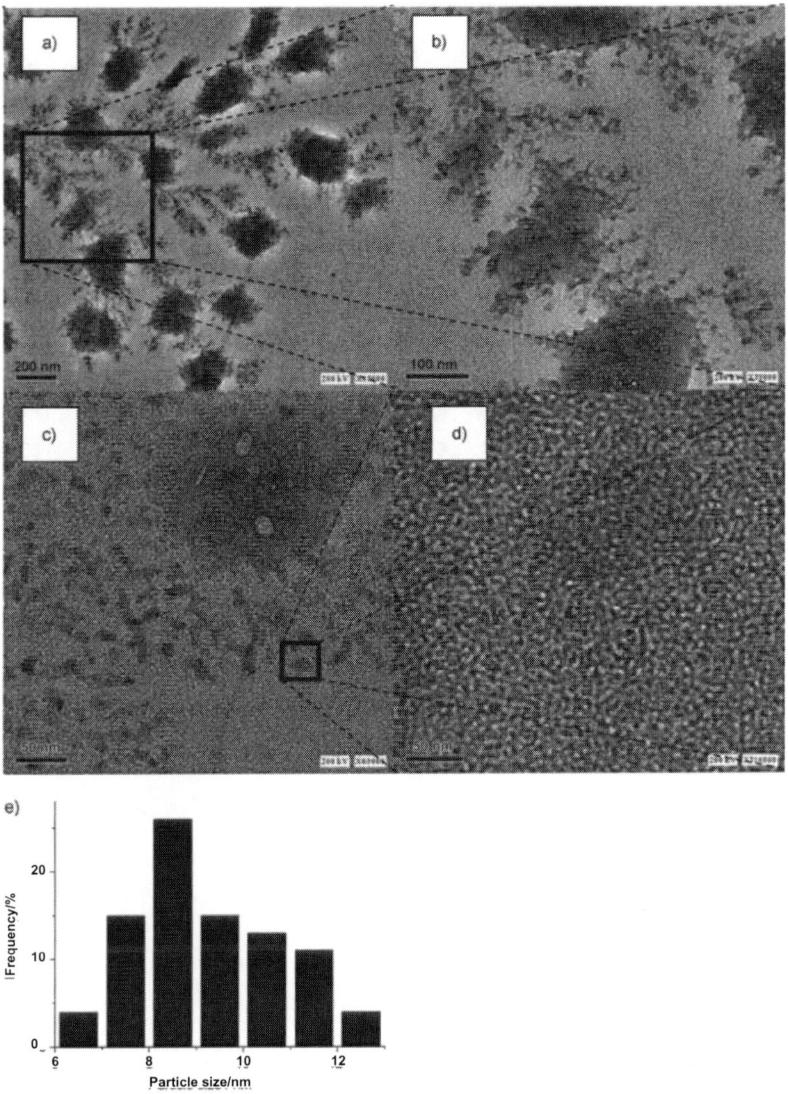

Figure 6.3 (a–c) TEM images of copper dendrites synthesized at a current pulse width of 300 ms, (d) fringes of an individual copper particle, and (e) a histogram of the sizes of individual copper particles. Reproduced from Haas et al. (2008) with permission from Wiley.

Other fascinating work has been explored by Haas for the case of copper ions in cetyltrimethylammonium bromide (CTAB) solutions that produced 95 wt % CuBr and 5% metallic copper instead of purely metallic copper, indicating a differing sonoelectrochemical mechanism (Haas et al., 2006). In terms of platinum, there are only a few reports on its sonoelectrochemical production. For example, Zin et al. (2009a) demonstrated that the electroreduction of Pt(II) to Pt(0) with average diameters from 11 nm to 15 nm was possible. However, as no capping agents are used, the platinum aggregates into larger particles in the range of 100–200 nm, which highlights a disadvantage of this approach compared with colloidal synthesis.

6.3.2 Magnesium

This material has potential as a hydrogen storage material due to its very low mass and there are very few approaches reported to fabricate magnesium nanoparticles (Connor et al., 1957; Leibenow, 1997; Aurbach et al., 2000; Sergeev, 2003; Kooi et al., 2006; Haas et al., 2008). Haas and Gedanken (2008) were the first to have reported the fabrication of nearly 4 nm magnesium nanoparticles using sonoelectrosynthesis. However, on transferring the fabricated nanoparticles outside of an inert atmosphere (such as a glove box), the nanoparticles readily transform to MgO. Note that sonoelectrosynthesis approach is beneficial, as metals, which cannot be reduced by chemical means, can be electrochemically driven to nucleate through the application of a large negative reduction potential where the applied voltage exceeds the standard reduction potential along with the overpotential.

6.3.3 Tungsten

Another example where sonoelectrosynthesis has benefits is in the synthesis of metals that are difficult to produce such as tungsten and molybdenum where very few fabrication options exist (Lei et al., 2007). Lei et al. (2007) report the synthesis of tungsten nanoparticles with four main reactions occurring in the electrolyte and at the cathode as follows:

Electrolyte

$$Fe^{2+} + (C_6H_5O_7)^{3-} \rightarrow (FeC_6H_5O_7)^- \qquad (6.2)$$

$$WO_2^{2+} + (C_6H_5O_7)^{3-} \rightarrow (WO_2C_6H_5O_7)^{-} \quad (6.3)$$

Cathode

$$(FeC_6H_5O_7)^{-} + 2e^{-} \rightarrow Fe + (C_6H_5O_7)^{3-} \quad (6.4)$$

$$(WO_2C_6H_5O_7)^{-} + 4H^{+} + 6e^{-} \rightarrow W + (C_6H_5O_7)^{3-} + 2H_2O \quad (6.5)$$

It was shown that 30 nm (diameter) tungsten nanoparticles consisting mainly of tungsten (with Fe impurities due to the above processes) in a body-centered cubic (bcc) structure can be advantageously produced.

6.3.4 Aluminum

Mahendiran et al. (2009) have reported for the first time the sonoelectrochemical synthesis of aluminum nanoparticles in the range 10–20 nm (diameter), which was found via X-ray diffraction (XRD) to correspond to face-centered cubic (fcc) metallic aluminum. Again, this approach is highly welcome, as aluminum plays a prominent role within material science due to its light weight, high strength, and corrosion resistance, and additionally, there are not many synthetic routes available.

6.3.5 Lead and Cadmium

Highly dispersed spherical and pure lead nanoparticles and dendritic structures in the presence of CTAB have been reported by Qiu et al. (2003), although the synthesis of 9 nm diameter cadmium selenide has also been reported (Masti et al. 2000), which has been extended to PbSe (Zhu et al., 2000) particles of size nearly 12 nm using a similar synthetic approach.

6.4 Alloy and Core–Shell Nanoparticles

Delplancke et al. (2000) first reported the sonoelectrochemical synthesis of magnetic nanopowders, including pure, binary and ternary iron, cobalt, and nickel. Work in this area has been extended to more exotic nanoparticles such as $Fe_{75}Co_{25}$ (Mancier et al., 2004), $Co_{65}Fe_{35}$ (Dabalà et al., 2008), FeCo (Zin et al., 2009b, 2010), and Fe-Cr (Zin and Dabalà, 2010).

Other work of note is the sonoelectrosynthesis of PbTe nanorods with diameters in the sub-10 nm region where the metal ions and ligands were used to control the nanostructures and stop coagulation/aggregation (Qiu et al., 2005). Related to these are core–shell nanoparticles where the particles consist of a core and a shell of two differing materials and can, in certain cases, have different chemical, magnetic, optical, and catalytic properties compared with the case of a single pure metal (Mancier et al., 2010).

Mancier et al. (2010) have reported Cu–Ag core–shell nanoparticles. Figure 6.4 shows an EELS spectrum wherein a Cu–Ag particle is evident, and through careful imaging, the Ag core and Cu-decorated nanoparticles are clearly evident.

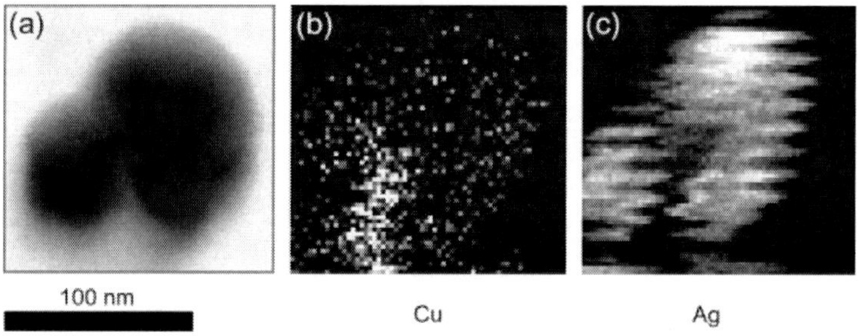

Figure 6.4 EELS spectrum-imaging of a Cu–Ag agglomerate: STEM-inverted dark field image (a), copper signal map (b), and silver signal map (c). Reproduced from Mancier et al. (2010) with permission from Elsevier.

6.5 Polymer Nanoparticles

Conducting polymers are of interest due to their applications in sensors, energy storage devices, conducting paints, and surface protection (Galal, 2006). Electrochemical methods have been used extensively to prepare such polymers and the application of ultrasound gives rise to different morphological structures that result in differing electrochemical and electronic characteristics. Usually in these approaches, the ultrasonic probe is placed directly opposite that of the working electrode (Galal, 2006), and recently, this has been simplified via the approach of sonoelectrosynthesis such as the inspiring work of Ganesan et al. (2008) and Atobe et al. (2009), where in the former 20–40 nm polyaniline nanoparticles have been fabricated, as well in the latter case poly (*N*-methylaniline) microspheres can be readily fabricated with precise control of the resultant sizes through electric pulses.

6.6 Sonoelectrochemical Nanoparticles Incorporated into New Nanomaterials

Graphene has received considerable attention in a plethora of areas, particularly in the area of electrochemistry (Brownson and Banks, 2010; Brownson et al., 2011). Pioneering work by Shi and Zhu (2011) have reported the sonoelectrochemical synthesis of palladium-graphene nanocomposite harnessing the power of the method to make palladium, as well as producing graphene from graphite via applying power ultrasound and a combination of these resulting in a highly novel composite active toward the sensing of chlorophenols.

Figure 6.5 shows the TEM, HRTEM, and XRD patterns of the graphene nanocomposites, where 30–40 nm (diameter) Pt is observed upon graphene sheets.

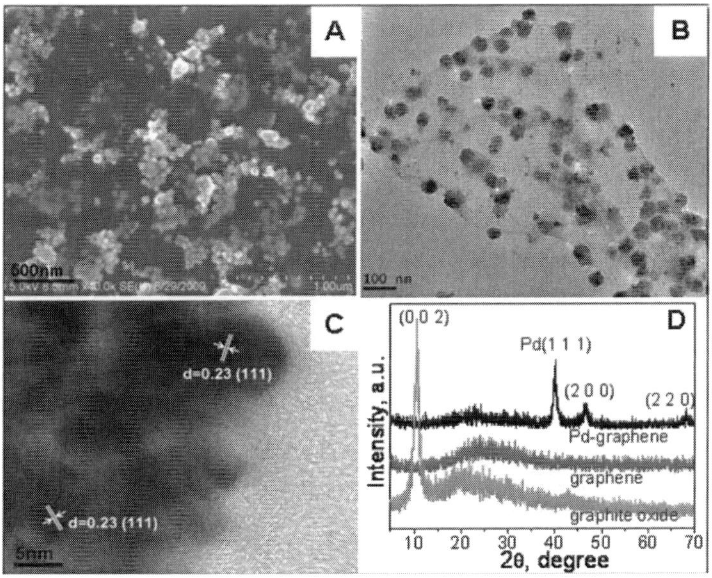

Figure 6.5 SEM (a), TEM (b), HRTEM (c) images, and XRD pattern (d) of Pd-graphene nanocomposite. Reproduced from Shi and Zhu (2011) with permission from Elsevier.

Figure 6.6 shows the Raman spectrum of the in situ synthesized graphene where the similarity of the Pd-graphene compared with graphene and graphene oxide strongly suggests that graphene has been successfully fabricated via the application of power ultrasound. The pertinent question is how is this composite formed? Figure 6.7 illustrates a four-step evolution driven under electric and ultrasonic fields, which can be explained as follows; firstly, the palladium complex ions diffused toward the surface of the sonoelectrode. Then, the palladium complex ions are reduced by the applied, controlled electric pulse to form the palladium primary particles at the electrode surface; these particles are then vibrated from the electrode surface by the applied ultrasound with the concomitant effects on the dispersion of graphene sheets. When the amount of primary palladium particles reaches a critical value, they possibly aggregate into larger palladium clusters spontaneously, and eventually lead to the formation of 3-D spherical palladium nanoparticles. Simultaneously, both the palladium nanoparticles and the residual primary palladium particles, which are dispersed under the ultrasonic field, are directly captured by graphene sheets through ultrasonic assembling processes coupled with van der Waals interactions between the PDDA-functionalized graphene and PDDA-capped palladium nanoparticles. Finally, a well dispersed aqueous solution with palladium–graphene nanocomposite can be obtained by repeating the above steps.

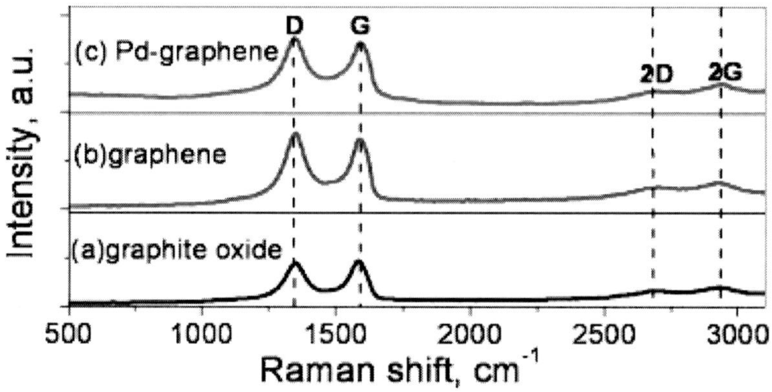

Figure 6.6 Raman spectrum of graphite oxide (a), graphene (b), and Pd-graphene nanocomposite (c). Reproduced from Shi and Zhu (2011) with permission from Elsevier.

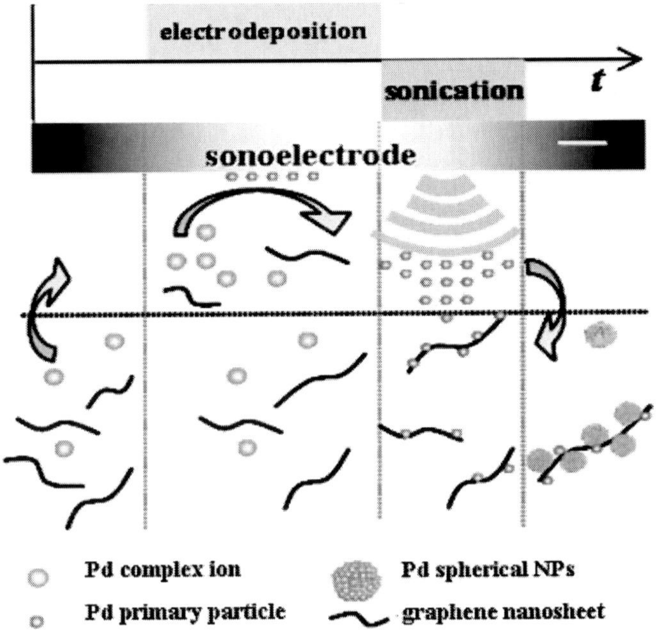

Figure 6.7 The formation of Pd-graphene nanocomposite via sonoelectrochemistry. Reproduced from Shi and Zhu (2011) with permission from Elsevier.

The unique composite was found to be beneficial toward the sensing of 2-chlorophenol at the low micromolar concentration range (Shi and Zhu, 2011). Clearly, this application is translational into forming a range of metal decorated graphene nanocomposites for a range of applications.

6.7 Sonoelectrochemical Production of Noble Metals and Fuel Cell Electrocatalysts

Few studies reported the use of the sonoelectrochemical method to produce noble metals and fuel cell electrocatalysts (Pollet, 2012). As stated previously, here the sonoelectrochemical technique involves either depositing a metal under continuous electrical current and ultrasound or producing nanosized metals at various currents and ultrasonic pulses (a few milliseconds) at a vibrating working electrode (Fig. 6.1).

6.7.1 Silver

Few reports investigated the sonoelectrochemical production of silver using various stabilizers and electrolytes (Zhu et al. 2002; Liu and Lin, 2004). For the first time, the Gedanken group produced silver nanoparticles of controlled shape by the pulse sonoelectrochemical method (Zhu et al., 2000). In their experimental work, they produced silver nanoparticles of various shapes, for example, spheres, rods, and dendrites from a silver nitrate solution ($AgNO_3$) in the presence of ethylenediaminetetraacetic acid (EDTA). They found that both the electrolyte solution and pulse time greatly affect the growth and shape of Ag nanoparticles. For example, in the absence of acetate, silver nanoparticles were not produced at all, but at low EDTA concentrations (<0.1 g/L), only Ag nanoparticles were formed of random phase. However, at high EDTA concentrations (up to 1 g/L), dendritic Ag nanoparticles were favored. They also observed that the silver nitrate concentration and reaction time played an important role in the shape of Ag nanoparticles, for example, spherical (approximately 20 nm), nanorod (10–20 nm), and dendritic Ag nanoparticles were obtained at 5 min, 12 min, and 25 min reaction time, respectively. They attributed the later structure to the excess of silver ions in solutions that favor growth and aggregation into highly ordered dendritic nanostructures. Following this, Socol et al. (2002) successfully synthesized silver nanoparticles again by pulse sonoelectrochemistry in a similar way to that of Zhu et al. (2000). In their study, they proposed a model to elucidate the formation of dendritic silver and found that this effect was mainly caused by intense acoustic streaming induced by ultrasound.

Zhu et al. (2002) found that it is possible to prepare silver nanowires without any template via the sonoelectrochemical route in aqueous solutions containing silver nitrate and EDTA under N_2 atmosphere. In their study, the sonoelectrolysis was performed under controlled-current and controlled-potential; they observed that almost uniform silver nanowires of single crystalline nature were formed by transmission electron microscopy (TEM) and selected-area electron diffraction (SEAD) (Fig. 6.8).

Liu et al. (2001) produced X-ray amorphous silver nanoparticles by the pulse sonoelectrochemical method from silver bromide (AgBr) solution in the presence of gelatine (used as a stabilizer for the precipitation of Ag nanoparticles). They found that gelatine plays an important role in preventing the X-ray amorphous silver from growing larger and helping easy removal of the produced nanoparticles. The silver nanoparticle size varied from 12 to over 30 nm and aggregation occurred with some particles with diameters larger than 50 nm. Liu and Lin (2004) also produced ultrafine silver nanoparticles with diameters less than 2 nm, which were first prepared in an aqueous solution of 0.1 M HCl without the addition of any stabilizer by the sonoelectrochemical methods.

Figure 6.8 The TEM image of multiple Ag nanowires by the controlled-current process (Zhu et al., 2002).

Jiang et al. (2004) reported the fabrication of monodispersed spherical silver nanoparticles with an average size of 20–25 nm with an fcc structure in a saturated solution of silver citrate ($C_6H_5Ag_3O_7$) and in the presence of a surfactant (polyvinylpyrrolidone, PVP).

Cheng and Yao (2005) also produced silver nanoparticles of different shape, that is, sphere, wire, and dendrite by the sonoelectrochemical method from an aqueous solution of $AgNO_3$

in the presence of EDTA disodium salt and PVP. In their work, they obtained spherical silver nanoparticles with the diameter of 30 nm and silver nanowires of 30 nm with the lengths of 200–900 nm. As observed by other researchers, they also found that the Ag dendrites were formed with increasing silver concentration and XRD patterns revealed that Ag nanoparticles were of fcc structure.

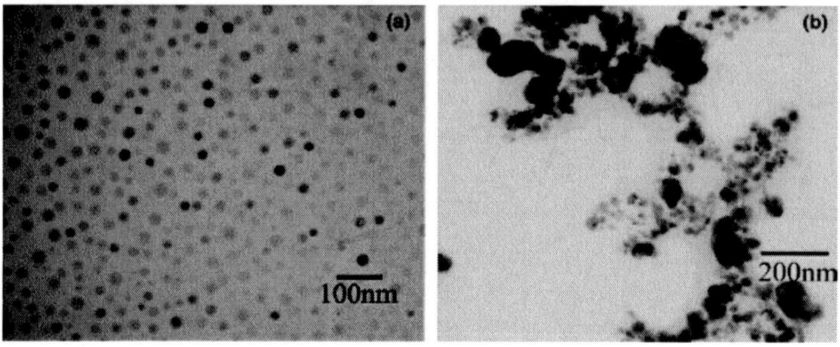

Figure 6.9 Silver nanoparticles prepared under different current density: (a) 70 and (b) 140 mA/cm^2. Reprinted from Jiang et al. (2004) with permission from Elsevier.

Tang et al. (2009) synthesized silver nanostructures with a control in shape via sonoelectrochemistry in an aqueous solution of $AgClO_4$ in the presence of poly(*N*-vinylpyrrolidone)-PVP. They observed that the morphology of silver nanostructure depended strongly upon the applied current and the concentration ratio of [PVP monomer]:[Ag^+]. By controlling the reaction conditions, not only 0-D silver nanostructure of spherical and oval shapes but also 3-D silver nanostructure of dendritic rod, dendritic sheet, and flower-like dendrites of staggered nanosheets was successfully produced. The authors proposed that the growth process depends upon the adjustable reduction current and the PVP concentration contributing to the finer structure in the shape-controlled synthesis.

Recently, Vu et al. (2009) produced silver nanoparticles of sphere and rod shapes by the pulse sonoelectrochemical method from an aqueous solution of $AgNO_3$ in the presence of a surfactant, for example, sodium dodecyl sulfate, $C_{12}H_{25}NaO_4S$ (SDS). By careful characterization (transmission electron microscope, TEM; scanning electron microscope, SEM; XRD; and ultraviolet-visible, UV-vis), they found that the concentrations of $AgNO_3$ and SDS affect the shape of the Ag nanoparticles. By controlling the various electrodeposition and ultrasonic parameters such as pulse duration and currents, the crystal size was varied from 5 nm up to 200 nm.

6.7.2 Gold

Similar to the sonoelectrochemical production of silver, there are few studies for the synthesis of gold nanoparticles. Liu et al. (2004) synthesized size-controlled gold nanoparticles, ranging from 2 nm to 15 nm in diameter sonoelectrochemically from gold substrates in aqueous solutions of 0.1 N HCl in the absence of any stabilizers. They first produced gold and chloride ions containing complexes by potentiodynamically cycling (100 cycles at a scan rate <100 mV/s) an Au electrode in a deoxygenated 0.1 N HCl solution. With the aim of synthesizing Au nanoparticles, the gold working electrode was then immediately replaced by a platinum electrode, and cathodic overpotentials and open circuit potentials (OCPs) were applied to the Pt working electrode under ultrasound. They have successfully produced Au nanoparticles from 2 nm to 15 nm in size and were able to control the size by adjusting the sonoelectrochemical reduction time.

Aqil et al. (2008) produced a high yield of stable suspensions of gold nanoparticles in water by either adding tailor-made polymers or mixtures of commercially available polymers (to minimize the aggregation). In their study, they have used α-methoxy-ω-mercapto-poly(ethylene oxide) and poly(vinyl pyrrolidone)/polyethylene oxide mixtures, which act as a coalescence barrier around the gold nanoparticles. The results showed that the size of the gold nanoparticles could be easily tuned in between 5 and 35 nm by simply controlling the electrochemical and ultrasonic parameters, for example, the deposition time (t_{ON}) from 10 to 20 ms. By comparing the properties of as-prepared gold nanoparticles with the ones of gold colloids prepared by precipitation method using chemical reductive agents, they concluded that the sonoelectrochemical method offers smaller and highly stable Au nanoparticles.

Yang et al. (2010) synthezised higher concentrations of pure Au nanoparticles in acidic electrolytes (0.1 N HCl and NaCl) mixed with chitosan (85% deacetylated or 1 g/L) in the absence of any stabilizers and reductants using the sonoelectrochemical (20 kHz, 100 W) pulse method. They observed that the process was pH-insensitive due to the contribution of the capping of chitosan whereby up to 100 ppm of Au nanoparticles of approximately 12 nm in size were obtained.

Recently, Shen et al. (2011) produced gold nanoparticles (Fig. 6.10) by the pulse sonoelectrochemical method in aqueous solutions and in the absence of any stabilizers using hydrogen tetrachloroaurate tetrahydrate ($HAuCl_4.4H_2O$), potassium nitrate (KNO_3), hydrochloric acid (HCl), and sodium hydroxide (NaOH). They found that the size and shape of the Au nanoparticles can be controlled by changing the current density, reaction time, and the pH of the precursor solution. Contrary to the work of Yang et al. (2010), they observed that the pH has a great effect on the size and dispersion of the produced gold nanoparticles. Characterization studies revealed that the Au nanoparticles (~20 nm) had cubic structure (fcc) with a high crystallinity.

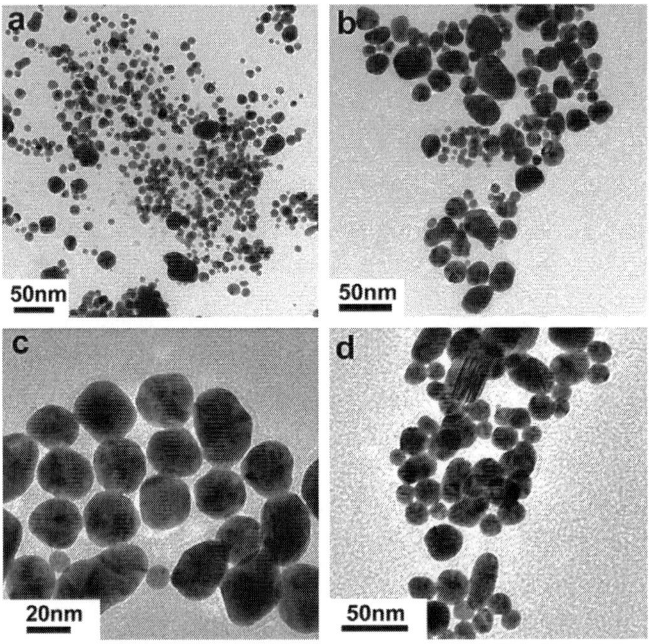

Figure 6.10 TEM images of the Au nanoparticles synthesized by the sonoelectrochemical method at different pulse times ($t_{current}/t_{sono}$). (a) 0.3 s/0.3 s, (b) 0.4 s/0.4 s, (c) 0.5 s/0.5 s, (d) 1.0 s/1.0 s (Shen et al. 2011).

6.7.3 Effect of Surfactants and Polymers

Shen et al. (2008) produced uniform and spherical-shaped three-dimensional dendritic Pt nanostructures (DPNs) with an average dimension of 2.5 ± 0.5 nm, at room temperature sonoelectrochemically (20 kHz; 20 W; pulse width of the current = 1.0 s, resting time of the current pulse = 0.5 s; duration of the ultrasonic pulse = 0.3 s) by using mixtures of hexachloroplatinum(IV) acid hydrate in the presence of various surfactants [PVP, poly(ethylene glycol)$_{20}$-poly(propylene glycol)$_{20}$-poly(ethyleneglycol)$_{20}$ (P123), SDS and poly-diallyl-dimethyl ammonium chloride (PDDA)] at current densities up to 40 mA/cm^2. They showed that at low current densities, the reduction rate of Pt was slower, the number of nuclei was small, and the rate of growth was faster than that of nucleation, which led to larger nanoparticles (5–7 nm). As the current density increased, the reduction rate of Pt^{4+} increased and the nucleation rate was higher than that of the growth leading to the generation of more nuclei and the formation of smaller primary Pt nanoparticles. They also showed that stabilized Pt nanoparticles do not aggregate in PVP due to stronger bonds between the Pt precursors with the C=O groups of the PVP; in other words, PVP adheres to the nanoparticles through a charge-transfer interaction between the pyrrolidone rings and Pt atoms. The DPNs showed improved electrocatalytic activity toward the methanol oxidation reaction (MOR) due to monodispersed Pt nanoparticles, and improved porosity structure leading to larger effective surface area. They proposed a mechanism (Fig. 6.11) whereby Pt ions are reduced by the electrical current and formed primary Pt nanoparticles, which are subsequently dislodged by the vibrating electrode. The primary nanoparticles in solution then spontaneously assembled together and formed smaller spherical DPNs. They showed that ultrasound leads to smaller primary nanoparticles in favor of the crystallite reorganization and growth of a stable near-single crystal.

Shen et al. (2009) also showed that by using the sonoelectrochemical (20 kHz; 20 W; pulse on time of the current = 0.5 s; pulse off time of the current = 0.5 s, duration of the ultrasonic pulse = 0.3 s) method, it is possible to realize the morphologically controlled synthesis of palladium nanostructures [spherical (SNP), multitwinned (MTP) and sponge-like (SSP)] at room temperature in the presence of various surfactants and polymers [cethyltrimethylammoniumbromide (CTAB), PVP, and PDDA] (Fig. 6.12).

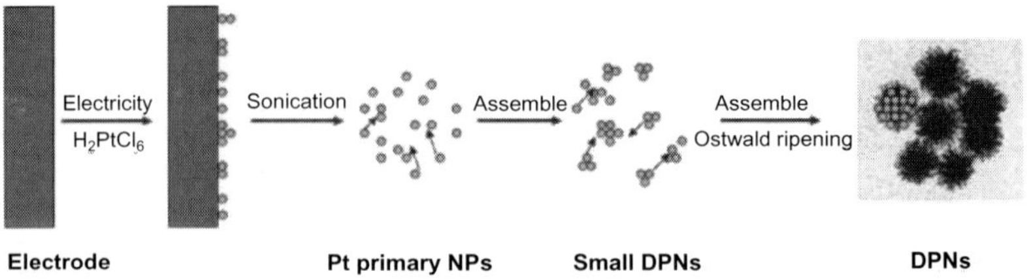

Figure 6.11 Schematic illustration of the formation of DPNs (Shen et al. 2008).

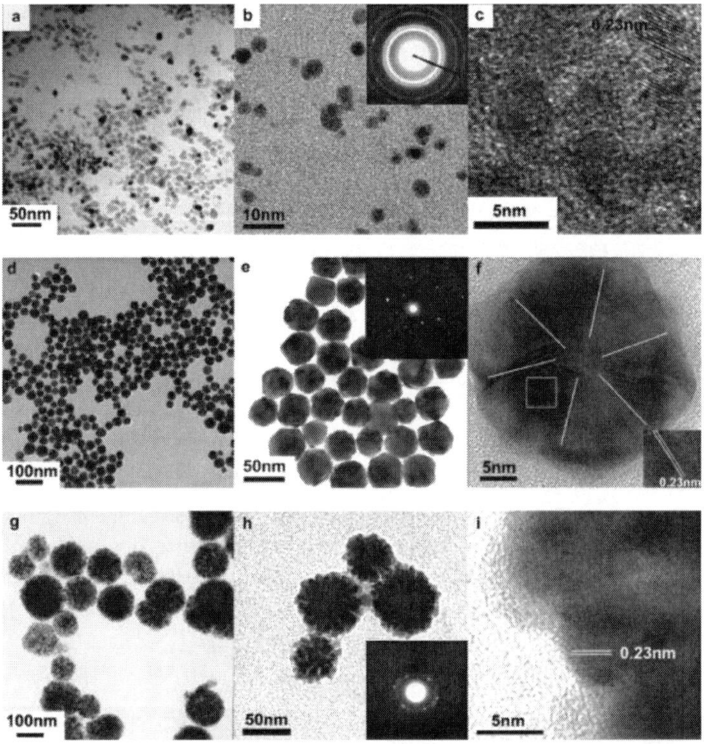

Figure 6.12 TEM and HR-TEM images of the Pd nanostructures prepared in different solutions. (a–c) Spherical nanoparticles (SNPs) prepared in CTAB solution. (d–f) Multitwinned particles (MTPs) prepared in PVP solution. (g–i) Spherical sponge-like particles (SSPs) prepared in PDDA solution. The insets in parts b, e, and h are SAED patterns of the corresponding Pd nanostructures. The inset in part (f) is the enlarged image of the square zone area. The white lines in part (f) indicate the twin boundaries separating the different adjacent facets (Shen et al. 2009).

They showed that the size and shape of the Pd nanostructures may be controlled by varying the current density and the precursor solution pH. Furthermore, the electrocatalytic activities of the produced sponge-like Pd nanostructures for direct alcohol oxidation in alkaline media showed higher electrochemical active surface than other Pd nanostructures. Qiu et al. (2003) also synthesized highly dispersed spherical Pd nanoparticles of a dendritic superstructure in the presence of CTAB by the pulse sonoelectrochemical (20 kHz) method at room temperature and by employing a reaction time above 2.5 h. They explained that the dendritic-structured Pd nanoparticles had a treelike structure and agreed with the diffusion-limited aggregation (DLA) model, involving cluster formation by the adhesion of particles to a selected seed on contact and allowing the particle to diffuse and stick to the growing structure. They stipulated that it is possible that Pd particles reach the anode and grow into a dendritic structure. They concluded that the shape and size of spherical nanocrystalline Pd may be controlled by varying the current density (8–13 mA/cm^2), the interval between continuous ultrasonic pulses, ultrasonic intensity (20–120 W/cm^2), and the CTAB concentration. For example, they observed that the shape of the nanoparticles appeared irregular and agglomerated in the region of 20 W/cm^2 and 120 W/cm^2.

6.7.4 Effect of Aqueous Solutions

As stated previously, recently Zin et al. (2010) produced platinum nanoparticles from aqueous chloroplatinic solutions by the pulse sonoelectrochemical (20 kHz, up to 118 W/cm^2) method on titanium alloy electrodes in the absence of any surfactants, alcohols, and polymers by producing short applied current pulses triggered and followed immediately by ultrasonic pulses at the working electrode (in this case the cathode). The time management sequence employed was as follows:

(1) A short current pulse of $|i|$ = 50 mA/cm^2 was applied to the sonoelectrode, whereby the titanium horn acted as an electrode only (t_{ON}); the time of this phase typically varied between 0.3 s and 0.5 s.

(2) Immediately after turning-off the electrochemical pulse, an ultrasonic pulse was sent to the sonoelectrode and here it

acted only as a vibrating ultrasonic horn (t_{US}); this second phase lasted no more than 0.5 s.

(3) A rest time, t_p, followed the two previous phases (this was useful to restore the initial electrolyte conditions close to the sonoelectrode).

They showed that the mean grain size of Pt ranging from 11 nm to 15 nm was produced and globular clusters had a mean size ranging between 100 nm and 200 nm, which in turn aggregated and built the complex structures (Fig. 6.13).

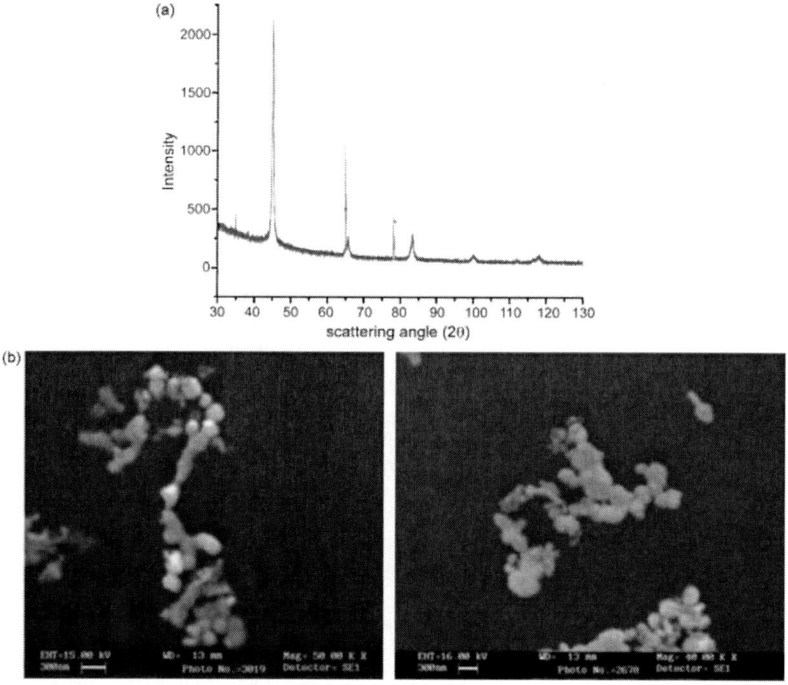

Figure 6.13 (a) XRD pattern of Pt nanopowders and (b) SEM micrograph of Pt nanoparticles (Zin et al. 2010).

For all the sonoelectrochemical production of nano-sized metals, a mechanism has been proposed, wherein metallic ions are reduced by a short current pulse to produce metallic nanoparticles on the sonoelectrode surface, which are then dislodged by the ultrasonic pulse. The metallic nanoparticles in solution tend to spontaneously assemble together whereby under insonation, the Ostwald ripening process, is accelerated, leading to smaller primary nanoparticles (Qiu et al., 2003; Shen et al., 2008, 2009).

6.8 Conclusions

The pulse sonoelectrochemical method, which combines ultrasound and electrochemistry, is an effective route for the shape-controlled production of various nanomaterials. In this chapter, it is shown that this method is simply accomplished by applying a short electrical current pulse (milliseconds) to create nucleation at the sonoelectrode surface, followed by a short burst of ultrasound (milliseconds) to remove the nanomaterial from the ultrasonic probe, which is acting as the cathode. It possesses the advantages of pulse sonoelectrodeposition, and the shape and size of the synthesized nanomaterials can be controlled by simply adjusting different parameters such as current, deposition time, sonication time, electrolyte temperature, pH, and surfactant/stabilizer concentrations.

Indeed, this novel method has been proven to be simple and fast for the preparation of polymeric, carbonaceous, metallic, noble, and fuel cell electrocatalyst nanoparticles. The pulse sonoelectrochemical technique has an advantage over many other methods due to the unusual experimental conditions caused by cavitation, water sonolysis, and enhanced mass-transport phenomenon. It is also clearly shown in this chapter that ultrasound can be used as a "reducing agent" to form materials of nanometer size. Overall, the sonoelectrochemical method is expected to be promising in materials chemistry and fuel cell technologies, as it is a cost-effective, easy-to-use, and less time-consuming than any other conventional methods.

References

Ahmadi, T. S., Wang, Z. L., Green, T. C., Henglein, A. and El-Sayed, M. A. (1996). Shape-controlled synthesis of colloidal platinum nanoparticles, *Science.* **272**, p. 1924.

Alivisatos, P. (2004). The use of nanocrystals in biological detection, *Nat. Biotechnol.*, **22**, p. 47.

Akiyama, T., Kawahara, T., Arakawa, T. and Yamada, S. (2008). Fabrication of densely packed gold nanoparticle films and their fluorescence enhancement effect, *Jpn. J. Appl. Phys.*, **47**, p. 3063.

Akkermans, R. P., Ball, J. C., Marken, F. and Compton, R. G. (1998). The use of sonotrodes for electroanalysis: sono-ASV detection of lead in aqueous solution, *Electroanal.*, **10**, p. 26.

Aqil, A., Serwas, H., Delplancke, J. L., Jerome, R., Jerome, C. and Canet, L. (2008). Preparation of stable suspensions of gold nanoparticles in water by sonoelectrochemistry. *Ultrason. Sonochem.*, **15**, p. 1055.

Atobe, M., Ishikawa, K., Asami, R. and Fuchigami, T. (2009). Size-controlled synthesis of conducting-polymer microspheres by pulsed sonoelectrochemical polymerization, *Angew. Chem. Int. Ed.*, **48**, p. 6069.

Aurbach, D., Lu, Z., Schechter, A., Gofer, Y., Gizbar, H., Turgeman, R., Cohen, Y., Moshkovich, M. and Levi, E. (2000). Prototype systems for rechargable magnesium batteries. *Nature*, **407**, p. 724.

Balamurugan, B., Aruna, I., Mehta, B. R. and Shivaprasad, S. M. (2004). Size-dependent conductivity-type inversion in Cu_2O nanoparticles. *Phys. Rev. B.*, **69**, p. 1654191.

Bang, J. H., Suslick, K. S. (2010) Applications of ultrasound to the synthesis of nanostructured materials. *Adv. Mater.*, **22**, pp. 1039–1059.

Banks, C. E., Compton, R. G. (2003). Voltammetric exploration and applications of ultrasonic cavitation, *Chem. Phys. Chem.*, **4**, p. 169.

Banks, C. E., Compton, R. G. (2004). Ultrasound: promoting electroanalysis in difficult real world media. *Analyst*, **129**, p. 678.

Banks, C. E., Compton, R. G. (2007), *Understanding Voltammetry*. Imperial College Press, London.

Brownson, D. A. C., Banks, C. E. (2010). Graphene electrochemistry: an overview of potential applications. *Analyst*, **135**, p. 2768.

Brownson, D. A. C., Kampouris, D. K. and Banks, C. E. (2011). An overview of graphene in energy production and storage applications, *J. Power Sourc.*, **196**, p. 4873.

Cheng, J., Suwei, Y. (2005). Synthesis and characterization of silver nanoparticles by sonoelectrodeposition. *Rare Metals*, **24**, p. 376.

Compton, R. G., Eklund, J. C., Marken, F., Rebbitt, T. O., Akkermans, R. P. and Waller, D. N. (1997). Dual activation: coupling ultrasound to electrochemistry—an overview. *Electrochim. Acta*, **42**, p. 2919.

Connor, J. H., Reid, W. E. and Wood, G. B. (1957). Electrodepostion of metals from organic solutions. *J. Electrochem. Soc.*, **104**, p. 38.

Dabala, M., Pollet, B. G., Zin, V., Campadello, E. and Mason, T. J. (2008). Sonoelectrochemical (20 kHz) production of $Co_{65}Fe_{35}$ alloy nanoparticles from aotani solutions. *J. Appl. Electrochem.*, **38**, p. 395.

Delplancke, J. L., Dille, J., Reisse, J., Long, G. J., Mohan, A. and Grandjean, F. (2000). Magnetic nanopowders: ultrasound-assisted electrochemical preparation and properties. *Chem. Mater.*, **12**, p. 946.

Eklund, J. C., Marken, F., Waller, D. N. and Compton, R. G. (1995). Voltammetry in the presence of ultrasound: a novel sono-electrode geometry. *Electrochim. Acta,* **41**, p. 1541.

Floate, S., Hardcastle, J. L., Cordemans, E. and Compton, R. G. (2002). A sonotrode for electroanalysis: the determination of copper in passivating media. *Analyst,* **127**, p. 1094.

Galal, A. (2006). Characterization of conducting poly(3-methylthiophene) films prepared under sono-electrochemical conditions. *J. Appl. Polym. Sci.*, **102**, p. 2416.

Ganesan, R., Shanmugam, S. and Gedanken, A. (2008). Pulsed sonoelectrochemical synthesis of polyaniline nanoparticles and their capacitance properties. *Synth. Met.*, **158**, p. 848.

Guzman, M., Delplancke, J. L., Long, G. J., Delwiche, J., Hubin-Franskin, M. J. and Grandjean, F. (2002). Morphologic and magnetic properties of Pd_{100}-xFe_x nanoparticles prepared by ultrasound assisted electrochemistry. *J. Appl. Phys.*, **92**, p. 2634.

Haas, I. Gedanken, A. (2006). Sonoelectrochemistry of Cu^{2+} in the presence of cetyltrimethylammonium bromide: obtaining CuBr instead of copper. *Chem. Mater.*, **18**, p. 1184.

Haas, I., Shanmugam, S. and Gedanken, A. (2006). Pulsed Sonoelectrochemical Synthesis of Size-Controlled Copper Nanoparticles Stabilized by Poly(N-vinylpyrrolidone). *J. Phys. Chem. B*, **110**, p. 16947.

Haas, I. Gedanken, A. (2008). Synthesis of metallic magnesium nanoparticles by sonoelectrochemistry. *Chem. Commun.*, **15**, p. 1795.

Handley, D. A. (1989). *Colloidal Gold Principles, Methods, and Applications, Vol. 1 (Ed.: M. A. Hayat).* Academic Press, New York, p. 13.

Jiang, L.P., Wang, A.N., Zhao, Y., Zhang, J.R., Zhu, J. J. (2004). A novel route for the preparation of monodisperse silver nanoparticles *via* a pulsed sonoelectrochemical technique. *Inorg. Chem. Commun.*, **7**, p. 506.

Kooi, B. J., Palasantzas, G., De Hosson, J. Th. M. (2006). Gas-phase synthesis of magnesium nanoparticles: a high-resolution transmission electron microscopy study *Appl. Phys. Lett.*, **89**, p. 161914.

Lei, H., Tang, Y.-J., Wei, J.-J., Li, J., Li, X.-B. and Shi, H.-L. (2007). Synthesis of tungsten nanoparticles by sonoelectrochemistry, *Ultrason. Sonochem.*, **14**, p. 81.

Liu, S., Huang, W., Chen, S., Avivi, S. and Gedanken, A. (2001). Synthesis of X-ray amorphous silver nanoparticles by the pulse sonoelectrochemical method. *J. Non-Cryst. Solids*, **283**, p. 231.

Liu, Y.-C., Lin, H.-L., and Chiu, W.-H. (2004). Size-controlled synthesis of gold nanoparticles from bulk gold substrates by sonoelectrochemical methods. *J. Phys. Chem. B.*, **108**, p. 19237.

Liu, Y.-C., Lin, L. (2004). New pathway for the synthesis of ultrafine silver nanoparticles from bulk silver substrates in aqueous solutions by sonoelectrochemical methods. *Electrochemistry Commun.*, **6**, p. 1163.

Mahendiran, C., Ganesan, R., Gedanken, A. (2009) Sonoelectrochemical synthesis of metallic. aluminum nanoparticles. *Eur. J. Inorg. Chem.*, **14**, p. 2050.

Mancier, V., Daltin, A.-L., Leclercq, D. (2008). Synthesis and characterization of copper oxide (I) nanoparticles produced by pulsed sonoelectrochemistry. *Ultrason. Sonochem.*, **15**, p. 157.

Mancier, V., Delplancke, J.-L., Delwiche, J., Hubin-Franskin, M.-J., Piquer, C., Rebbouh, L. and Grandjean, F. (2004). Morphologic, magnetic, and mössbauer spectral properties of $Fe_{75}Co_{25}$ nanoparticles prepared by ultrasound-assisted electrochemistry. *J. Magn. Magn. Mater.*, **281**, p. 27.

Mancier, V., Rousse-Bertrand, C., Dille, J., Michel, J., Fricoteaux, P. (2010). Sono and electrochemical synthesis and characterization of copper core–silver shell nanoparticles. *Ultrason. Sonochem.*, **17**, p. 690.

Masti, V., Polsky, V., Koltypin, V., Gedanken, A. (2000). Pulsed sonoelectrochemical synthesis of cadmium selenide nanoparticles. *J. Am. Chem. Soc.*, **121**, p. 10047.

Panyam, J., Labhasetwar, V. (2003). Biodegradable nanoparticles for drug and gene delivery to cells and tissue. *Adv. Drug Delivery Rev.*, **55**, p. 329.

Pollet, B. G., Hihn, J. Y. (2011). Sonoelectrochemistry – from theory to applications, Chapter 24 in *Handbook on Applications of Ultrasound – Sonochemistry for Sustainability*, Eds: D. Chen, S. K. Sharma, A. Mudhoo, CRC Press, p. 623.

Pollet, B. G. (2010) The use of ultrasound for the fabrication of fuel cell materials. *Int J Hydrogen Energy*, **35**(21), p. 11986.

Pollet, B. G. (2012), *Power Ultrasound in Electrochemistry: From Versatile Laboratory Tool to Engineering Solution*, Ed. B.G. Pollet, John Wiley & Sons.

Pollet, B. G., Valzer, E. F., Curnik, O. J. (2011). Platinum sonoelectrodeposition on glassy carbon and gas diffusion layer electrodes. *Int. J. Hydrogen Energy*, **36**, p. 6248.

Qiu, X.-F., Lou, Y., Samia, A. C. S., Devadoss, A., Burgess, J. D., Dayal, S., Burda, C. (2005). PbTe nanorods by sonoelectrochemistry. *Angew. Chem. Int. Ed.*, **44**, p. 5855.

Qiu, X.-F., Xu, J.-Z., Zhu, J.-M., Zhu, J.-J., Zu, S., Chen, H.-Y. (2003). Controllable synthesis of palladium nanoparticles via a simple sonoelectrochemical method. *J. Mater. Res.*, **18**, p. 1399.

Rabani, E., Reichman, D. R., Geissler, P. L., Brus, L. E. (2003). Drying-mediated self-assembly of nanoparticles. *Nature*, **426**, p. 271.

Reisse, J., Francois, H., Vandercammen, J., Fabre, O., Kirsch-de Mesmaeker, A., Maerschalk, C., Delplancke, J. L. (1994). Sonoelectrochemistry in aqueous electrolyte: a new type of sonoelectroreactor. *Electrochim. Acta*, **39**, p. 37.

Sergeev, G. B. (2003). Cryochemistry of metal nanoparticles. *J. Nanopart. Res.*, **5**, 529.

Shen, Q., Jiang, L., Zhang, H., Min, Q., Hou, W., Zhu, J.-J. (2008). Three-dimensional dendritic Pt nanostructures: sonoelectrochemical synthesis and electrochemical applications. *J. Phys. Chem. C*, **112**, p. 16385.

Shen, Q., Min, Q., Shi, J., Jiang, L., Hou, W., Zhu, J.-J. (2011). Synthesis of stabilizer-free gold nanoparticles by pulse sonoelectrochemical method, *Ultrason. Sonochem.*, **18,** p. 231.

Shen, Q., Min, Q., Shi, J., Jiang, L., Zhang, J.-R., Hou, W., Zhu, J.-J. (2009). Morphology-controlled synthesis of palladium nanostructures by sonoelectrochemical method and their application in direct alcohol oxidation. *J. Phys. Chem. C*, **113**, p. 1267.

Shi, J.-J., Zhu, J.-J. (2011). Sonoelectrochemical fabrication of Pd-graphene nanocomposite and its application in the determination of chlorophenols. *Electrochim. Acta*, **56**, p. 6008.

Shipway, A. N., Katz, E., Willner, I. (2000) Nanoparticle arrays on surfaces for electronic, optical and sensor applications. *Chem. Phys. Chem.*, **1**, p. 18.

Socol, Y., Abramson, O., Gedanken, A., Meshorer, Y., Berenstein, L., Zaban, A. (2002). Suspensive electrode formation in pulsed sonoelectrochemical synthesis of silver nanoparticles. *Langmuir*, **18**, p. 4736.

Sztrum, C. G., Hod, O., Rabani, E. (2005). Self-assembly of nanoparticles in three-dimensions: formation of stalagmites. *J. Phys. Chem. B*, **109**, p. 6741.

Sztrum, C. G., Rabani, E. (2006). Out-of-equilibrium self-assembly of binary mixtures of nanoparticles. *Adv. Mater.*, **18**, p. 565.

Tang, S., Meng, X., Lu, H., Zhu, S. (2009). PVP-assisted sonoelectrochemical growth of silver nanostructures with various shapes. *Mater. Chem. Phys.*, **116**, p. 464.

Turkevich, J., Stevenson, P. C., Hillier, J. (1951). A study of the nucleation and growth processes in the synthesis of colloidal gold. *Discuss. Faraday Soc.*, **11**, p. 55.

Vu, L. V., Long, N. N., Doanh, S., Trung, B. Q. (2009). Preparation of silver nanoparticles by pulse sonoelectrochemical method and studying their characteristics. *J. Phys.: Conf. Ser.* **187**, p. 012077.

Wang, J. (2006), *Analytical Electrochemistry*. Wiley and Sons.

Yang, K. H., Liu, Y. C., Hsu, T. C. and Tsai, H. I. (2010). pH-insensitive fabrication of gold nanoparticles with high concentration by ultrasound-assisted electrochemical process via aid of chitosan. *Mater. Res. Bull.*, **45**, p. 63.

Yee, C., Scotti, M., Ulman, White, H., Rafailovich, M. and Sokolov, J. (1999). One-phase synthesis of thiol-functionalized platinum nanoparticles. *Langmuir*, **15**, p. 4314.

Yosef, G., Rabani, E. (2006). Self-assembly of nanoparticles into rings: a lattice-gas model. *J. Phys. Chem. B*, **110**, p. 20965.

Zhu, J., Aruna, S. T., Koltypin, Y. and Gedanken, A. (2000). A novel method for the preparation of lead selenide: pulse sonoelectrochemical synthesis of lead selenide nanoparticles. *Chem. Mater.*, **12**, p. 143.

Zhu, J.-J., Qiu, Q.-F., Wang, H., Zhang, J.-R., Zhu, J.-M. and Chen, Z.-M. (2002). Synthesis of silver nanowires by a sonoelectrochemical method, *Inorganic Chem. Commun.*, **5**, p. 242.

Zin, V., Campadello, E., Zanella, A., Brunelli, K., Dabalà, M. (2010). Synthesis and characterization of metal nanoparticles produced by sonoelectrochemistry. *La Metallurgia Italiana*, **1**, p. 29

Zin, V., Dabalà, M. (2010). Iron–chromium alloy nanoparticles produced by pulsed sonoelectrochemistry: synthesis and characterization. *Acta Mater.*, **58**, p. 311.

Zin, V., Pollet, B. G., Dabalà, M. (2009a). Sonoelectrochemical (20 kHz) production of platinum nanoparticles from aqueous solutions. *Electrochim. Acta*, **54**, p. 7201.

Zin, V., Zanella, A., Agnoli, A., Brunelli, K., Dabalà, M. (2009b). Sonoelectrochemical synthesis of FeCo nanoparticles: study of the effects of bath's composition on process efficiency and particles features. *Curr. Nanosci.*, **5**, p. 232.

Chapter 7

Preparation of Nanomaterials Under Combined Ultrasound/Microwave Irradiation

Giancarlo Cravotto and Luisa Boffa
Dipartimento di Scienza e Tecnologia del Farmaco,
University of Turin, via Pietro Giuria n. 9, 10125 Torino, Italy
giancarlo.cravotto@unito.it

The aim of this chapter is to shed light on the recent combined use of power ultrasound (US) and microwaves (MW) in the preparation of nanomaterials. After a general overview of this hybrid technique and applications, its specific additive or synergic effects will be thoroughly described. Maeda and Amemiya first described the surprising synergic effects in sono-luminescence and chemi-luminescence experiments carried out under simultaneous US/MW irradiation (Maeda and Amemiya, 1995). Although sequential or simultaneous irradiation with US and MW sources entails technical and safety considerations, their outstanding potential in chemico-physical modifications is now well established (Cravotto and Cintas, 2006, 2007).

Cavitation: A Novel Energy-Efficient Technique for the Generation of Nanomaterials
Edited by Sivakumar Manickam and Muthupandian Ashokkumar

ISBN 978-981-4411-54-7 (Hardcover), 978-981-4411-55-4 (eBook)
www.panstanford.com

7.1 Introduction

Combined ultrasound (US)/microwave (MW) irradiation can be performed simultaneously or otherwise sequentially by circulating the reacting mixture. US can be conveyed inside a modified MW oven by inserting a non-metallic horn made of quartz, pyrex®, or ceramic through its wall and down into the reaction vessel. Cravotto's group experimented on various non-metallic horns together with Danacamerini (Torino, Italy) and found that a particular type of PEEK® containing a small amount of glass fiber was a good solution. Simultaneous US/MW irradiation usually requires a cooling system to keep the temperature under strict control so as to avoid superheating that would decrease the acoustic cavitation. Efficient cooling can be achieved by circulating a refrigerating fluid that is transparent to MW, for example, silicone oil or Galden® (Solvay-Solexis), a perfluoropolyether with a high boiling point and low viscosity (Fig. 7.1). Combined irradiation can otherwise be achieved through the use of flow loop-reactors for sequential US/MW treatment that allow the use of commercially available metallic horns. In these reactors, a pump circulates the reacting mixture through two separate reaction cells, one placed inside the MW oven and the other (fitted with a US probe) standing outside it (Cravotto et al., 2005) (Fig. 7.2).

Figure 7.1 Simultaneous US/MW irradiation with a cooling system.

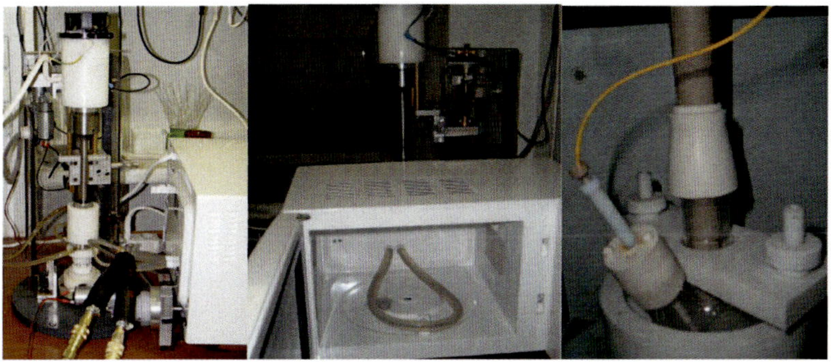

Figure 7.2 US/MW loop reactor with a peristaltic pump and cooling by coaxial tubings.

Separate US and MW irradiation have been widely used for the preparation of nanomaterials as documented by several reviews (Kerner et al., 2001; Gedanken, 2004; Dahl et al., 2007). Both techniques contribute to greener nanosyntheses, shorter reaction times, reduced energy consumption, and better product yields. Such remarkable improvements provide a unique platform for the growth of novel nanostructures (Nemamcha et al., 2006). The possibility of scaling up the production of nanoparticles using these methodologies for industrial applications is truly noteworthy (Gerbec et al., 2005; Leonelli and Mason, 2010). Many studies have explored the effects of temperature, solvent, reactants, and reaction times on the formation of nanoparticles. Although MW and US irradiation have proven to be a useful tool in synthesizing nanomaterials, they have some limitations. Both have in fact limited penetration depth and, in some cases, poor experimental repeatability (Nikolai, 2002).

7.2 Combined Use of Ultrasound and Microwaves

In recent years, a few studies have documented the favorable combination of US and MW for the preparation of nanomaterials. Ishikawa et al. (2008) described the preparation of platinum nanoparticles in a heterogeneous solid-liquid system by simultaneous US and MW irradiation. The average size of Pt nanoparticles was particularly fine with an excellent homogeneous distribution.

Tai and Guo (2008) developed a US/MW-assisted process for the controlled synthesis of cadmium sulfide 3D nanostructures (flower-like hexagonal nanopyramids and/or nanoplates) of high purity. This efficient method was fast, seedless, template-free, and environmentally friendly. A powder mixture with $CdCl_2$, thiourea (Tu), or thioacetamide (TAA) and sulfur was placed in a combined US/MW reactor (Fig. 7.3) and irradiated at 140°C for 45 min (MW and US power 400 W and 60 W/cm^2, respectively).

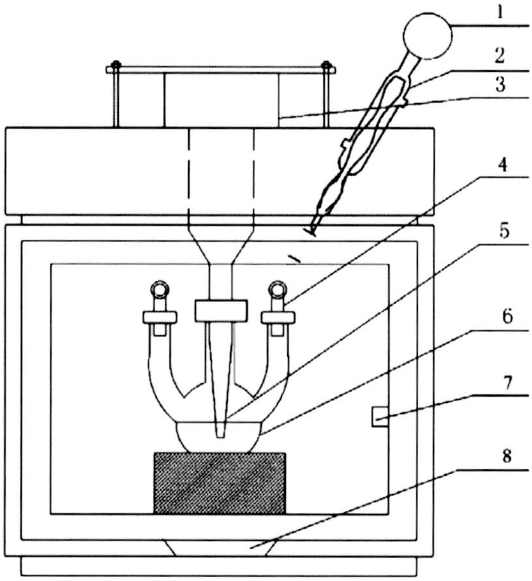

Figure 7.3 US/MW reactor. (1) Balloon, (2) condenser, (3) US transducer, (4) glass pipe, (5) US probe [from Xinzhi Co., China, JY92-2D, with a 10 mm diameter horn (20 kHz) working in a pulsed mode with a duty cycle of 0.5 s], (6) reactor, (7) infrared thermometer and (8) MW generator. Reprinted from *Ultrason. Sonochem.*, 15, Tai, G. and Guo, W., Sonochemistry-assisted microwave synthesis and optical study of single-crystalline CdS nanoflowers, 350-356, Copyright (2008), with permission from Elsevier.

The treatment time was a critical point for the production of well-defined CdS nanoflowers. After 15 min of treatment, the nanopyramids in the nanoflowers are changing progressively (Fig. 7.4a), and after 30 min they become thin, nearly hexagonal flakes (Fig. 7.4b).

The optimal irradiation time was 45 min (Fig. 7.4d,e), while prolonged treatment (60 min, Fig. 7.4c) gave a mixture of hexagonal nanopyramids and nanoplates.

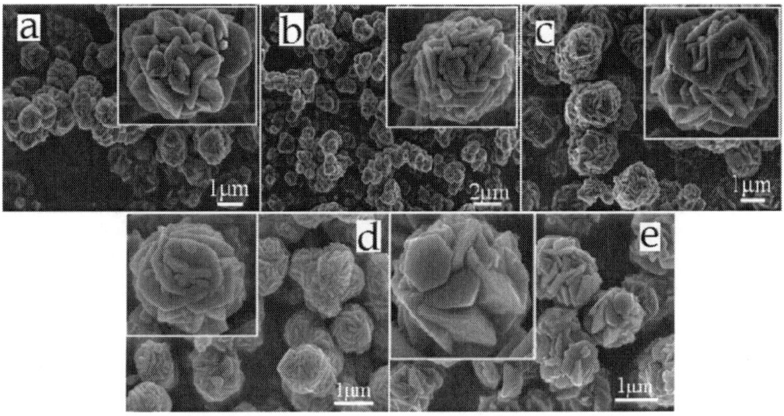

Figure 7.4 FESEM images of CdS nanoflower clusters after 15 (a), 30 (b), 60 (c) and 45 min (d, e) US/MW irradiation at 140°C with $CdCl_2$, sulfur powder and other different sulfur sources [Tu (a,b,c,d) and TAA (e)]. Reprinted from *Ultrason. Sonochem.*, 15, Tai, G. and Guo, W., Sonochemistry-assisted microwave synthesis and optical study of single-crystalline CdS nanoflowers, 350–356, Copyright (2008), with permission from Elsevier.

The authors compared these products with those obtained by the application of MW irradiation only. The sole MW heating generated a mixture of triangular and hexagonal pyramid nanostructures and a large amount of residual sulfur powder was detected in the resulting products. The combined US/MW synthesis method could fully convert sulfur powders into CdS, as shown by the XRD patterns (Fig. 7.5).

The transient cavitation produced by US irradiation reduced sulfur powder aggregation and promoted the effective reaction between the sulfur powders and cadmium ions. Moreover, US could avoid or reduce the localized overheating that may be caused by MW heating and produce more uniform nanostructures. Such synergistic effects provided the optimal conditions to the complete conversion of sulfur sources into CdS hexagonal nanopyramids and/or nanoplates and to their regular stacking into 3D nanoflowers.

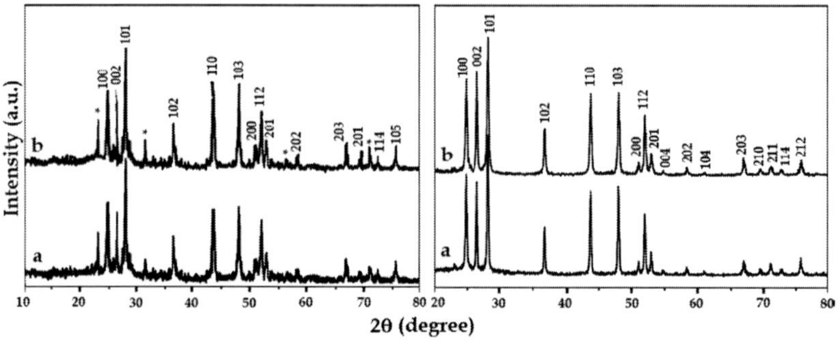

Figure 7.5 XRD patterns of CdS nanostructures prepared by only MW (on the left) or by US/MW irradiation (on the right) at 140°C for 45 min with reacting $CdCl_2$ and different sulfur sources: (a) TAA and sulfur powder; (b) Tu and sulfur powder. The peaks marked * are from sulfur powders. Reprinted from *Ultrason. Sonochem.*, 15, Tai, G. and Guo, W., Sonochemistry-assisted microwave synthesis and optical study of single-crystalline CdS nanoflowers, 350–356, Copyright (2008), with permission from Elsevier.

Ag-doped CdS nanoparticles have been successfully synthesized by Ma et al. (2010) under US/MW irradiation using the combined reactor as shown in Fig. 7.3. Cadmium acetate, thiourea, and sulfur powders in the presence of $AgNO_3$ were irradiated for 15 min (140°C). The authors investigated the contributions of US and MW, carrying out blank tests: MW alone, US alone, and mechanical stirring. Although desired products could be obtained in small amounts under simple MW irradiation, they were highly contaminated by unreacted sulfur powder. Much longer reaction times were required under simple US irradiation or mechanical stirring affording quite inhomogeneous products. Combined US/MW irradiation was considered the method of choice for the production of noble metal doped semiconductor nanostructures. Thanks to fast and homogeneous volumetric MW heating and simultaneous pulsed sonication, the sulfur powder was finely dispersed and Ag easily penetrated the CdS matrix.

Recently, Tai et al. (2010) published a further application on the combined US/MW irradiation in the synthesis of CdS nanoparticles from a zinc-blend to wurtzite. Shen studied the US/MW combination for the rapid synthesis of Pb(OH)Br nanowires (Shen, 2009). The ionic liquid 1-butyl-3-methylimidazolium bromide ([BMIM]Br) was employed both as a reactant and as a structure-directing agent in this reaction.

A mixture of $Pb(OAc)_2$ and [BMIM]Br was subjected to US/MW irradiation (50 W, 10 min) (Fig. 7.6). For comparison, the synthesis was also performed under the same conditions using conventional heating and MW or US separately. The synthesis under conventional heating was carried out in a water bath at 70°C for 24 h (yield 23%). The wires produced (Fig. 7.7a, 7.7b) typically crystallized as multi-angular prisms of 20–30 μm in diameter and 2–3 mm in length.

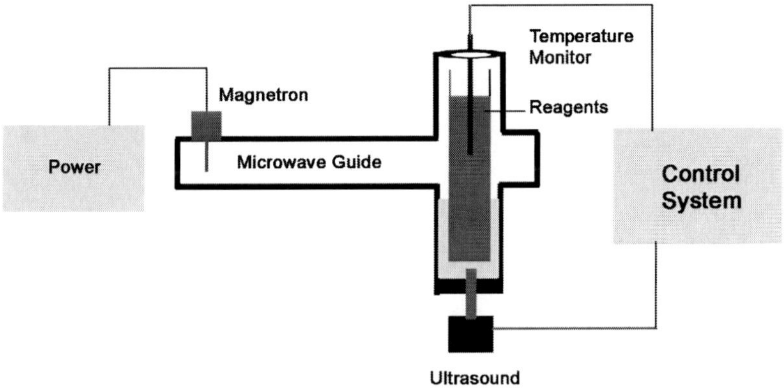

Figure 7.6 Schematic diagram of the laboratory-made US/MW system (MW power range from 0 to 300 W, US from 0 to 50 W at a frequency of 38.4 kHz). Reprinted from *J. Chem. Techn. Biotech.*, 84, Shen, X.-F., Combining microwave and ultrasound irradiation for the rapid synthesis of nanowires: a case study on Pb(OH)Br, 1811–1817, Copyright (2009), with permission from Wiley.

MW irradiation (50 W, 10 min, see Table 7.1) gave a yield of 35.1%. The resultant wires showed that uniform morphology was about 1800–2000 nm in diameter and 60–80 μm in length (Fig. 7.7c). Applying US irradiation alone to the reactant mixture (50 W, 10 min) resulted in a lower yield (32.9%). Most of the wires (Fig. 7.7d) were 700–1500 nm in diameter and 20–40 μm in length. The resulting 1D wires were significantly less uniform than those obtained under MW irradiation. Under combined US/MW irradiation (both 50 W, 10 min), the product yield was higher (45%) than those obtained under MW or US alone.

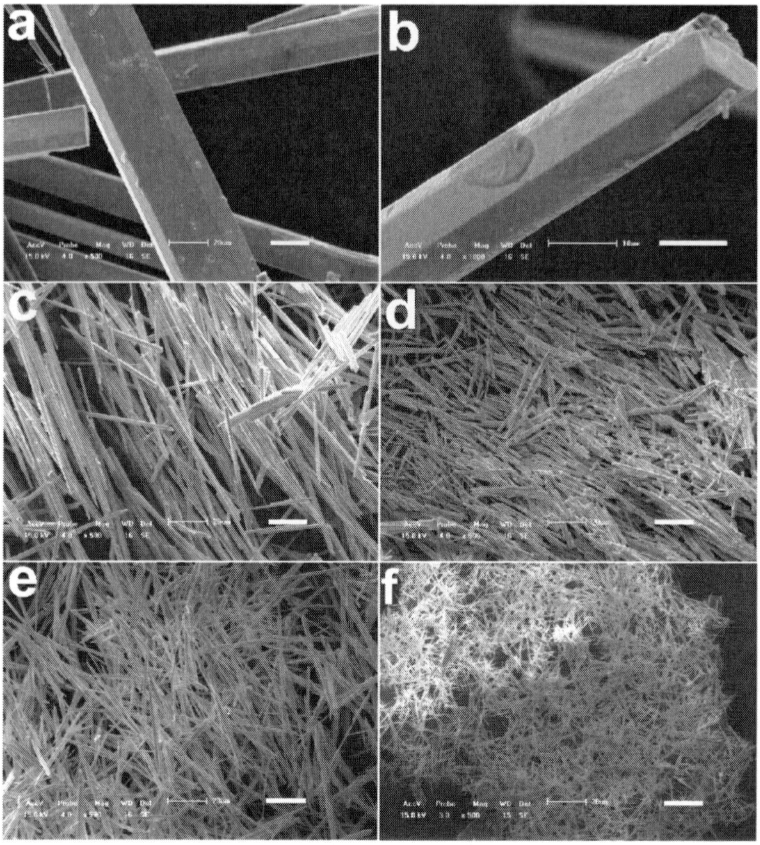

Figure 7.7 SEM images of Pb(OH)Br synthesized by (a) and (b) conventional heating at 70°C for 24 h; (c) MW irradiation at 50 W for 10 min; (d) US irradiation at 50 W for 10 min; (e) simultaneous US/MW irradiation at 50 W/50 W for 10 min, and (f) at 250 W/50 W for 80 s. The scale bar in (a) and (c–f) is 20 μm and in (b) is 10 μm. Reprinted from *J. Chem. Techn. Biotech.*, 84, Shen, X.-F., Combining microwave and ultrasound irradiation for rapid synthesis of nanowires: a case study on Pb(OH)Br, 1811–1817, Copyright (2009), with permission from Wiley.

The image of synthesized nanowires (Fig. 7.7e) showed straight, smooth wires with diameter ranging from 80 nm to 800 nm and length from 50 μm to 100 μm. Further increase in irradiation power under combined US/MW irradiation (250 W/50 W, respectively) led to a dramatic reduction in the reaction time (80 s) with a slightly increased yield (48.2%) (Fig. 7.7f). Higher power irradiation also resulted in the formation of nanowires with much shorter diameter and length. Most of the 1D product ranged from 100 nm to 500 nm in diameter and from 10 μm to 30 μm in length.

Table 7.1 Synthesis of Pb(OH)Br nanowires: comparison of conventional heating, MW irradiation, US irradiation, and simultaneous US/MW irradiation

	Reaction time	Temperature (°C)	Yield (%)	Diameter (nm)	Length (μm)
Conventional heating	24 h	70	23.0	20000–30000	2000–3000
MW (50 W)	10 min	89	35.1	1800–2000	60–80
US (50 W)	10 min	34	32.9	700–1500	20–40
US/MW (50 W-50 W)	10 min	66	45.0	80–800	50–100
US/MW (250 W-50 W)	80 s	95	48.2	100–500	10–30

These results demonstrate the favorable effect of simultaneous US/MW irradiation on the synthesis of 1D nanostructures with a marked improvement in the product yield, reaction rate, and an enlarged 1D structure aspect ratio.

7.3 Sequential Use of Ultrasound and Microwaves

In the literature, several synthetic applications for sequential US/MW irradiation have been described. US and MW can be applied separately in two steps or sequentially by means of loop reactors that circulate the reacting mixture through the two irradiating cells.

7.3.1 US and MW Sequential Irradiation in the Same Reaction

US nozzle spray processes generate a fine, low-velocity mist from the precursors solution, a technique that has been exploited for the preparation of multi-component metal oxides and highly activated catalysts (Espinal et al., 2007).

Nyutu et al. (2008) developed a coprecipitation technique to obtain spinel metal oxides (nickel ferrites and zinc aluminates). This method combines in situ mixing (I), US nozzle treatment (N), and MW heating (M) in a so-called INM process (Fig. 7.8). The authors investigated the influence of MW heating and US nozzle mixing on the interfacial properties of the obtained materials and on the composition, particle size, purity, and surface area of the resulting powders.

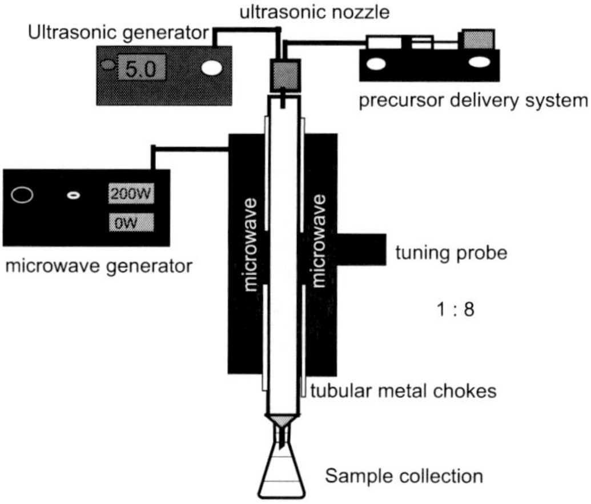

Figure 7.8 Scheme of continuous flow US-nozzle MW (INM) reactor, composed of a US nozzle (Sono-Tek Corp., 48 kHz and 120 kHz) and a tunable, cylindrical single-mode cavity (Wavemat MW Processing System model CMPR 250, continuous power ranging from 0 to 1250 W). Reprinted from *J. Phys. Chem. C*, 112, Nyutu, E. K., Conner, W. C., Auerbach S. M., Chen, C.-H. and Suib, S. L., Ultrasonic nozzle spray in situ mixing and microwave-assisted preparation of nanocrystalline spinel metal oxides: Nickel ferrite and zinc aluminate, 1407–1414, Copyright (2008), with permission from ACS.

When the nozzle was silent and the starting materials were simply flown into the test tube, the product obtained after calcination was impure (multiple phase) regardless of MW irradiation and MW power used (samples 1–3, 9; Table 7.2).

Similar results were achieved when the US nozzle was used without MW irradiation (sample 5, Table 7.2).

Table 7.2 Experimental conditions of Nickel ferrite preparation and results from XRD and Raman spectroscopy

Entry	Nozzle (kHz)	MW (W)/temp. (°C)	Calcination temp. (°C)	XRD crystallite size (nm)	Phases present
1/9	-	-	800/600	39[a], 22[b]	$NiFe_2O_3$ + impurities
2	-	100 W	800	46[b]	$NiFe_2O_3$ + impurities
3	-	300 W	800	13[a]	$NiFe_2O_3$ + impurities
4	120	100 W	800	29[a], 18[b]	$NiFe_2O_3$ + impurities
5	120	-	800	7[a], 6[b]	$NiFe_2O_3$ + impurities
6	120	300 W	600	6[b]	$NiFe_2O_3$
7	120	300 W	800	9[a], 13[b]	$NiFe_2O_3$
8	120	400 W	800	19[b]	$NiFe_2O_3$
10	120	100°C	600	11	$NiFe_2O_3$
11	120	150°C	600	10	$NiFe_2O_3$
12	48	300 W	600	8	$NiFe_2O_3$
13	120	300 W	600	12	$NiFe_2O_3$

[a]CO_3^{2-} as a coprecipitating agent

[b]OH^- as a coprecipitating agent

When the nozzle was used in combination with MW using a power less than 300 W, no single-phase products were obtained (sample 4, Table 7.2). Pure products were obtained with an MW power setting of 300–400 W combined with 120 kHz nozzle (samples 6–8, 12–13, Table 7.2). The FESEM image of the control experiment (no MW, no US nozzle) (sample 9, Fig. 7.9a) shows a large amount of agglomerated particles.

The high degree of precursor mixing and MW heating guaranteed optimal particle purity and morphology. High-frequency US nozzle mixing (120 kHz) resulted in a higher degree of particle agglomeration (Fig. 7.9c, entry 13) than the low-frequency nozzle use (48 kHz) (Fig. 7.9b, entry 12). This could be due to the formation of smaller particles at 120 kHz that tend to agglomerate. The median water droplet diameter at 120 kHz and 48 kHz are 18 and 38 nm, respectively.

Figure 7.9 FESEM micrographs of (a) conventional mixing of nickel ferrite precursors (no nozzle and MW, sample 9 in Table 7.2). (b) INM 300 W MW, 48 kHz nozzle (entry 12). (c) INM 300 W MW, 120 kHz nozzle (entry 13). (d) INC 100°C, 120 kHz nozzle (sample 10). Reprinted from *J. Phys. Chem. C*, **112**, Nyutu, E. K., Conner, W. C., Auerbach S. M., Chen, C.-H. and Suib, S. L., Ultrasonic nozzle spray in situ mixing and microwave-assisted preparation of nanocrystalline spinel metal oxides: Nickel ferrite and zinc aluminate, 1407–1414, Copyright (2008), with permission from ACS.

The authors also compared MW and conventional heating, which was called as INC: in situ (I), nozzle (N), conventional heating (C). Even under conventional heating (entries 10, 11, Table 7.2), pure $NiFe_2O_3$ was obtained, but the highly aggregated particles were less uniform and had lower surface areas than those obtained from the INM process (Fig. 7.9d). These results show that a suitable combination of both MW irradiation and the ultrasonic nozzle acts synergistically affording single-phase products. In fact, the use of nozzle or MW separately did not result in a pure phase material.

A recent work described the formation of $CoFe_2O_4$ nanocrystals from $CoCl_2$ and $FeCl_3$ in the presence of NaOH under hydrothermal conditions (MW or conventional heating) and the effects of preliminary US treatment (Kuznetsova et al., 2009).

The hydrothermal synthesis of $CoFe_2O_4$ was carried out at 130°C for a time ranging from 1 min to 45 h in a MARS-5 MW oven (heating rate of 10°C/min) for the MW process or in a steel autoclave (heating rate of 2°C/min) for the conventional method. Nanocrystalline $CoFe_2O_4$ ferrite was formed after 24 h under the conventional treatment, whereas under MW heating, a large amount of product was found after 1 min of irradiation and the complete conversion of starting materials was also obtained in 2 h.

In the case of conventional heating, presonication (3 min, in the presence of NaOH) did not affect the particle formation and growth, while in the case of MW heating, the reaction time was strongly reduced with a much higher conversion (64% in 1 min, 40% without presonication; Table 7.3, lines 2–3).

When the sonicated sample was left 2–4 h aging, the yield was reduced by half (Table 7.3, lines 4–5), because clusters formed under US irradiation underwent degradation due to instability. US alone did not lead to $CoFe_2O_4$ ferrite formation (Table 7.3, line 1).

Hydrothermal MW synthesis is much more advantageous than conventional heating, because it offers a much faster and homogeneous heating. Therefore, the reaction rate can be highly increased. Furthermore, the rate and the effectiveness of the hydrothermal synthesis can be significantly increased by subjecting it to ultrasonic treatment, as cavitation bubbles produce an efficient disaggregation of particles and favor the formation of nucleating centers of the new phase.

Table 7.3 Influence of the hydroxide aging time (after US treatment) on the crystallization degree and particle size upon hydrothermal MW treatment (1 min)

Sonication (3 min)	MW hydrothermal treatment (1 min)	Aging (h)	Conversion (%)	Particle size (nm)
+	–	–	0	–
–	+	–	40	7
+	+	0	64	9
+	+	2	50	6
+	+	4	49	6

MW heating has been applied in the sonocatalyzed hydrothermal preparation of tetragonal phase-pure lead titanate nanopowders with a stoichiometric chemical composition (Tapala et al., 2008). The reaction mixtures [$Pb(NO_3)_2$, TiO_2, and KOH] were sonicated at 70(±5)°C for 3 h using a general US bath (Bandelin Electronic RK255H, 160/320 W, 35 kHz), and then placed in a 95(±5)°C water bath, located in a domestic MW oven (Whirlpool XT-25ES/S, 900 W), for 1–3.5 h applying a continuous MW power (720 W).

The effects of MW heating on the synthetic process and powder characteristics were a reduction in the reaction temperature from 180°C for conventional heating (Rujiwatra et al., 2006) to 95(±5)°C, a reduction in the tetragonal distortion and an improved regularity of the crystal habit. MW heating did not affect the particle size and aggregation, which may be governed mainly by the US irradiation step.

Kabiri et al. (2007) applied sequential US/MW irradiation in the preparation of chitosan-modified nanoclays. The authors described a novel approach for the rapid preparation of chitosan-intercalated bentonite in green conditions (Fig. 7.10). This technique replaces the conventional time-consuming method (up to 48 h). Different conditions were used to treat a mixture of bentonite and chitosan (Table 7.4). A domestic MW oven (Panasonic) and a classic US cleaning bath were used.

The interlayer distance was increased from 12.34 Å (2θ° 7.15) to 14.12 Å (2θ° 6.26) for the clay modified by MW irradiation. This was further increased to 15.72 Å (2θ° 5.63) for the sample prepared by stirring for 4 h at room temperature followed by MW irradiation for

3 min. The highest interlayer distance (16.45 Å) was obtained in the case in which US and MW irradiation were alternatively used for the intercalation process in a total reaction time of 33 min.

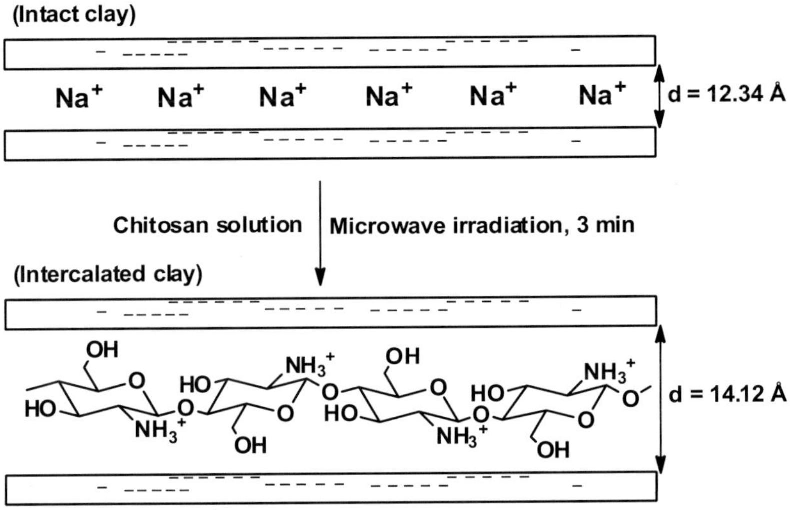

Figure 7.10 Schematic of unmodified clay (bentonite) and chitosan-intercalated bentonite prepared via MW irradiation (one-step) (Kabiri et al., 2007, copyright Springer).

Table 7.4 Intercalation conditions and interlayer distance for the unmodified and bio-modified clay samples

Intercalation conditions	2θ°([a])	Interlayer distance (Å)
Pristine clay	7.15	12.34
MW irradiation (3 min)	6.26	14.12
Stirring at r.t. (4 h) +MW (3 min)	5.63	15.72
Conventional heating at 60°C (48 h)	5.59	15.78
US (30 min)	5.82	15.17
US (30 min) + MW (3 min)	5.36	16.45

[a]θ is the angle of diffraction.

CuO nanofluid was synthesized by transforming an unstable $Cu(OH)_2$ precursor into CuO in water under sonication, followed by MW irradiation (Zhu et al., 2007). Nanofluids, prepared by suspending solid nanoparticles in the conventional fluids, usually show excellent heat transfer properties. The suspension of $Cu(OH)_2$ in the presence of ammonium citrate as dispersant was sonicated for 30 min in a US disrupter (Autoscience, AS3120A, 120 W) and then irradiated at low power with MW (3 min) (Galanz, WP750) to afford a black CuO nanofluid.

The samples synthesized without US or MW irradiation gave unsatisfactory results. In particular, the US treatment alone did not give a complete transformation of the precursor, while when only MW irradiation was used, the CuO particles obtained were irregular flakes in large agglomerates.

Hsin et al. (2006) presented a facile process to rapidly convert iron-filled multi-walled carbon nanotubes (Fe-MWCNTs) to multiple functional materials by a combination of ultrasonication and MW-induced radical co-polymerization.

Starting from both hydrophobic (ex. styrenes, Fig. 7.11d) and hydrophilic (e.g., acrylamide, Fig. 7.11c) olefinic monomers, Fe-MWCNTs could be rapidly surface functionalized to become soluble in organic solvents or aqueous solutions, reserving their magnetic properties crucial for biomedical applications (MRI contrast agents). The monomer and nanotubes suspension in water, THF, or toluene in the presence of benzoyl peroxide was sonicated for 2 min and then irradiated with MW for 10–20 s; the sequence was repeated several times.

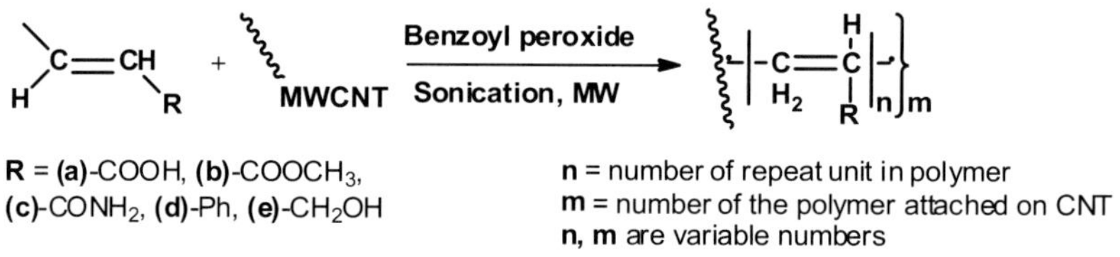

Figure 7.11 Process of surface functionalization of MWCNTs in the presence of THF and benzoyl peroxide carried out under US irradiation (2 min), followed by a few seconds (10–20 s, 600 W) of MW irradiation. Reprinted from *Carbon*, 44, Hsin, Y.-L., Lai, J.-Y., Hwang, K. C., Lo, S.-C., Chen, F.-R. and Kai, J. J., Rapid surface functionalization of iron-filled multi-walled carbon nanotubes, 3328–3335, Copyright (2006), with permission from Elsevier.

The current combined US/MW process was easy, rapid (within tens of min), and homogeneous if compared with the literature using reflux conditions (several hours) and can be used to easily attach a large number of different types of polymer chains onto MWCNTs (Fig. 7.11). The ultrasonication could overcome the aggregation (and thus inhomogeneous functionalization) problem of carbon nanotubes encountered in most of the functionalization processes.

7.3.2 Application of US and MW in Sequential Reactions

Recently, Rangari et al. (2010) studied the effect of MW on the curing behavior of an epoxy resin system in the presence and absence of MWCNTs. US irradiation was used for the reinforcement of the EPON 862 Part A resin with 0.1–0.3 wt% of MWCNTs (Ti-horn, 20 kHz, 100 W/cm^2) at an amplitude of 50% for 15 min. The mixture obtained was cured in an MW oven (Sharp BP210) for only 10 min instead of 8 h, which would be the case for conventional heating (curing process is 4 h at 120°C and post-curing at 170°C for 4 h). Owing to the high capacity of graphite carbon to absorb MW radiation (1100°C in 2 min), this process led to the rapid curing of EPON 862 resin in the presence of CNTs.

An MW-assisted protocol followed by US irradiation was used for the synthesis of Pt/H-ZSM5 catalysts (Mediavilla et al., 2010). Platinum salt solutions in ethylene glycol were irradiated at different power (184 W and 461 W) for the time periods of 30 s and 60 s (Samsung MW1050 oven) to obtain Pt nanoparticle suspensions, which were then mixed with the selected supports, irradiated with US for 30 min and finally reduced under H2 current at 500°C for 6 h to obtain the desired platinum catalysts. Temperature-programmed reduction studies indicated time and power of MW irradiation affect on the reducibility of the catalysts. There was a strong (13-fold) increase in the activity when the MW irradiation time was increased from 30 s to 60 s. However, a further increase in the irradiation time to 90 s resulted in a total loss of the catalytic activity.

The catalyst prepared from suspensions irradiated at 461 W for 60 s showed the highest activity, which seems to be related to the lower particle size and increased Pt dispersion (Fig. 7.12). As higher reaction temperatures produce smaller size of nanoparticles in the polyol method, the catalyst synthesized at high power (higher temperature) resulted in more activity.

This sequential US/MW protocol strongly improved the synthesis and the activity of the catalysts.

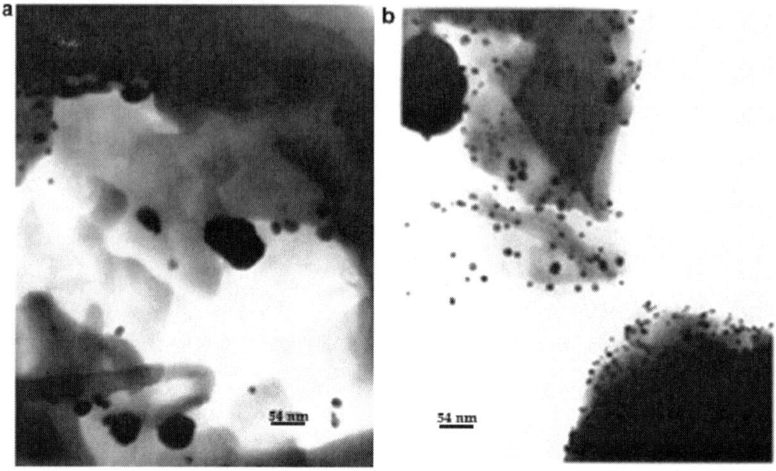

Figure 7.12 TEM images of Pt/H-ZSM5 solid synthesized from suspension irradiated with MW at 461 W for 30 s (a) and 60 s (b), and then with US (30 min). Reprinted from *Microporous and Mesoporous Materials*, **131**, Mediavilla, M., Morales, H., Melo, L., Sifontes, A. B., Albornoz, A., Llanos, A., Moronta, D., Solano, R. and Brito, J. L., Microwave-assisted polyol synthesis of Pt/H-ZSM5 catalysts, 342–349, Copyright (2010), with permission from Elsevier.

Sequential US/MW irradiation was applied in several cases for the preparation of nanofluids. Chandrasekar et al. (2010) prepared nanocrystalline alumina (Al_2O_3) powder by an MW-assisted chemical precipitation method (Fig. 7.13). Nanofluid was then prepared by dispersing a specified amount of Al_2O_3 nanoparticles in water using a US device (Toshiba, India, US pulses of 100 W, at 36 ± 3 kHz). The US treatment was prolonged to 6 h to obtain a uniform dispersion and stable suspension, which determined the final properties of the nanofluids.

In another example, SnO_2 nanoparticles were synthesized via an MW-induced combustion route (Habibzadeh et al., 2010). As they form clusters within the fluids, the mixtures were sonicated for 2 h to improve the dispersion and stability of the nanoparticles (Bandelin, power 200 W, frequency 20 kHz).

Ultrasonication provides the mechanical energy that helps the particles to overcome the attractive van der Waals forces at contact. Even though sonication cannot break up nanoparticles individually, the nanoparticle cluster breaks into smaller clusters resulting into higher stability.

Figure 7.13 SEM photograph of Al_2O_3 particles obtained under MW irradiation. Reprinted from *Experimental Thermal and Fluid Science* 34, Chandrasekar, M., Suresh, S. and Chandra Bose, A., Experimental investigations and theoretical determination of thermal conductivity and viscosity of Al_2O_3/water nanofluid, 210–216, Copyright (2010), with permission from Elsevier.

Tian et al. (2004) reported the use of ultrasonication in the synthesis of Pt/C catalysts simply as a strong and efficient means of stirring the reaction mixture (Pt salt and carbon in acetone), which was then heated to dryness in an MW oven in a repeated 5 s on/off cycle. The platinum oxide (on charcoal) formed was reduced with formic acid and dried again in MW in a repeated 20 s on/15 s off cycle under Ar, affording a uniformly dispersed Pt/C catalyst (Fig. 7.14).

Compared with conventional drying methods, MW drying heats the sample bulk evenly, making every particle to reach the same temperature immediately. In this case, the fast evaporation of moisture broke the lumped sample into powder thanks to the internal pressure developed by the expansion of the moisture, resulting in smaller particles and subsequently a higher surface area.

Moller et al. (2007) investigated the influence of high-power US (20 kHz, 750 W, Vibracell) on the particle size distribution of nanoscaled ZrO_2:Pr_3^+ powders obtained via a hydrothermal MW-driven process. Preliminary tests showed that extended treatment with high-power US heated the samples in such a way as to lead to an increased agglomeration of the particles. Therefore, the pulsed mode was used and the sample was placed in a water cooling bath.

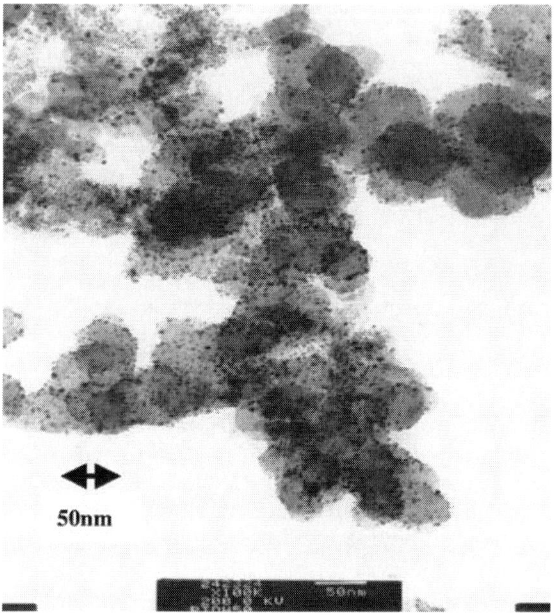

Figure 7.14 TEM image of a 50% Pt/C catalyst. Reprinted from *Journal of Materials Science*, **39**, Tian, Z. Q., Xie, F. Y. and Shen, P. K., Preparation of high loading Pt supported on carbon by on-site reduction, 1507–1509, Copyright (2004), with permission from Kluwer Academic Publishers.

Through a set of experiments with a 65% amplitude and pulsed mode (3 s on, 3 s off) US setting, the authors established that an application time of 10 min was the optimum time to obtain dispersions with smaller agglomeration tendency.

7.4 Conclusions

The application of combined US/MW irradiation in the preparation of nanomaterials shows favorable additive and even synergic effects. US irradiation promotes an efficient and powerful mixing both on the homogenous and heterogeneous mixtures, which is a fundamental factor in the synthesis of nanoparticles. Moreover, cavitation bubbles produce an efficient disaggregation of particles, favoring the formation of nucleating centers of the new phases.

MW irradiation offers several advantages such as fast, efficient, and homogeneous heating, a significant improvement in the product yield and reaction rates. US irradiation applied during MW heating can avoid or reduce overheating in a local area and produce more uniform nanostructures. The combined US/MW irradiation has been shown to be the method of choice for the synthesis of noble metal-doped semiconductor nanostructures and in general metal nanoparticles and nanowires.

With low boiling solvents or when the reaction temperature is close to the boiling point of the reaction solvent, US cannot be applied throughout MW heating, as the boiling bubbles destroy the cavitation bubbles countering US effects. In this case, the sequential application of US and MW can be chosen in order to benefit from the distinct US and MW irradiation. This method can be useful also when different steps of a reaction or different reactions in sequence need different treatments (e.g., heating or just mixing). The sequential US/MW irradiation has been demonstrated to be convenient for the preparation of multi-component metal oxides, highly activated catalysts, functionalized MWCNTs, and nanofluids. This hybrid technique will surely find a wide range of practical applications in the nanoscience and nanotechnology.

References

Chandrasekar, M., Suresh, S. and Chandra Bose, A. (2010). Experimental investigations and theoretical determination of thermal conductivity and viscosity of Al_2O_3/water nanofluid, *Exp. Thermal Fluid Sci.* **34**, pp. 210–216.

Cravotto, G., Beggiato, M., Penoni, A., Palmisano, G., Levêque, J.M., Bonrath, W. (2005). High-intensity ultrasound and microwave, alone or combined, promote Pd/C-catalyzed aryl–aryl couplings *Tetrahedron Lett.* **46**, pp. 2267–2271.

Cravotto, G. and Cintas, P. (2006). Power ultrasound in organic synthesis: moving cavitational chemistry from academia to innovative and large-scale applications, *Chem. Soc. Rev.* **35**, pp. 180–196.

Cravotto, G. and Cintas, P. (2007). The combined use of microwaves and ultrasound: new tools in process chemistry and organic synthesis, *Chem. Eur. J.* **13**, pp 1902–1909.

Cravotto, G., Di Carlo, S., Curini, M., Tumiatti, V. and Roggero, C. (2007). A new flow reactor for the treatment of polluted water with microwave and ultrasound, *J. Chem. Tech. Biotech.* **82**, pp. 205–208.

Cravotto, G., Boffa, L., Mantegna, S., Perego, P., Avogadro, M. and Cintas, P. (2008). Improved extraction of natural matrices under high-intensity ultrasound and microwave, alone or combined, *Ultrason. Sonochem.* **15**, pp. 898–902.

Dahl, J. A., Maddux, B. L. S. and Hutchison, J. E. (2007). Toward greener nanosynthesis, *Chem. Rev.* **107**, pp. 2228–2269.

Espinal, L., Malinger, K., Espinal, A., Gaffney, A. M. and Suib, S. L. (2007). Preparation of multicomponent metal oxides using nozzle spray and microwaves, *Adv. Funct. Mater.* **17**, pp. 2572–2579.

Gedanken, A. (2004). Using sonochemistry for the fabrication of nanomaterials, *Ultrason. Sonochem.* **11**, pp. 47–55.

Gerbec, J. A., Magana, D., Washington, A., and Strouse, G. F. (2005). Microwave-enhanced reaction rates for nanoparticle synthesis, *J. Am. Chem. Soc.* **127**, pp. 15791–15800.

Habibzadeh, S., Kazemi-Beydokhti, A., Khodadadi, A. A., Mortazavi, Y., Omanovic, S. and Shariat-Niassar, M. (2010). Stability and thermal conductivity of nanofluids of tin dioxide synthesized via microwave-induced combustion route, *Chem. Eng. J.* **156**, pp. 471–478.

Hsin, Y.-L., Lai, J.-Y., Hwang, K. C., Lo, S.-C., Chen, F.-R. and Kai, J. J. (2006). Rapid surface functionalization of iron-filled multi-walled carbon nanotubes, *Carbon* **44**, pp. 3328–3335.

Hu, Y., Wang, T., Wang, M.-X., Han, S.-F., Wan, P.-Y., Fan, M.-H. (2008). Extraction of isoflavonoids from Pueraria by combining ultrasound with microwave vacuum, *Chem. Eng. Process.* **47**, pp. 2256–2261.

Ishikawa, D., Hayashi, Y. and Takizawa, H. J. (2008). Preparation of platinum nanoparticles in heterogeneous solid-liquid system by ultrasound and microwave irradiation, *Nanosci. Nanotech.* **8**, pp. 4482–4487.

Kabiri, K., Mirzadeh, H. and Zohuriaan-Mehr, M. J. (2007). Highly rapid preparation of a bio-modified nanoclay with chitosan, *Iranian Polymer J.* **16**, pp. 147–151.

Kerner, R., Palchik, O. and Gedanken, A. (2001). Sonochemical and microwave-assisted preparations of PbTe and PbSe. A comparative study, *Chem. Mater.* **13**, pp. 1413–1419.

Kuznetsova, V. A., Almjasheva, O. V. and Gusarov, V. V. (2009). Influence of microwave and ultrasonic treatment on the formation of $CoFe_2O_4$ under hydrothermal conditions, *Glass Phys. Chem.* **35**, pp. 205–209.

Lévêque, J.-M. and Cravotto, G. (2006). Microwaves, power ultrasound and ionic liquids. A new synergy in green organic synthesis, *Chimia* **60**, pp. 313–320.

Lianfu, Z. and Zelong, L. (2008). Optimization and comparison of ultrasound/microwave assisted extraction (UMAE) and ultrasonic assisted extraction (UAE) of lycopene from tomatoes, *Ultrason. Sonochem.* **15**, pp. 731–737.

Leonelli, C., Mason, T. J. (2010). Microwave and ultrasonic processing: Now a realistic option for industry, *Chem. Eng. Proc.* **49**, pp. 885–900.

Lou, Z., Wang, H., Wang, D. and Zhang, Y. (2009). Preparation of inulin and phenols-rich dietary fibre powder from burdock root, *Carbohydr. Polym.* **78**, pp. 666–671.

Ma, J., Tai, G. and Guo, W. (2010). Ultrasound-assisted microwave preparation of Ag-doped CdS nanoparticles, *Ultrason. Sonochem.* **17**, pp. 534–540.

Maeda, M. and Amemiya, H. (1995). Chemical effects under simultaneous irradiation by microwaves and ultrasound, *New J. Chem.* **19**, pp. 1023–1028.

Mediavilla, M., Morales, H., Melo, L., Sifontes, A. B., Albornoz, A., Llanos, A., Moronta, D., Solano, R. and Brito, J. L. (2010). Microwave-assisted polyol synthesis of Pt/H-ZSM5 catalysts. *Microporous Mesoporous Mater.* **131**, pp. 342–349.

Möller, S., Fidelus, J. D. and Lojkowski, W. (2007). Particle size distribution of ZrO_2:Pr^{3+}– Influences of pH, high power ultrasound, surfactant and dopant quantity, *Solid State Phenomena* **128**, pp. 97–100.

Nemamcha, A., Rehspringer, J. L. and Khatmi, D. (2006). Synthesis of palladium nanoparticles by sonochemical reduction of palladium(II) nitrate in aqueous solution, *J. Phys. Chem. B* **110**, pp. 383–387.

Nyutu, E. K., Conner, W. C., Auerbach S. M., Chen, C.-H. and Suib, S. L. (2008). Ultrasonic nozzle spray in situ mixing and microwave-assisted preparation of nanocrystalline spinel metal oxides: nickel ferrite and zinc aluminate, *J. Phys. Chem. C* **112**, pp. 1407–1414.

Rangari, V. K., Bhuyan, M. S. and Jeelani, S. (2010). Microwave processing and characterization of EPON 862/CNT nanocomposites, *Mater. Sci. Eng. B* **168**, pp. 117–121.

Rujiwatra, A., Thammajak, N., Sarakonsri, T., Wongmaneerung, R., Ananta, S. (2006). Influence of alkali reagents on phase formation and crystal morphology of hydrothermally derived lead titanate, *J. Cryst. Growth* **289**, pp. 224–230.

Shen, X.-F. (2009). Combining microwave and ultrasound irradiation for rapid synthesis of nanowires: a case study on Pb(OH)Br, *J. Chem. Techn. Biotech.* **84**, pp. 1811–1817.

Tai, G. and Guo, W. (2008). Sonochemistry-assisted microwave synthesis and optical study of single-crystalline CdS nanoflowers, *Ultrason. Sonochem.* **15**, pp. 350–356.

Tai, G., Zhou, J. and Guo, W. (2010). Inorganic salt-induced phase control and optical characterization of cadmium sulfide nanoparticles, *Nanotechnology*, **21**, pp. 175601/1-175601/7.

Tapala, S., Thammajak, N., Laorattanakul, P. and Rujiwatra, A. (2008). Effects of microwave heating on sonocatalyzed hydrothermal preparation of lead titanate nanopowders, *Mater. Lett.* **62**, pp. 3685–3687

Tian, Z. Q., Xie, F. Y. and Shen, P. K. (2004). Preparation of high loading Pt supported on carbon by on-site reduction, *J. Mater. Sci.* **39**, pp. 1507–1509.

Zhu, H. T., Zhang, C. Y., Tang, Y. M. and Wang, J. X. (2007). Novel synthesis and thermal conductivity of CuO nanofluid, *J. Phys. Chem. C* **111**, pp. 1646–1650.

Chapter 8

Ultrasound-Assisted Preparation of Nanopolymeric and Micropolymeric Materials for the Encapsulation of Bioactive Agents

Francesca Cavalieri,[a,b] Meifang Zhou,[b] Mariarosaria Tortora,[a] and Muthupandian Ashokkumar[b,c]

[a]*Dipartimento di Scienze e Tecnologie Chimiche, Universita di Roma Tor Vergata, Roma, Italy*
[b]*School of Chemistry, The University of Melbourne, 1-100 Grattan Street, Parkville VIC 3010, Australia*
[c]*Chemistry Department, King Abdulaziz University, Abdullah Sulayman, Jeddah 22254,Saudi Arabia*
francesca.cavalieri@uniroma2.it

The use of nanopolymeric materials (i.e., micro-nanocontainers) for microencapsulation of active agents provides a simple and cost-effective way to encapsulate drugs, cells, pesticides, and flavors within a semi-permeable polymeric membrane for the purpose of protecting and releasing the enclosed substances in a controlled fashion. The effective encapsulation of bioactive molecules may require the synthesis of a tailor-made polymer and the use of specific

Cavitation: A Novel Energy-Efficient Technique for the Generation of Nanomaterials
Edited by Sivakumar Manickam and Muthupandian Ashokkumar

ISBN 978-981-4411-54-7 (Hardcover), 978-981-4411-55-4 (eBook)
www.panstanford.com

microencapsulation techniques. A number of colloidal nanosystems, such as polymersomes, dendrimers, and polymeric micelles, can also be suitable platforms for encapsulation of therapeutic agents. The use of ultrasound technique to synthesize and functionalize micro-nanopolymeric materials has been discussed in this chapter.

8.1 Introduction

The concept of microencapsulation was first proposed by Chang to entrap bioactive materials such as enzymes, proteins, and cells in a semi-permeable membrane in the 1960s (Chang, 1964). Since then, tremendous efforts in microencapsulation have been made for applications in biomedicine such as drug delivery, cell implantation, and gene therapy, as well as applications in biotechnology such as fermentation and large-scale cell culture involving a variety of materials and preparation technologies (Liu et al., 2000; Xue et al., 2003; Chang, 2004). Nowadays, encapsulation technologies are utilized in medicine, pharmaceutical, food, and the cosmetic industries for the development of controlled-release delivery systems.

Encapsulation is the technique by which one material or a mixture of materials is coated with or entrapped within another material or system. The coated material is typically an active agent, whereas the coating material is a colloidal carrier that stabilizes the core by forming a thin or thick shell. Carrier systems of colloidal dimensions, such as liposomes, polymeric particles, and microemulsion droplets, are used for the sustained release of drugs, pesticides, flavors, and other substances. The use of nanopolymeric materials (i.e., micro-nanocontainers) for microencapsulation of active agents provides a simple and cost-effective way to enclose drugs, cells, pesticides, and flavors within a semi-permeable polymeric membrane for the purpose of protecting and realeasing the enclosed substances in a controlled fashion.

In a relatively simplistic form, a polymer-based micro-nanocapsule is a vesicular system in which a bioactive agent is confined in a reservoir with a shell (Fig. 8.1). Alternatively, the active agents can be embedded into homogeneous polymeric micro-nanoparticles. The core may contain an active lipophilic or hydrophilic substance in either liquid or solid form (i.e., flavors and drugs) depending on the preparation method. In addition, micro-nanocapsules can also carry the active substance on their surface or imbibed in the polymeric film. The coated material is called core or internal phase and the coating material is called shell, carrier, or encapsulant.

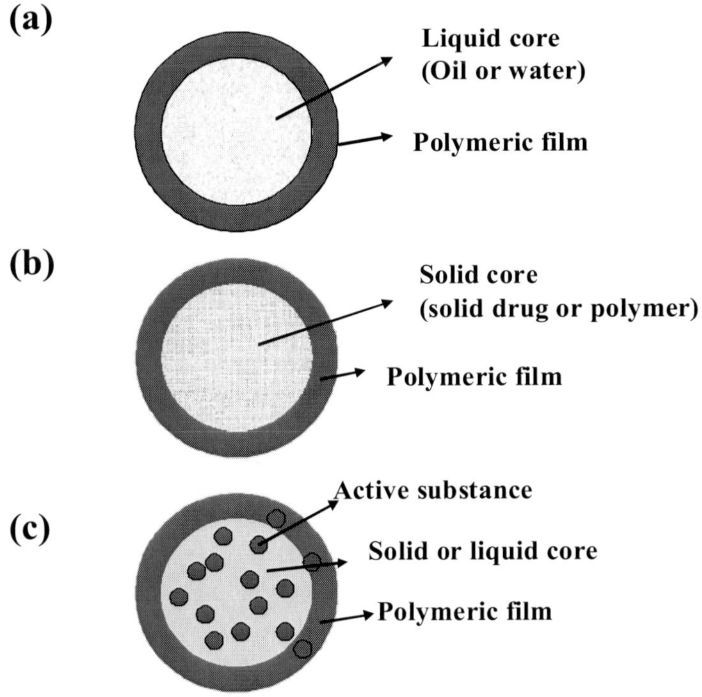

Figure 8.1 Polymeric microcapsules or nanocapsules containing a liquid core (a) or a solid core (b) or a bioactive agent (c).

In food processing, fats and oils, aroma compounds and oleoresins, vitamins, minerals, colorants, and enzymes have been encapsulated (Dziezak, 1988; Jackson et al., 1991; Shahidi et al., 1993). In the case of flavor encapsulation, microcapsules can be employed to retain aroma in a food product during storage, to protect the flavor from undesirable interactions with food, to minimize flavor–flavor interactions, to guard against light-induced reactions and/or oxidation, and to increase flavors shelf-life and/or allow a controlled release.

Research in the production of polymeric drug delivery micro-nanodevices has been focused on the microencapsulation of both hydrophobic and hydrophilic therapeutic agents. The possibility to realize injectable reservoirs for controlled release (e.g., drug-loaded micro-nanoparticles) may provide a strategy for the administration of chemical drugs with poor solubility. The low oral bioavailability of hydrophobic drugs, their broad therapeutic windows, the necessity for a low daily dose, and long-time treatment of disease are the items that support the requirement of reservoir-type polymeric microdevices. In addition, the advantages of using nanoencapsulated systems include high drug encapsulation efficiency and protection of therapeutic agents against degradation factors such as pH and light. For instance, the protective action of orally administered drugs from the attack of acids in the stomach before they can be absorbed in the intestine is exerted by gelatin microcapsules, which are insoluble at the low pH values found in the stomach allowing the desired release later in the intestinal tract (Wasan, 2001).

A more challenging task is the sustained and targeted release of drugs in the body. In principle, the drug should be carried to the target organ without any loss and then released by applying a specific internal or external stimulus. External and internal stimuli to trigger the delivery of drugs can be due to a magnetic field, electric field, ultrasound field, temperature, light, or mechanical forces. For the potential use of micro-nanocapsules in this therapeutic application, the chemical nature, mechanical and surface properties, and biofunctionality of the shell are crucial features. The effective encapsulation of biocative molecules may require the synthesis of a tailored-made polymer and the use of specific microencapsulation techniques. A number of colloidal nanosystems, such as polymersomes, dendrimers, and polymeric micelles, can be also suitable platforms for the encapsulation of therapeutic agents.

8.2 Micro-nanoencapsulation Techniques

The encapsulation of flavors and drugs has been accomplished by a variety of methods. The choice of the method is determined by the nature of the specific core and/or shell materials and the end use of microcapsules. The most common microencapsulation procedures (Park et al., 2005) are nanoprecipitation, emulsion–diffusion, double emulsification, emulsion–coacervation, layer-by-layer deposition technique, and ultrasonic encapsulation.

8.2.1 Microemulsion and Nanoemulsion Techniques

The "emulsion diffusion" is a two-step process based on a conventional emulsification step followed by the removal of part of the oil phase as a second step (Fig. 8.2). First, an oil-in-water emulsion is fabricated with "oil phase" containing the polymer in an organic solvent. The emulsion is usually prepared by stirring with an Ultra-Turrax® T25 rotor–stator shearing device, a high-pressure homogenizer, or using an ultrasonic dispersion apparatus. It has been shown that nanocapsule size is related to the shear rate used in the emulsification process, chemical composition of the organic phase, polymer concentration, oil-to-polymer ratio, and the droplet size of the primary emulsion. Smaller nanocapsules were produced when the stirring process was changed from high speed stirring mode to ultrasonic cavitation mode. Therefore, it is possible to decrease the micro-nanocapsule size by a suitable choice of the emulsification technology. Indeed, the typical size range of microcapsules is 10–100 μm and the polymer membrane is about a few micrometers in thickness. It is worth noting that the ultrasonic treatment is used to reduce the oil droplet size.

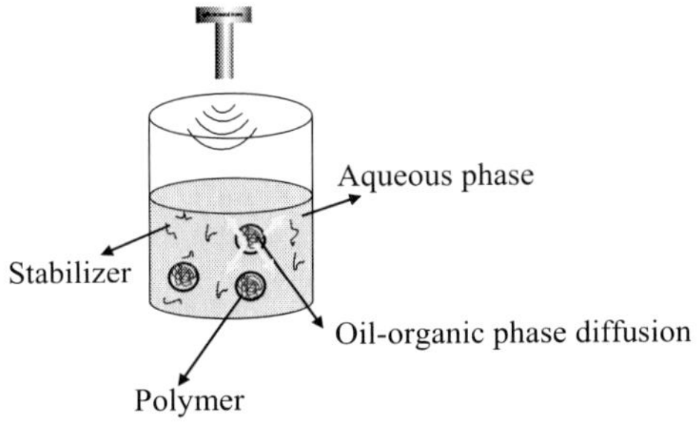

Figure 8.2 Preparation of micro-nanocapsule using the "emulsion–diffusion" method.

The formation of nanocapsules occurs at the diffusion step when the organic solvent (dichloromethane, hexane, and so on) diffuses out of the oil droplets, leaving the immiscible polymer and oil inside. The elimination of the organic solvent contained in the oil phase causes the separation of the polymer and the oil and a reduction in the particle size. The mild removal of organic solvent is made by means of its diffusion into the aqueous phase caused by water addition (diffusion step); then, the organic solvent can be safely evaporated under reduced pressure. The organic solvent is selected such that both the oil and polymer are soluble in it, and it is partially soluble in water for the diffusion to be possible by dilution. Alternatively, direct evaporation of the organic solvent is often performed. The latter method is not suitable for nanocapsules preparation, as the nanocapsules do not resist a direct evaporation of the solvent, possibly because of the mechanical stress when gas bubbles form inside the aqueous suspension. The nanoprecipitation method (Fig. 8.3) is also known as solvent displacement or interfacial deposition. A solvent phase essentially consists of an organic medium or a mixture of solvents, a film-forming polymer soluble in the solvent, and an active substance (capric/caprylic triglyceride types, oleic acid, ethyl oleate, argan oil, sunflower seed oil, and soybean oil).

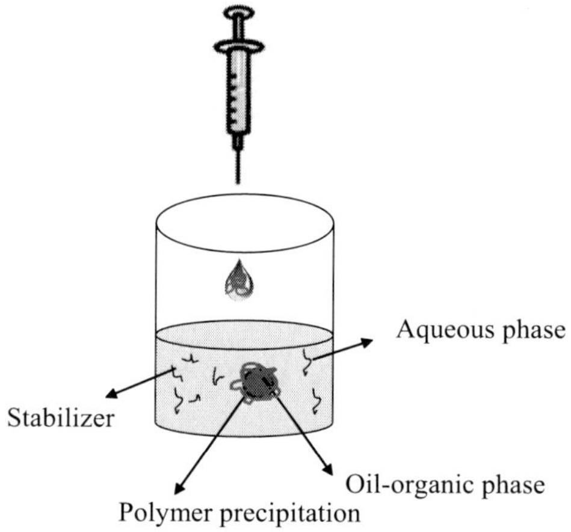

Figure 8.3 Preparation of microcapsule or nanocapsule using the nanoprecipitation method.

The non-solvent phase (i.e., water) for the film-forming substance is supplemented by a stabilizer (poloxamer 188 or polysorbate 80).

In the nanoprecipitation method, the nanocapsules (100–200 nm) are obtained as a colloidal suspension when the organic phase is slowly added to the aqueous phase with moderate stirring. The polymers commonly used are biodegradable polyesters, especially poly-e-caprolactone (PCL), poly(lactide) (PLA), and poly(lactide-co-glicolide) (PLGA).

The emulsion–coacervation process (Fig. 8.4) is mainly presented as a strategy for nanocapsules preparation from naturally occurring polymeric materials. Polysaccharides (sodium alginate) and protein (gelatin) have been used, though synthetic polymeric materials could also be used for this purpose. Nanocapsules formation by the emulsion–coacervation method uses the oil (nonsolvent) in water emulsion as a template phase and the formation of a coacervate phase that causes polymer precipitation from the continuous emulsion-phase to form a film on the template forming the nanocapsule. In addition, nanocapsules can be hardened by physical intermolecular or covalent cross-linking, which typically can be achieved by altering pH or temperature, or by adding cross-linking agents or salts. The use of electrolytes for polymer desolvation is known as salting-out and the electrolytic efficiency for this process follows the Hofmeister series, which arranges ions in an increasing order according to their capacity to immobilize water molecules in solvation in the ternary polymer–water–salt system.

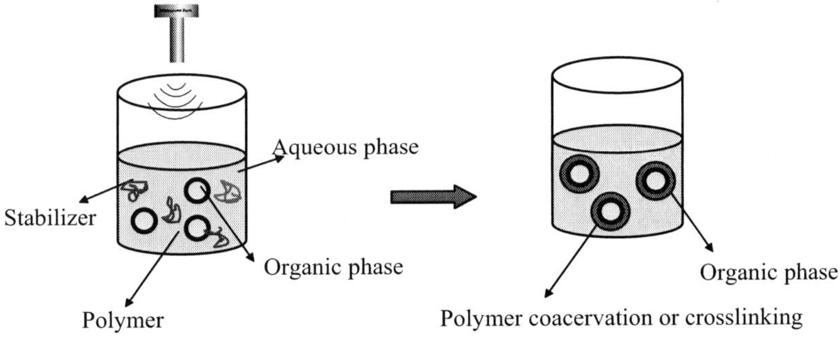

Figure 8.4 Preparation of microcapsules and nanocapsules using the nanocoacervation process.

In a typical procedure for preparation of micro-nanocapsules by double emulsification, the primary emulsion is formed by ultrasound horns or high shearing devices and a hydrophobic surfactant stabilizes the interface of the w/o internal emulsion (Fig. 8.5). The second emulsion is also formed by ultrasound and a hydrophilic surfactant stabilizes the external interface of the oil globules for w/o/w emulsions. Finally, the solvents are removed by evaporation or extraction by vacuum, leaving hardened nanocapsules in the aqueous medium. The mechanism involved in the double-emulsion micro-nanocapsules formation (size about 150–200 nm) is associated with the principles of both nanoprecipitation and emulsion–diffusion methods. In this case, in the primary w/o emulsion, the oil is changed by an organic phase containing a solvent that is partially miscible in water, the film-formed polymer, and a w/o surfactant. Then, water containing a stabilizing agent is added to the system to obtain water in organic solvent in the water emulsion. Particle hardening is obtained through solvent diffusion or evaporation and polymer precipitation. Water is usually added to the double emulsion in order to achieve full solvent diffusion. The inner aqueous phase is composed of an aqueous solution of the active substances, insulin (Zhu et al., 2005), penicillin G (Khoee and Yaghoobian, 2009), and plasmid DNA (Perez et al., 2001).

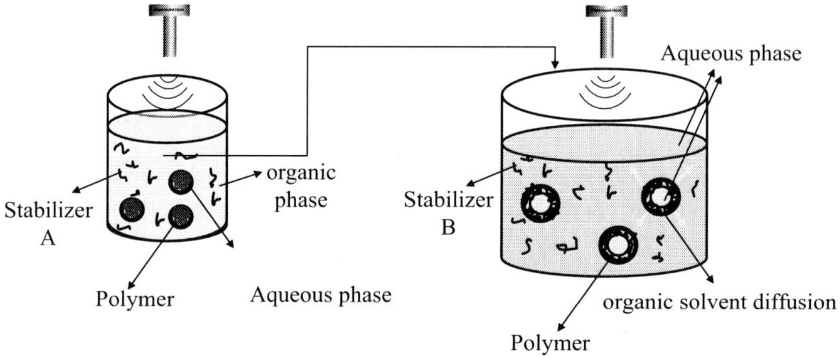

Figure 8.5 Preparation of micro-nanocapsules using the double emulsification technique.

As organic phase, ethyl acetate, methylene chloride, and dichloromethane have been used as solvents and biodegradable polyesters, such as PCL, PLA, and PLGA have been frequently used as film-forming polymers. Among o/w surfactants, sorbitan esters are preferred. PVA and polysorbates are frequently used as a stabilizer of the external aqueous phase.

8.2.2 Layer-by-Layer Encapsulation Technique

The layer-by-layer (LbL) assembly process, developed by Decher (Decher and Hong, 1991; Decher, 1997) and later extended to colloidal particle preparation (Caruso et al., 1998; Donath et al.,

1998), makes it possible to obtain vesicular particles with well-defined chemical and structural properties. The versatility of this method is the ability to create highly tailored polymer, thin caspules with a nearly unlimited range of functional groups incorporated in the film structure. The mechanism of microcapsule and nanocapsule formation is based on irreversible electrostatic attraction that leads to polyelectrolyte adsorption and complexation.

This method requires a template onto which a polymer layer is adsorbed by incubation in the polymer solution subsequently washed. This procedure is then repeated with a second polymer and multiple polymer layers are deposited sequentially as schematically presented in Fig. 8.6.

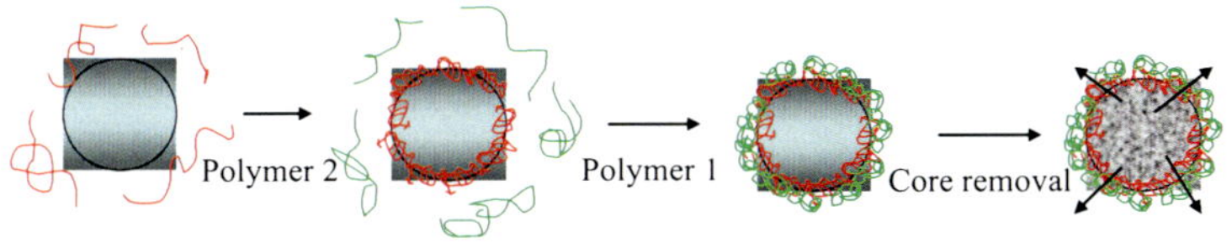

Figure 8.6 Preparation of hollow micro-nanocapsules using the LbL technique.

The adsorption of oppositely charged polyelectrolytes or charged nanoparticles can be achieved on the surface of colloidal particles (silicon dioxide, $CaCO_3$, or $MnCO_3$, melamine–formaldehyde or polystyrene latices) with subsequent core dissolution. An active substance or living cells can be used directly as a template (Krol, 2004).

The LbL encapsulation technique was originally applied to polyelectrolytes, in which assembly is driven by electrostatic interactions; however, electrostatically assembled films are typically highly charged, which can lead to biofouling and aspecific interactions in biological systems. More recent studies have exploited hydrogen bonding (Zelikin et al., 2006), DNA hybridization (Cavalieri et al., 2009; Lee et al., 2010), and sequential chemical reactions (Ochs et al., 2010). A wide range of molecules (Sivakumar et al., 2009; Becker et al., 2010) including small anticancer drugs, oligonucleotides, plasmids, peptides and whole proteins, have been encapsulated within LbL microcapsules.

The LbL encapsulation techniques fall broadly into three categories: incorporating the cargo within the film during film assembly; postloading preformed films/capsules; or preloading the cargo onto the template (either planar or colloidal). Incorporating cargo during the assembly of LbL films has been widely exploited in electrostatically associated LbL films, as charged biopolymers can readily interact with conventional polyelectrolyte materials. Proteins (Mehrotra, 2010), oligonucleotide (Becker et al., 2009), and hydrophobic drug (Kim et al., 2008) can be incorporated within multilayers by exploiting the electrostatic interactions.

Biocolloid particles with tailored enzymatic activities can be produced by means of protein multilayer architectures on colloidal particles. Multilayer films of bovine serum albumin, immunoglobulin alone (Caruso and Mohwald, 1999), or in combination with a fluorescent label (Yang et al., 2001), β-glucosidase (Caruso et al., 2000), glucose oxidase (Caruso and Schuler, 2000, Caruso et al., 2000), horseradish peroxidase (Perazzini et al., 2011), and urease have been assembled on polystyrene microspheres. The simplest method of encapsulation consists of the core itself as substance to be encapsulated. The problem is to obtain colloidal crystals of low dispersity in the desired size range.

The encapsulation of catalase (Caruso and Schuler, 2000; Caruso et al., 2000) by the controlled polymer multilayer coating of biocrystals was achieved by the sequential adsorption of oppositely charged polyelectrolytes onto enzyme crystal templates. An extremely high enzyme loading in each polymer capsule was obtained, and the activity of the encapsulated enzyme was preserved. The polymer-encapsulated enzyme was stable against protease degradation. The polymer-coated enzyme retained 100% of its activity after incubation for 100 min with protease, whereas uncoated, solubilized catalase lost more than 90% of its initial activity within 100 min under the same conditions. The post-loading approach (Fig. 8.7a) for encapsulation consists of loading preformed capsules by switching the shell-wall permeability through variation of environmental conditions, such as pH and ionic strength. The pre-loading encapsulation method has two stages (Fig. 8.7b). First, polymer is infiltrated into a porous sacrificial core. Second, stable classical LbL polyelectrolyte multilayers are assembled. The core is removed, and the inside of the capsule remains filled with the polymer.

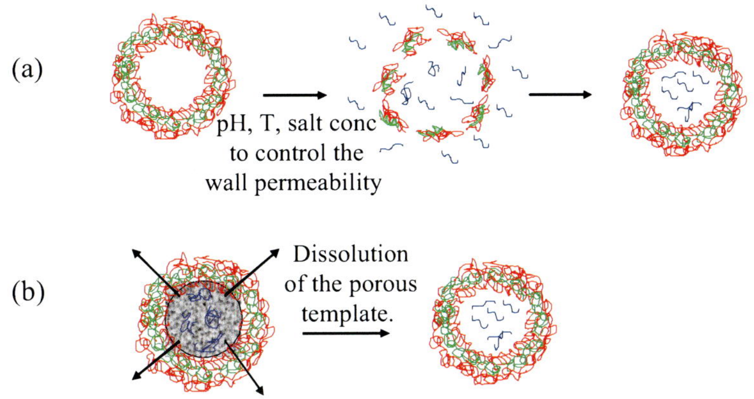

Figure 8.7 Post-loading encapsulation by changing the capsules permeability (a). Pre-loading encapsulation by infiltration of the biactive agent into a porous template (b).

8.3 Ultrasonic Synthesis of Biopolymer-Based Microcapsules

8.3.1 Mechanism of Microcapsules Formation

The fabrication of proteinaceous and protein hybrid microspheres using the sonochemical method was widely reported in the last decade. Protein microspheres, air-filled and oil-filled, have attracted much research interest in the biomedical field because of their biodegradability and biocompatibility. Applications such as ultrasound contrast agents, magnetic resonance imaging, optical coherence tomography, oxygen, and drug delivery have been considered (Langer, 1990; Webb et al., 1996; Lee et al., 2003). Several types of protein (Lee et al., 1981; Heidebach et al., 2009) hollow protein microbubbles (Grinstaff and Suslick, 1991; Suslick, 1995), and core–shell protein microspheres (Toublan, 2006) have been prepared. The loading capability of these microspheres can be improved by modifying the outer surface of the protein microspheres, the interior core, or the matrix structure.

The mechanism responsible for the formation of protein-based microcapsules is a combination of two ultrasound-induced phenomena: emulsification and cavitation. Hydrophobic interaction between the protein subunits is one of the requirements for the formation of microbubbles, but it is not the main mechanism responsible for the stabilization of microbubbles. Suslick proposed that for sulfur containing proteins (e.g., bovine serum albumin, BSA) (Suslick and Grinstaff, 1990), human serum albumin (HAS), and hemoglobin (Hb), the mechanism of the sonochemical formation of stable oil and gas-filled microspheres involves the cross-linking of cysteine residues through disulfide bond formation by the superoxide radicals generated under sonication conditions. For sulfhydryl-free polypeptides (e.g., polyglutamate), hydrogen bonding or ion pairing between carboxylates (Dibbern, 2006) provided the mechanism for stabilization of oil-filled microparticles. Avivi and Gedanken (Avivi, 2002) suggested that hydrophobic interactions in acidic medium are responsible for the fabrication of poly(glutamic acid) and streptavidin microspheres; –SH functional groups are absent in both of these systems. It has been claimed that the one-step sonochemical encapsulation process is extremely effective in producing a core–shell structure with high encapsulation efficiency. BSA is the most commonly used shell material for entrapping a variety of oils. Suslick also reported the ultrasonic synthesis of proteinaceous microcapsules of BSA filled with n-dodecane, n-hexane, and cyclohexane. Another protein that recently was used to synthesize both microbubbles and oil-filled microcapsules is lysozyme (Fig. 8.8). Hen egg white lysozyme contains four intramolecular disulfide bonds that can be partly broken using reducing agents, such as dithiothreitol. This enhances the probability of intermolecular disulfide bond formation leading to cross-linking between lysozyme molecules. On the basis of this strategy, lysozyme air-filled microspheres (Fig. 8.9a, b) have been successfully synthesized using chemically denatured lysozyme under high-intensity ultrasound (Cavalieri et al., 2008). Upon sonication, new disulfide linkages are formed enabling interprotein cross-linkage to occur leading to the formation of lysozyme microspheres with a thick (0.1–0.8 μm) protein shell that are stable for several months. The lysozyme microbubbles represent a multifunctional ultrasound-responsive system endowed with antimicrobial and biodegradability features.

When tailored with targeting moieties, the lysozyme-shelled microspheres are promising candidates for controlled drug delivery applications. However, loading a sufficient amount of therapeutic

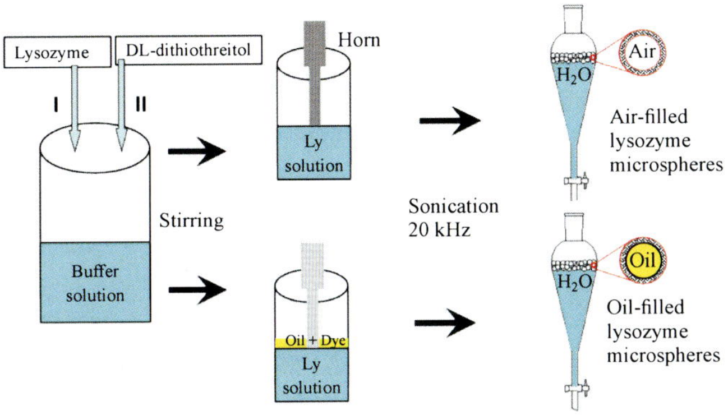

Figure 8.8 Scheme of lysozyme microbubble and microsphere preparation.

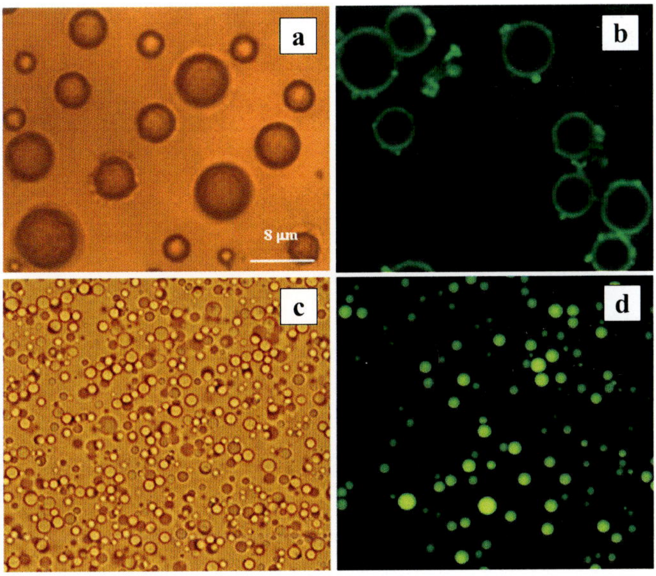

Figure 8.9 Optical microscopy images of lysozyme microbubbles (a), confocal microscopy image of lysozyme microbubbles (b), tetradecane-filled lysozyme microspheres (c); and tetradecane-filled lysozyme microspheres (d).

agent onto the micro bubbles is a current limitation for these ultrasound contrast agent based drug delivery systems. A better alternative is to encapsulate an oil-phase containing the drug that increases their lipophilic drug-loading capacity. We have explored the possibility of using crosslinked lysozyme as the shell material to ultrasonically encapsulate a vegetable oil and some non-polar organic liquids (Fig. 8.9c,d), which make them suitable reservoirs for lipophilic compounds.

The morphologies of lysozyme microspheres filled with oil and organic liquids have also been studied. The effective encapsulation of the liquid materials within the microspheres has been confirmed by ultrasonically breaking the microspheres loaded with liquids containing an organic dye and studying the fluorescence properties of the dye (Meifang et al., 2010). To further demonstrate the core–shell nature of the microspheres, hexane-filled lysozyme microspheres were imaged using SEM (Fig. 8.10a) and broken by the electron beam. The oil is released (Fig. 8.10b) from the microsphere, but the protein shell retains its structural integrity (Fig. 8.10c).

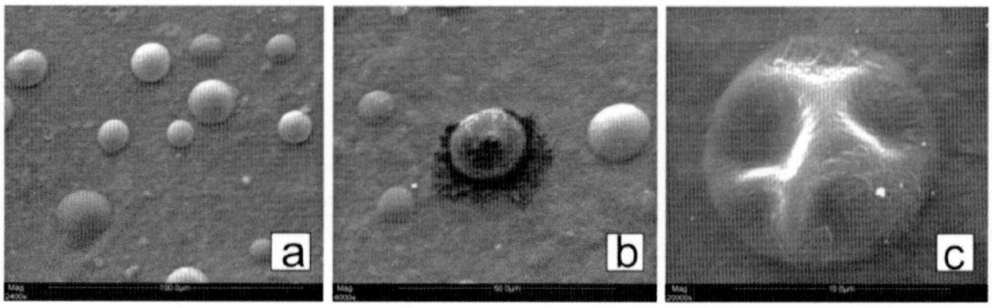

Figure 8.10 SEM images of (a) hexane-filled lysozyme microspheres; (b) broken hexane-filled lysozyme microspheres; and (c) hexane-evaporated lysozyme microcapsule.

8.3.2 Controlling the Yield and Size of Proteinaceous Microspheres

Controlling the yield and size of proteinaceous micropheres can be achieved by optimizing experimental parameters. A model system, BSA, was used to study the effect of few experimental variables on the yield and size of the proteinaceous microspheres. The results can be extended to other type of proteins (Makino et al., 2001). Makino and coworkers varied the following parameters: concentration of BSA; percentage of toluene in the mixture of toluene and BSA solution; acoustic frequency and sonication time; and the core materials (toluene, chloroform, peanut oil, or soybean oil) to investigate the effect of these parameters on the yield of proteinaceous microspheres and the size distribution of proteinaceous microspheres. They found that as the concentration of BSA increases from 0.005 to 0.01% (w/v), the micro encapsulation yield increases. Small changes in the yield between 0.02% and 1% (w/v) were observed. At concentrations below 0.01% (w/v), gradient changes in the microsphere formation yield appeared. In dilute solutions, the frequency of collision of BSA molecules is low and the density of the microcapsule membrane seems to be low, which in turn decreases yield. The influence of toluene volume fraction on the microsphere yield becomes remarkable only above a value of 0.3; up to this value, the yield was constant at a level of 80%. The microsphere yield decreases when the volume fraction of toluene becomes higher than 0.3. In addition, when the volume fraction of toluene is over 0.8, the phase inversion from o/w emulsion to w/o emulsion is clearly observed. The microsphere formation yield was studied as a function of sonication time and acoustic frequency. The yield of BSA microspheres was higher at 28 kHz than at 45 kHz, and no microcapsule was formed at 100 kHz. As the sonication time increases, the microencapsulation yield increases at 45 kHz, but then remains almost constant after 5 min. Also, the dependence of the microencapsulation yield upon the sonication time was more clearly observed at 45 kHz than at 28 kHz, as at 28 kHz, the microencapsulation yield reaches almost 80% within 30 s.

The size distributions of microcapsules were narrower at 45 kHz than at 28 kHz. The diameters of microcapsules are about 3.8 and 2.7 μm at 45 and 28 kHz, respectively. The authors studied the influence of the type of organic liquids on the size distribution of the microspheres as well (Grinstaff and Suslick, 1991). Using chloroform as the organic liquid resulted in capsules with average diameters of 5.1 and 3.6 μm at 28 and 45 kHz, respectively. Chloroform-filled BSA microcapsules show a wider size distribution and a larger average size than toluene-filled BSA microcapsules. Toluene is less miscible with water than chloroform. Such a difference of solubility of two organic solvents to water may make the size distribution of microcapsules containing each organic solvent different. The size distributions of BSA microcapsules containing soybean oil and peanut oil were narrower than those obtained for toluene and chloroform. The volume-averaged diameters are 1.48 μm and 1.67 μm for toluene and chlorform, respectively. In conclusion, smaller microcapsules were obtained when the encapsulated organic solvent has a higher viscosity.

To reduce the microbubbles size to the nanometer range, Avivi (Avivi and Gedanken, 2002; Gedanken, 2008) has investigated the influence of the addition of a surfactant to the protein solution. Two proteins were examined, BSA and α-amylase. The surfactants used in these experiments were SDS (sodium dodecyl sulphate), CTAB (cetyltrimethyl ammonium bromide), oleic acid, Tween 40, and Tween 80. The concentration of the BSA was 3% w/v, and that of the α-amylase 0.025% w/v. For all surfactants, a bimodal distribution was obtained. The distribution consisted of a major part of micron-sized microcapsules and only a small percentage of proteinaceous nanobubbles (500–800 nm). However, the addition of Tween 80 had a dramatic effect on increasing the percentages of the nanospheres, and with larger amounts of this surfactant, a situation was reached in which only nanobubbles are obtained. The same effect was obtained for the two proteins. In general, smaller nanospheres are obtained for amylase than for the BSA. The use of a surfactant in stabilizing polymeric microspheres is well known (Wu, 1994).

Magnetic protein microspheres containing nanoparticle iron oxide have been also synthesized by combining the sonochemical synthesis of nanostructured iron oxide (Cao, 1997) with sonochemical synthesis of proteinaceous microspheres. Selective targeting of protein microspheres to tumor region was demonstrated. Suslick and co-workers (Toublan, 2006) reported on a noncovalent, electrostatic LbL modification that successfully targets protein microspheres to the integrin receptors that are overexpressed in several tumor types. These core–shell microspheres are highly charged due to the numerous ionizable groups present in the BSA shell. Under physiological conditions, BSA typically has 185 counterions resulting from both acidic and basic surface residues, with a net charge of −17 at pH 7. For a typical 2 μm microsphere, there are nearly 10^6 BSA molecules per shell. Thus, the protein microspheres are sufficiently charged for the electrostatic adhesion of polyelectrolytes onto their surface. Suslick has found that these negatively charged vesicles are excellent templates for LbL electrostatic adhesion. It was demonstrated that protein microspheres can be selectively targeted to human tumor cells by using an LbL approach to modify their surface with integrin receptor–specific peptide ligands. Integrin receptors are overexpressed in several tumor types, and the RGD tripeptide motif has been used as a label for these tumor cells and their vasculature.

8.3.3 One-Step Preparation of Polysaccharides Microspheres by a Sonochemical Method

A fast and one-step process for the preparation of stable chitosan soybean oil microspheres using the sonochemical method was reported by Skirtenko et al. (2010). Chitosan is obtained from the deacetylation of chitin, a naturally occurring and abundantly available (in marine crustaceans) biocompatible polysaccharide. Chitosan is a biodegradable polymer with great potential for pharmaceutical applications due to its biocompatibility, high charge density, nontoxicity, and mucoadhesion properties. The mechanism of formation of chitosan microsphere by high intensity ultrasound treatment was however not clearly elucidated and it is ascribed to a combination of chemical and physical interactions involving imine bond and hydrogen bonds. The major advantage of the sonochemical method, compared with other techniques used for the preparation of chitosan microspheres, is the ease and shorter reaction time. For example, the preparation of chitosan microspheres using other techniques such as emulsification and ionotropic gelation (Lim et al., 1997; Daly and Knorr, 1998; Shu and Zhu, 2001; Mi et al., 2002) requires about 3 h, whereas in the sonochemical process, the microspherization of chitosan takes only 3 min. Another advantage of the sonochemical process is that cross-linking chemicals such as tripolyphosphate or glutaraldehyde are not required for the spherization process. The average microsphere size of approximately 1 μm with a narrow size distribution is much smaller than the 20–70 μm for the chitosan microspheres prepared by emulsification and ionotropic gelation.

8.4 The Structural and Functional Features of the Nanopolymeric Encapsulant

The functional properties of a polymer carrier should fulfill a number of requirements such as water solubility, the capacity for the encapsulation of poorly water-soluble molecules, biodegradability, and non-toxicity. For the encapsulation of flavor compounds, the carrier material must have no reactivity with the core molecules, allowing complete elimination of organic solvents after microcapsules production, protecting the active ingredients, ensuring good emulsion-stabilization properties, and high efficiency in releasing the flavor at the time and the place desired. In order to play these roles, the wall materials are requested to have excellent barrier against functional oxidation for microencapsulated molecules, solubility, viscosity, emulsification, and film-forming properties. An enormous range of different biomaterials are used for the flavor encapsulation, the most common are carbohydrates (starch, maltodextrins and β-cyclodextrin, cellulose), lipids, proteins (sodium caseinate, whey protein, and soy protein), gums, and cellulose. The encapsulating bioagents are typically endowed with amphiphilic properties and ability to self-associate, and during emulsion, they become rapidly adsorbed at the newly formed oil–water interface. The resulting steric-stabilizing layer immediately protects the oil droplets against coalescence and Ostwald ripening providing physical stability to the emulsion during processing and storage. More stringent requirements for the polymeric matrixes have to be considered for biomedical application. Crucial properties are biocompatibility, non-immunogenicity and stealth, lack of long-term accumulation in host, and in vivo stability. The carrier is also required to carry out a selective delivery to target site and prevent drug release before reaching the site of action. In an injectable biomacromolecular depot, drug release should be faster than the complete degradation of the respective polymer; thus, one of the criteria to adopt for the selection of a biocompatible carrier polymer is its degradation time.

8.4.1 Polymeric Micro-Nanoparticles for Encapsulation of Bioactive Agents

The majority of biodegradable polymers have been used in the form of micro-nanoparticles, from which the incorporated drug is released to the environment in a controlled manner. Among the biocompatible polymers, hydrophobic polylactic acid and poly(lactic-co-glycolic) acid, poly-caprolactone, and poly(butyl cyanoacrylate) are largely used to prepare micro-nanoparticles because of their approval in numerous biomedical and pharmaceutical products in the US and Europe for use in human, their availability, and a good solubility in organic solvents. The PLA-based device shows a slow degradation and the drug leuprolide acetate (antineoplastic agent) can be delivered over months (Trenantone®), whereas PLGA degrades faster and can release drugs in 2–4 weeks (Enantone®) (Hutchinson and Furr, 1989). The degradation time and the water uptake can be tuned by the selected molecular weight, the polymer end-group, the lactide/glycolide ratio, and, for PLA, the crystallinity of polymer (Witschi and Doelker, 1998; Li, 1999).

A higher lactide/glycolide ratio induces slower release, while the presence of free carboxylic groups causes a faster degradation of free-acid end-group PLGA if compared with that of the capped polymer bearing an aliphatic group at the end of the chain. The size of microparticles, pH, and temperature of the medium are other impact factors that influence polymer degradation (Dunne et al., 2000). If a faster release is required, a PLA-PEG-PLA block polymer could be used instead of PLGA copolymers (Ruan and Feng, 2003). The use of blends of high molecular weight (MW) polymer with a small portion of a low MW polymer or different polymers with a similar low MW can also be the precondition for fast matrix degradation and drug release. The presence of low MW polymer allows a faster degradation due to an easier break-up that involves the fragments to be removed from the matrix, leading to pores that open the access of the particle core to the water, thus the drug being released (Witschi and Doelker, 1998).

Considerable research efforts have been directed toward developing safe and efficient chitosan-based particulate drug delivery systems. Many reports are available on the preparation of chitosan microspheres using a number of methods such as the emulsion cross-linking technique and coacervation–precipitation method (Sunil, 2004). Selection of any of the methods depends upon factors such as particle size requirement, thermal and chemical stability of the active agent, reproducibility of the release kinetic profiles, stability of the final product, and residual toxicity associated with the final product. The emulsion cross-linking method is based on the use of the reactive functional amine group of chitosan to cross-link with aldehyde groups of the cross-linking agent. Water-in-oil (w/o) emulsion is prepared by emulsifying the chitosan aqueous solution in the oil phase. The stable emulsion is cross-linked by using an appropriate cross-linking agent such as glutaraldehyde to harden the droplets. The emulsion cross-linking method has few drawbacks such as the harsh cross-linking agents, which might possibly induce chemical reactions with the active agent. In addtion, the complete removal of the unreacted and toxic glutaraldehyde agent may be difficult in this process. However, more biocompatible crosslinker, such as genipin, can be used to crosslink chitosan microspheres. In addition, reversible physical cross-linking exploiting the electrostatic interaction, with tripolyphosphate instead of chemical cross-linking, has been applied to avoid the possible toxicity of reagents and other undesirable effects. Coacervation–precipitation method was already described that relies on the physicochemical property of chitosan that is insoluble in alkaline pH medium, but precipitates/coacervates when it comes in contact with alkaline solution.

8.4.2 Polymersomes

Recently, block copolymer-based vesicles, also called polymer-based liposomes or polymersomes, appear to be promising for drug encapsulation, as their double layer resembles the structure of lipids in membrane cells that could facilitate their biological performance and the design of targeted nanoparticles. They can be obtained from amphiphilic di-block, triblock, graft, or charged copolymers by means of self-assembled or covalently-assembled strategies. The production of polymersomes using poly(butadiene)-poly(ethylene glycol) (PBD-PEO) amphiphilic diblock copolymers and the biocompatible and biodegradable block-copolymer PEO-b-PLA, based on PB-b-PLGA, has been reported (Motornov, 2010).

Polymersome preparation can be carried out using both solvent-free and solvent displacement techniques. In the first method, dried amphiphile polymer is brought in contact with the aqueous medium and then is hydrated to form vesicles. In the second method, the block copolymer is dissolved in organic solvents, then water is added, and subsequently the organic solvent is eliminated by evaporation. The encapsulation of active substances inside the polymer vesicles is obtained by incubation-based techniques. Polymersomes are mechanically stronger and possess superior camouflage ability than lipid-based liposomes. Moreover, polymersomes offer more flexibility in the choice of vesicle size, bilayer thickness, and camouflaging ability via appropriate selection of the molecular weight of the hydrophilic/hydrophobic block. Bovine hemoglobin (Hb) was encapsulated inside polymer vesicles (polymersomes) to form polymersome-encapsulated Hb (PEH) dispersions (Dian and Palmer, 2005). The Hb-loading capacities of these polymersomes were also higher than lipogel-encapsulated hemoglobin. In addition, unlike liposomes composed of natural lipids, polymersomes did not induce Hb oxidation.

8.4.3 Polymeric Micelles

Among different drug carriers used for controlled drug delivery, there has been a rising interest in polymeric micelles that are nanosized, usually spontaneously formed by self-assembly in a liquid. Micelles are formed when amphiphiles are placed in water. They consist of an inner core of assembled hydrophobic segments capable of solubilizing lipophilic substances and an outer hydrophilic corona serving as a stabilizing interface between the hydrophobic core and the external aqueous environment (Fig. 8.11). The most important feature of micellar delivery systems, which distinguishes them from other particulate carriers, lies in their small size (10–30 nm). Their dimension is similar to natural carriers (viruses and serum lipoproteins), thus preventing the micelles uptake by the reticuloendothelial system. At the same time, an easy extravasation at leaky sites of capillaries can lead to passive accumulation in certain tissue (Yuan et al., 1994). The decoration of the micelles surfaces with recognizable moieties or the development of thermo or pH-sensitive block copolymers has been used as active targeting to increase the cellular interaction (Lavasanifar et al., 2002).

The drug loading can be pursued by physical encapsulation of the hydrophobic drug molecules in the core, or by chemical conjugation and complex formation between block copolymers and charged therapeutics. For drug physically encapsulated in stable polymeric micelles, release is controlled by the rate of drug diffusion in the micellar core or break up of the micelles. In the former case, the diffusion rate may be quite low if a strong interaction exists between the drug molecule and the core.

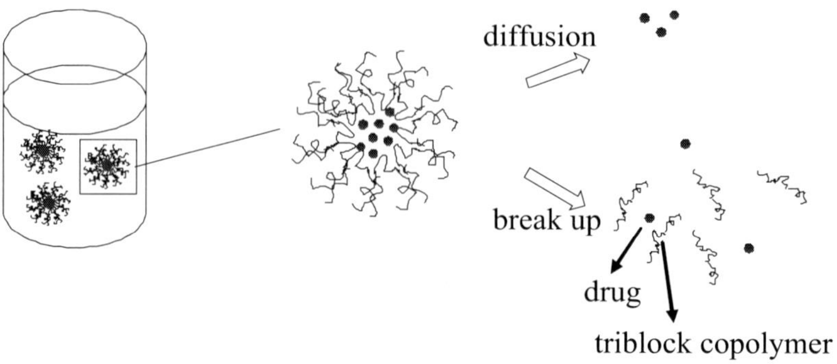

Figure 8.11 Encapsulation of drugs into the hydrophobic core of micelles.

In the case of drug conjugates, the covalent bond between therapeutic molecule and the polymer has to be cleaved for drug release. Water penetration and hydrolysis of the bonds in the micellar core, followed by drug diffusion, may occur in relatively hydrophilic liquid like core structures. Water diffusion into hydrophobic and rigid cores may be restricted. Therefore, in this case, release may be dependent on the rate of micellar dissociation. The slow dissociation of the micellar structure to single polymeric chains and further hydrolysis of the liable bonds may result in a sustained drug release. With such a system, a delay in the therapeutic release is achieved, thus allowing the reaching of the target cells by the carrier.

Micelles made of nonionic surfactants are widely used as adjuvants and drug carrier systems in many areas of pharmaceutical technology and controlled drug delivery. Surfactant micelles form only above a critical concentration, the critical micelle concentration (CMC), and rapidly break apart upon dilution (milliseconds), which can result in premature leakage of the drug and its precipitation in situ. For the polymeric micelles, the concentration at which the association first takes place is lower by several orders of magnitudes than typical surfactant CMC values, thus resulting in a slow break up (s, min, to h).

A wide range of hydrophobic blocks have been explored, resulting in different micellar systems with distinct physicochemical properties. Toxicological studies on Pluronic® triblock copolymers [also termed Poloxamer; poly(ethylene oxide)x-b-poly(propylene oxide)y-b-poly(ethylene oxide)x; PEOx-b-PPOy-b-PEOx] have demonstrated that they are safe, particularly those with a high PEO content. Moreover, the stealth properties of the PEO corona result in increased blood circulation times and allow drugs to be administered over prolonged periods of time. The biocompatibility studies on block copolymer composed of polyesters and poly(L-amino acids) (PLAA) have shown that their hydrolysis and/or enzymatic degradation produce biocompatible monomers.

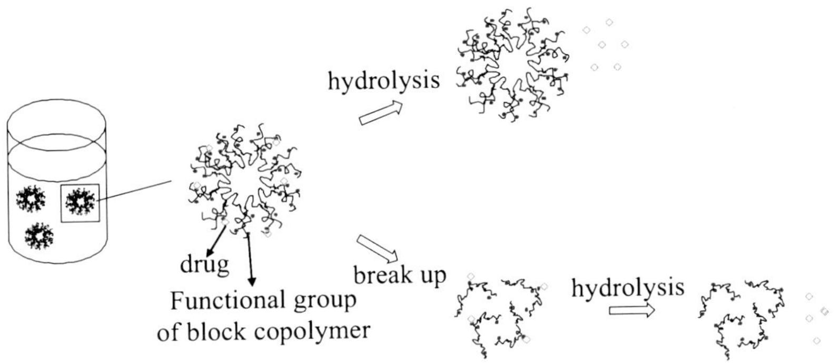

Figure 8.12 Mechanism of drug release promoted by either the hydrolysis of the conjugated drug or the dissolution of micelles.

Generally, PEOs with molecular weight ranging from 1000 to 20000 g/mol are used as a hydrophilic block in the micelle forming block copolymers. The length of the core-forming block is usually equal or shorter than the PEO block to maintain water solubility and form spherical micelle structure. Pluronics have been studied extensively as a delivery medium for injectable drug formulations of poorly water-soluble drugs such as paclitaxel, haloperidol, indomethacin, amphotericin B, doxorubicin, and epirubicin (Lavasanifar et al., 2002). Overall, they proved to be highly effective drug delivery vehicles. By balancing the ratio of PEO to PPO content, the properties of the carriers in terms of drug loading and stability were tuned. Biodegradable block copolymers with polyester core-forming structures such as poly(lactic acid), poly(glycolic acid), poly-(caprolactone), and their copolymers have been developed and used for drug delivery. PEO-b-poly(D,L lactic acid), PEO-b-PDLLA, and block copolymer have been shown to form a micellar fraction, which is able to solubilize model hydrophobic drugs. PEO-b-poly(lactic acid), PEO-b-PLA, and block copolymers with different lengths of PLA block have been used to encapsulate water-soluble drug. An increase of PLA length caused an increase in the size of nanoparticles but did not affect their encapsulation efficiency. Micelles of PEO and PDLLA, poly(D,L-lactic acid-co-caprolactone) (PDLLACL), or poly(glycolic-co-caprolactone) (PGACL) have been used to encapsulate taxol. The taxol-loaded PEO-b-PDLLA micelles revealed high stability due to the high hydrophobicity and glass transition temperature, *Tg*, of PDLLA block. Moreover, higher content of PDLLA determined higher taxol solubility and loading content up to 25%, which is in contrast with the loading of 0.5% for the same drug in Pluronic micelles.

PEO-b-PDLLA micelles with aldehyde groups on their surface were produced to conjugate peptidyl ligands through Schiff base formation and further reduction. Also, PEO blocks with site-specific sugar groups at their chain ends have been obtained through ring opening polymerization using D,L-lactide, in such a way that the micelle surfaces have glucose or galactose moieties. The galactose residues bind to the sugar-bind receptor RCA1-lecetine molecule, thus realizing an active targeting by receptor-mediated endocytosis. Another versatile system for micelles self-assembly is constituted from block copolymers of PEO-b-poly(L-amino acids) (PLAA). In particular, three different types of drug delivery systems based on PEO-b-PLLA block copolymers have been investigated: micelle-forming block copolymer-drug conjugates; micellar nanocontainers; and polyion complex micelles. Micelle-forming block copolymer-drug conjugates are based on the capability of self-assembling PEO-b-PLLA block copolymers to attach drugs, drug-compatible moieties, genes, or intelligent vectors in the micellar core through free carboxylic and amine groups of the amino acid chain. The physical entrapment of drugs within polymeric micelles can be pursued when drug molecules do not bear reactive functional carboxyl, hydroxyl, or amino groups for chemical conjugation or the free functional site may be required for the pharmacological effectiveness of the drug. The micellar nanocontainers are usually prepared by dialysis or o/w emulsion. Their principal drawbacks are the possibility of low encapsulation capacity or the rapid release of encapsulated drugs. Through the attachment of drug-compatible moieties to the core-forming block of PEO-b-PLLA, the encapsulation and the release of the therapeutics can be tuned for each particular drug or class of drugs. Finally, the positive or negative charges associated with PEO-b-PLLA block copolymers may form polyion complex with oppositely charged macromolecules such as DNA or peptides, the neutralization of the charge inducing the required amphiphilicity for micellization of the complex. The polyion complex micelles are salt-sensitive, so burst release can be achieved if the salt concentration increases above a certain value.

8.4.4 Dendrimers

There are various approaches in delivering a therapeutic substance to the target site in a controlled release manner. An alternative approach to microcapsules is the use of dendrimers as carrier for drugs. Dendrimers are synthetic, monodisperse macromolecules with highly branched structure and globular shape. Because molecular size and generation dendrimers are increased by iterative synthetic methodology, their size and structure are highly controllable. They also have modifiable surface functionality that can be used to attach several drugs and targeting groups. Due to the large number of terminal groups to which drug molecules can be conjugated, dedrimers can carry drugs at high density. On the other end, internal cavities are capable of encapsulating various molecules. It has been shown that small guest molecules can be physically entrapped in the interior of dendrimers with rigid shell (Jansen et al., 1994). Also, dendrimers with a hydrophobic interior and hydrophilic chain ends are able to solubilize various hydrophobic guest molecules in aqueous solution (Newkome et al., 1991).

Dendrimers can be prepared using either a divergent growth approach or a convergent one (Klajnert and Bryszewska, 2001). In the divergent methods, dendrimers grow outward from a multifunctional core molecule; in the convergent approach, the dendrimer is constructed stepwise, starting from the end groups and progressing inward. The divergent process allows the production of large quantities of dendrimers, but problems occur from side reactions and incomplete reactions of the end groups that lead to structure defects. Moreover, the necessity to use large excess of reagents to avoid side reactions and to force reactions to completion entails some difficulties in the purification of the final products. The convergent growth process does not have the disadvantages of the divergent one, but it does not allow the formation of high generations, because of the steric problems that occur in the reaction of dendrons and the core molecule. Architectures of different dendrimers have been shown in Fig. 8.13.

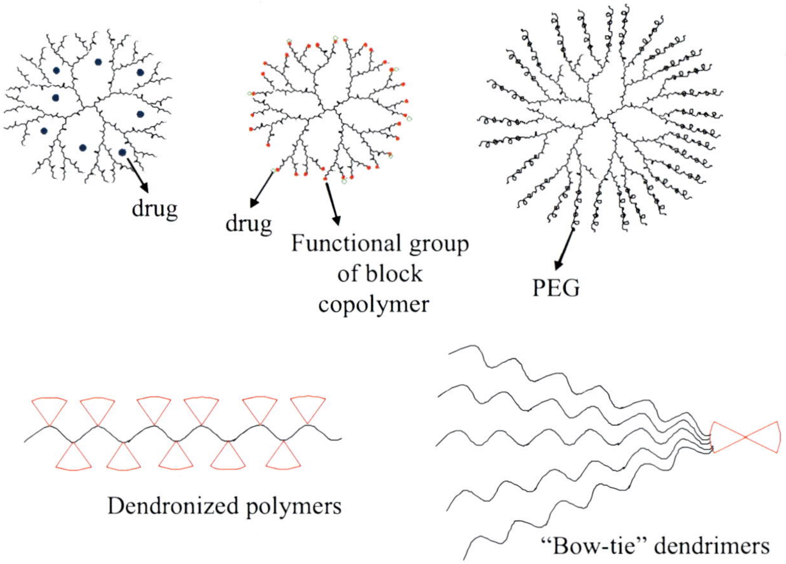

Figure 8.13 Different dendrimers architectures.

The family of dendrimers most investigated for drug delivery is the polyamidoamine (PAMAM) drendimers that bear primary amino group termini. It has been demonstrated that they are quite cytotoxic (Roberts et al., 1996), and to avoid the toxicity and liver accumulation associated with their polycationinc surface, it is generally necessary to modify the surface (Malik, 2000). To overcome the toxicity connected to PAMAM dendrimers, PEG can be incorporated to the dendrimers (Kojima et al., 2000).

An advantage of using dendritic unimolecular micelles instead of polymeric micelles is the high stability of dendrimers at all concentration, because the segments are covalently connected. For polymeric micelles below a CMC, the polymeric aggregates dissociate into free chains, leading to the rapid release of the drug. The drawback of the unimolecular micelles is the difficulty to control the release of the molecules; according to the drug characteristics, harsh condition can be requested or a rapid release can be observed (Gillies and Fréchet, 2005).

PAMAM dendrimers with PEG grafts have been used for the physical encapsulation of anticancer drugs, such as adriamycin and methotrexate (Kojima et al., 2000). A promising new approach to control the release is the use of pH-sensitive hydrophobic acetal groups on the dendrimer periphery. At acidic pH, the loss of the hydrophobic groups triggers disruption of micelles and release of the entrapped molecules (Gillies et al., 2005). An alternative strategy for drug delivery is the covalent conjugation of the tumor agent to the dendrimers. In the literature, there are examples of PAMAM dendrimers conjugated with drugs as cisplatin or Ara-c and folic acid for targeting to tumor cells (Patri et al., 2002; Gillies and Fréchet, 2005). Frechet et al. (2005) developed novel polyaryl ether dendrimers containing dual functionality on the surface. One type of functionality was used to attach PEG units on the surface to increase the water solubility of dendrimers and decrease the immunogenicity. The other functionality was utilized to attach hydrophobic drug molecule (Patri et al., 2002). Carriers of anticancer drug doxorubicin have been prepared by conjugation of poly[*N*-(2-hydroxypropyl) methacrylamide macromonomers to PAMAM dendrimers, followed by attachment of the drug to the polymer arms by a biodegradable peptide spacer (Gillies and Fréchet, 2005).

Another alternative approach is the preparation of hybrids of polyester dendrimers and PEO star polymers. The hybrid polymers were then conjugated to doxorubicin via a hydrazone linkage that undergoes hydrolysis at the mildly acidic subcellular organelles. These drug conjugates show an increased circulation time of more than 1 h. "Bow tie" hybrids of polyester dendrimers and PEO have recently been prepared. It has been demonstrated that these derivates are generally long circulating with half-lives greater than 24 h. Finally, a new class of macromolecules termed "dendronized polymers" has been recently prepared, in which linear polymer bears dendrons at each repeat unit. Due to the result of steric interactions, at high generation numbers, dendronized polymers attain rigidified conformation (Gillies and Fréchet, 2005). Application of dendrimers to boron neutron capture therapy is also studied. The strategy used consists in the preparation of boronated dendrimers that are conjugated to a monoclonal antibody, which in turn is directed to the tumor cells. An in vivo study of a PAMAM dendrimer carrying 1100 boron atoms, conjugated to cetuximab, a monoclonal antibody specific for the EGF receptor, revealed a 10-fold greater concentration in brain tumors than in normal brain tissue. Biodistribution study of these systems has demonstrated their accumulation in the liver and the spleen. In order to reduce the liver uptake of the boronated dendrimers, another strategy introduced PEO chains into the dendrimer periphery. When compared with naked dendrimers, a lower uptake in the liver was found for PEO-conjugated dendrimers; on the other hand dendrimers with a higher PEO content and a lower PEO molecular weight showed a higher uptake. The photodynamic therapy is another pursued field of dendrimers application. Conjugates of a dendrimer with 5-aminolevulinic acid (ALA), a natural precursor of the photosensitizer proporphyrin (PIX), have been prepared. An increased production of PIX relative to free ALA and a higher toxicity after irradiation was demonstrated for this kind of delivery vehicles. Polyaryl ether dendrimer porphyrins with either 32 quaternary ammonium groups or 32 carboxylic acid groups on the periphery were prepared and compared with the PIX activity. An increased toxicity of the cationic dendrimer porphyrin was explained as a result from the electrostatic association with cell membranes. PEO-polyaspartic acid block polymers were electrostatically assembled with cationic porphyrin dendrimers, in order to prolong the circulation of the carriers in the blood and to enhance tumor accumulation. It was found out that a cellular uptake three times slower when compared with porphyrin, but when normalized to the porphyrin content, the conjugate showed about 40 times greater photodynamic efficacy. This has been attributed to the decreased tendency of the dendrimers to aggregate under physiological conditions. A potential disadvantage of this system is that only one drug molecule can be loaded for each micelle. The dendrimer porphyrin used has an absorption maximum at 559 nm. Because an increased permeability of tissue is observed to near-infrared and infrared light, efforts have been carried out for the preparation of water-soluble polymer conjugates of metallo-phthalocyanines with an absorption maximum at 675 nm. In recent studies, a water-soluble dendrimer porphyrin incorporating several two-photon absorbing chromophore was prepared and it has been demonstrated that it generates singlet oxygen efficiently by irradiation at 780 nm (Gillies and Fréchet, 2005).

8.5 Conclusion

Herein, numerous examples involving ultrasound for the preparation of nanopolymeric devices for the encapsulation and conjugation of therapeutic agents have been discussed. Active substances encapsulated are mainly anticancer drugs such as adriamycin, paclitaxel, and doxorubicin and therapeutic proteins and antisense molecules for gene therapy. Polymeric capsules have the ability to protect the encapsulated agents from enzymatic degradation and bacteria. The polymeric shell allows the control of permeability toward small substrates, which can enter and leave the capsules to react with the biomolecules in the interior. Apart from the physicochemical properties of the polymeric matrix, the active molecule's solubility, stability, and the interactions with the polymer are properties that altogether should be considered for the indentification, development, and optimization of new microparticles formulation. The microencapsulation technique is primarly used for molecules that exhibit poor solubility in water but that are typically soluble in various organic solvents. Micro-nanoparticles are almost produced by emulsion techniques, which generally require the use of high-energy ultrasound equipments. The ultrasound-assisted microencapsulation compared with other reported methods offers the advantage of being effective in controlling the size distribution, the thickness of the encapsulant micro-nanoreservoir, and the dose of the encapsulated bioactive agents. We reported on the use of ultrasound techniques as reactive processing methods to modify and functionalize micro-nanopolymeric materials. Radicals, reducing and oxidizing species generated by acoustic cavitation are a powerful tool to functionalize both the synthetic and natural polymeric materials at the nano scale. The LbL deposition method also appears to be a versatile and feasible approach for the encapsulation of hydrophilic and labile bioactive molecules.

References

Avivi, S., Gedanken, A. (2002), S-S bonds are not required for the sonochemical formation of proteinaceous microspheres: the case of streptavidin, *Biochem. J.*, **366**, pp. 705–707.

Becker, A. L., Johnston, A. P. R., Caruso, F. (2010), Layer-by-layer-assembled capsules and films for therapeutic delivery, *Small*, **6**, pp. 1836–1852.

Becker, A. L., Zelikin, A. N., Johnston, A. P. R., Caruso, F. (2009) Langmuir, tuning the formation and degradation of layer-by-layer assembled polymer hydrogel *Microcapsules*, **25**, pp. 14079–14085.

Cao, X., Prozorov, R., Koltypin, Y., Kataby, G., Felner, I., Gedanken, A. (1997) Synthesis of pure amorphous Fe_2O_3 *J. Mater. Res.*, **12**, pp. 402–406.

Caruso, F., Caruso, R. A., Mohwald, H. (1998), Nanoengineering of inorganic and hybrid hollow spheres by colloidal templating *Science*, **282**, pp. 1111–1114.

Caruso, F., Fiedler, H. Haage, K. (2000), Assembly of β-glucosidase multilayers on spherical colloidal particles and their use as active catalysts, *Colloids Surf. A*, **169**, pp. 287–293.

Caruso, F., Mohwald, H. (1999), Protein multilayer formation on colloids through a stepwise self-assembly technique, *J. Am. Chem. Soc.*, **121**, pp. 6039–6046.

Caruso, F., Schuler, C. (2000), Enzyme multilayers on colloid particles: Assembly, stability, and enzymatic activity, *Langmuir*, **16**, 9595–9603

Caruso, F., Trau, D., Möhwald, H., Rennenberg, R. (2000) Enzyme encapsulation in layer-by-layer engineered polymer multilayer capsules, *Langmuir*, **16**, pp. 1485–1488.

Cavalieri, F., Ashokkumar, M., Grieser, F., Caruso, F. (2008), Ultrasonic synthesis of stable, functional lysozyme microbubbles *Langmuir*, **24**, pp. 10078–10083.

Cavalieri, F., Postma, A., Lee, L., Caruso, F. (2009), Assembly and functionalization of DNA–polymer microcapsules, *ACS Nano,* **3**, pp. 234–240.

Chang, T. M. S. (1964). Semipermeable microcapsuels, *Science*, **146**, pp. 524–525.

Chang, T. M. S. (2004), Artificial cells for cell and organ replacements, *Artif. Organs*, **28**, pp. 265–270.

Daly, M. M., Knorr, D. (1998), Chitosan-alginate complex coacervate capsules, *Biotechnol. Prog.*, **4**, pp. 76–81.

Decher, G. (1997), Fuzzy nanoassemblies: Toward layered polymeric multicomposites, *Science*, **277**, pp.1232–1237.

Decher, G., Hong, J. D. (1991), Buildup of ultrathin multilayer films by a self-assembly process. I. Consecutive adsorption of anionic and cationic bipolar amphiphiles *Makromol. Chem. Macromol. Symp.*, **46**, pp. 321–327.

Dian, R., Palmer, A. F. (2005), Polymersome encapsulated hemoglobin: a novel type of oxygen carrier, *Biomacromolecules*, **6**, pp. 2172–2181.

Dibbern, E. M., Toublan, F. J., Suslick, K. S. (2006), Formation and characterization of polyglutamate core-shell microspheres, *J. Am. Chem. Soc.*, **128**, pp. 6540–6541.

Donath, E., Sukhorukov, G. B., Caruso, F., Davis, S. A., Möhwald, H. (1998), Novel hollow polymer shells by colloid-templated assembly of polyelectrolytes *Angew. Chem. Int. Ed. Engl.*, **37**, pp. 2202–2205.

Dunne, M., Corrigan, O. I., Ramtoola, Z. (2000), Influence of particle size and dissolution conditions on the degradation properties of polylactide-co-glycolide particles, *Biomaterials*, **21**, pp. 1659–1668.

Dziezak, J. D. (1988), Microencapsulation and encapsulated ingredients. *Food Technol.*, **42**, pp. 136–151.

Gedanken, A. (2008) Preparation and properties of proteinaceous microspheres made sonochemically. *Chem. Eur. J.*, **14**, pp. 3840–3853.

Gillies, E. R., Fréchet, J. M. J. (2005), Dendrimers and dendritic polymers in drug delivery, *Rev. Drug Disco. Today*, **10**, pp. 35–43.

Grinstaff, M. W., Suslick, K. S. (1991), Air-filled proteinaceous microbubbles: synthesis of an echo-contrast agent, *Proc. Natl. Acad. Sci. USA*, **88**, pp. 7708–7710.

Heidebach, T. Forest, P., Kulozik, U. (2009), Transglutaminase-induced caseinate gelation for the microencapsulation of probiotic cells. *Int. Dairy J.* **19**, pp. 77–84.

Hutchinson, F. G., Furr, B. J.A. (1989), Biodegradable polymers for controlled release of peptides and proteins, *Biochem. Biophys.*, **9**, pp. 111–129.

Jackson L. S., Lee, K. (1991), Microencapsulation in the food industry, *Lebensmittel-Wissenschaft und –Technologie,* **24**, pp. 289–297.

Jansen, J. F. G. A, de Brabander-van den Berg, E. M. M, Meijer, E. W. (1994), Encapsulation of guest molecules into a dendritic box, *Science*, **266**, pp. 1226–1229.

Khoee, S., Yaghoobian, M. (2009), An investigation into the role of surfactants in controlling particle size of polymeric nanocapsules containing penicillin-G in double emulsion, *Eur. J. Med. Chem.*, **44**, pp. 2392–2399.

Kim, B. S., Park, S. W., Hammond, P. T. (2008), Hydrogen-bonding layer-by-layer-assembled biodegradable polymeric micelles as drug delivery vehicles from surfaces, *ACS Nano*, **2**, pp. 386–392.

Klajnert, B., Bryszewska. M. (2001), Dendrimers: properties and applications. *Acta Biochimica Polonica*, **48,** pp. 199–208.

Kojima, C., Kono, K., Maruyama, K., Takagishi, T. (2000), Synthesis of polyamidoamine dendrimers having polyethylene glycol grafts and their ability to encapsulate anticancer drugs, *Bioconjugate Chem.* **11**, pp. 910–917.

Krol, S., Diaspro, A., Cavalleri, O., Cavanna, D., Ballario, P., Grimaldi, B., Filetici, P., Ornaghi, P., Gliozzi, A. (2004). Nanocapsules: a novel tool for medicine and science. In: Buzaneva, E., Scharff, P. (Eds.), *Frontiers of Multifunctional Integrated Nanosystems*. Academic Publishers, The Netherlands, pp. 439–446.

Langer, R. (1990), New methods of drug delivery, *Science*, **249**, pp. 1527–1533.

Lavasanifar, A., Samuel, J., Kwon, G. S. (2002), Poly(ethylene oxide)-block-poly(L-amino acid) micelles for drug delivery, *Adv. Drug Del. Rev.*, **54**, pp. 169 –190.

Lee, L., Cavalieri, F., Johnston, A. P. R., Caruso., F. (2010), Influence of salt concentration on the assembly of DNA multilayer films, *Langmuir*, **6** pp. 3415–3422.

Lee, T. K., Sokoloski, T. D., Royer, G. P. (1981), Science, serum albumin beads: an injectable, biodegradable system for the sustained release of drugs, **213**, pp. 233–235.

Lee, T. M., Oldenburg, A. L., Sitafalwalla, S., Marks, D. L., Luo, W., Toublan, F. J., Suslick, K. S., Boppart, S. A. (2003), Engineered microsphere contrast agents for optical coherence tomography, *Opt. Lett.*, **28**, pp. 1546–1548.

Li, S. (1999), Hydrolytic degradation characteristics of aliphatic polyesters derived from lactic and glycolic acids, *J. Biomed. Mater. Res. (Appl. Biomater.)*, **48**, pp. 342–353.

Lim, L. Y., Wan, L. S. C., Thai, P. Y. (1997),Chitosan microparticles prepared by emulsification and ionotropic gelation, *Drug Dev. Ind. Pharm.*, **23**, pp. 981–985.

Liu, X. D., He, Y., Liu, Q., Xiong, Y., Fu, Y. L., Ma, X. J., Yu, X. J., Yuan, Q. (2000), Microcapsule and its applications in biomedical field. *Chin. Sci. Bull.*, **45**, pp. 2476–2485.

Makino, K., Mizorogi, T., Ando, S., Tsukamoto, T., Ohshima, H. (2001), Sonochemically prepared bovine serum albumin microcapsules: factors affecting the size distribution and the microencapsulation yield, *Colloids Surf. B*, **22**, pp. 251–255.

Malik, N., Wiwattanapatapee, N. R., Klopsch, R, Lorenz, K., Frey, H., et al. (2000), Dendrimers: Relationship between structure and biocompatibility in vitro, and preliminary studies on the biodistribution of 125I-labelled polyamidoamine dendrimers in vivo *J. Control. Release*, **65**, pp. 133–148.

Mehrotra, S., Lynam, D., Maloney, R., Pawelec, K. M., Tuszynski, M. H., Lee, I., Chan C., Sakamoto, J. (2010), Time controlled protein release from layer-by-layer assembled multilayer functionalized agarose hydrogels, *Adv. Funct. Mater.*, **20**, pp. 247–258.

Meifang, Z., Leong, T. S. H., Melino, S. Cavalieri, F. Kentish, S. and Ashokkumar, M. (2010), Sonochemical synthesis of liquid-encapsulated lysozyme microspheres, *Ultrason. Sonochem.*, **17**, pp. 333–337.

Mi, F. L., Tan, Y. C., Liang, H. F., Sung, H. W. (2002), In vivo biocompatibility and degradability of a novel injectable-chitosan-based implant, *Biomaterials*, **23**, pp. 181–191.

Motornov, M., Roiter, Y., Tokarev, I., Minko, S. (2010), Stimuli-responsive nanoparticles, nanogels and capsules for integrated multifunctional intelligent systems, *Prog. Polym. Sci.*, **35**, pp. 174–211.

Newkome, G. R., Moorefield, C. N., Baker, G. R., Saunders, M. J., Grossman, S. H. (1991), Alkane cascade polymers possessing micellar topology: Micellanoic acid derivatives, *Angew. Chem. Int. Ed. Engl.* **30**, pp. 1176–1178.

Ochs, C. J., Such, G.K., Yan, Y., van Koeverden, M. P., Caruso, F. (2010) Biodegradable click capsules with engineered drug-loaded multilayers, *ACS Nano*, **4**, pp. 1653–1663.

Park, J. H., Ye, M., Park K. (2005), Biodegradable polymers for microencapsulation of drugs, *Molecules*, **10**, pp. 146-161.

Patri, A. K., Majoros, I. J., Baker, J. R. (2002), Dendritic polymer macromolecular carriers for drug delivery, *Cur. Opin. Chem. Biol.,* **6,** pp. 466–471.

Perazzini, R., Saladino, R., Guazzaroni, M., Crestini, C. (2011), A novel and efficient oxidative functionalization of lignin by layer-by-layer immobilised Horseradish peroxidase, *Bio. Med.Chem.* **61**, pp. 440–447.

Perez, C., Sanchez, A., Putnam, D., Ting, D., Langer, R., Alonso, M. J. (2001), Poly(lactic acid)-poly(ethylene glycol) nanoparticles as new carriers for the delivery of plasmid DNA *J. Control. Release*, **75**, pp. 211–224.

Roberts, J. C., Bhalgat, M. K., Zera, R. T. (1996), Preliminary biological evaluation of polyamidoamine (PAMAM) Starburst dendrimers. *J. Biomed. Mater.* **30**, pp. 53–65.

Ruan, G., Feng, S. S. (2003), Preparation and characterization of poly (lactic acid)–poly(ethylene glycol)–poly(lactic acid) (PLA–PEG–PLA) microspheres forcontrolled release of paclitaxel, *Biomaterials*, **24**, pp. 5037–5044.

Shahidi, F., Han, X. Q. (1993), Encapsulation of food ingredients, *Crit. Rev. Food Sci.*, **33**, pp. 501–547.

Shu, X. Z., Zhu, K. J. (2001), Chitosan/gelatin microspheres prepared by modified emulsification and ionotropic gelation, *J. Microencapsulation*, **18**, pp. 237–245.

Sivakumar, S., Bansal, V., Cortez, C., Chong, S.F., Zelikin, A. N. Caruso, F. (2009), Degradable, surfactant-free, monodisperse polymer-encapsulated emulsions as anticancer drug carriers, *Adv. Mater. (Weinheim, Ger.)*, **21**, pp. 1820–1824.

Skirtenko, N., Tzanov, T., Gedanken, A., Rahimipour., S. (2010), One-step preparation of multifunctional chitosan microspheres by a simple sonochemical method, *Chem. Eur. J.*, **16**, pp. 562–567.

Sunil, A. A., Nadagouda. N. M., Tejraj, N. M. A. (2004), Recent advances on chitosan-based micro- and nanoparticles in drug delivery, *J. Controlled Release*, **100**, pp. 5–28.

Suslick, K. S. (1995), Sonochemically produced hemoglobin microbubbles, *Mater. Res. Soc. Symp. Proc.*, **372**, pp. 89–94.

Suslick, K. S., Grinstaff, M. W. (1990), Protein microencapsulation of nonaqueous liquids, *J. Am. Chem. Soc.*, **112**, pp. 7807–7809.

Toublan, F. J., Boppart, S., Suslick, K. S. (2006), Tumor targeting by surface-modified protein microspheres, *J. Am. Chem. Soc.*, **128**, pp. 3472–3473.

Toublan, F.J-J., Boppart, S. Suslick, K. S. (2006) Tumor targeting by surface modified protein microspheres, *J. Am. Chem. Soc.*, **128**, pp. 3472–3473.

Wasan K. M. (2001), Formulation and physiological and biopharmaceutical issues in the development of oral lipid-based drug delivery systems, *Drug. Dev. Ind. Pharm.*, **27**, pp. 267–276.

Webb, A. G., Wong, M., Kolbeck, K. J., Magin, R. L., Wilmes, L. J., Suslick, K. S. (1996), Sonochemically produced fluorocarbon microspheres: a new class of magnetic resonance imaging agent, *J. Magn. Reson. Imag.*, **6**, pp. 675–683.

Witschi, C., Doelker, C., E.(1998), Influence of the microencapsulation method and peptide loading on poly(lactic acid) and poly(lactic-co-glycolic acid) degradation during in vitro testing, *J. Control. Release*, **51**, pp. 327–341.

Witschi, C., Doelker, E. (1998), Influence of the microencapsulation method and peptide loading on poly(lactic acid) and poly(lactic-co-glycolic acid) degradation during in vitro testing, *J. Control. Release*, **51**, pp. 327–341.

Wu, C. (1994), Laser light scattering determination of the surfactant interface thickness of spherical polystyrene microlatices, *Macromolecules*, **27**, pp. 7099–7102.

Xue, W. M., Yu, W. T., Wang, W., Liu, X. D., Ma, X. J. (2003), Materials and fabricating methods of microcapsules for cell transplantation, *Chem. Online*, **66**, w12 (1–7).

Yang, W. J., Trau, D., Renneberg, R., Yu, N. T., Caruso, F. (2001), Layer-by-layer construction of novel biofunctional fluorescent microparticles for immunoassay applications, *J. Colloid Interface Sci.*, **234**, pp. 356–362.

Yuan, F., Leuning, M., Huang, S. K., Berk, D. A., Papahadjopoulos, D. Jain, R. K. (1994), Microvascular permeability and interstitial penetration of sterically stabilized (stealth) liposomes in a human tumor xenograft, *Cancer Res.*, **54**, pp. 3352–3356.

Zelikin, A. N., Li, Q., Caruso, F. (2006), Degradable polyelectrolyte capsules filled with oligonucleotide sequences, *Angew. Chem., Int. Ed.*, **45**, pp. 7743–7745.

Zhu, Y., Zhang, G., Yang, H., Hong, X. (2005), Influence of surfactants on the parameters of polylactide nanocapsules containing insulin, *J. Surfact. Deterg.*, **8**, pp. 353–358.

Chapter 9

Innovative Inorganic Nanoparticles with Antimicrobial Properties Attached to Textiles by Sonochemistry

Aharon Gedanken, Nina Perkas, Ilana Perelshtein, Guy Applerot, Anat Lipovsky, Yeshayahu Nitzan, and Rachel Lubart
Department of Chemistry and Kanbar Laboratory for Nanomaterials,
Bar-Ilan Centre for Advanced Materials and Nanotechnology,
Bar-Ilan University, Ramat-Gan 52900, Israel
gedanken@mail.biu.ac.il

This chapter is devoted to research on the functionalization of textiles with antimicrobial inorganic nanoparticles by a sonochemical method. Nanosized materials can be finely spread on fabrics due to the large surface-to-volume ratio and high surface energy that provides a strong affinity to the textile fibers. Over the last two decades, progress in nanotechnology has discovered many synthetic methods for the fabrication of various kinds of nanomaterials. The Introduction section emphasizes the necessity for new, effective, and environmentally friendly methods for the antibacterial treatment of textiles. The second section explains the principles behind the sonochemical process of embedding nanoparticles

Cavitation: A Novel Energy-Efficient Technique for the Generation of Nanomaterials
Edited by Sivakumar Manickam and Muthupandian Ashokkumar

ISBN 978-981-4411-54-7 (Hardcover), 978-981-4411-55-4 (eBook)
www.panstanford.com

into solids, particularly on the textiles. It also presents the work on the sonochemical deposition of nanometal oxides on various substrates. The third section of the review follows publications on antimicrobial properties obtained by functionalizing the textiles with nano silver and metal oxide nanoparticles [ZnO, CuO, TiO_2, MgO]. In the fourth section, the advantages of the sonochemical approach for the antimicrobial finishing of textiles with nanometric metal oxides, in comparison with other coating methods, are demonstrated. The mechanism of antimicrobial activity of metal nano-oxides is discussed in Section 9.5. Our continuing research on the functionalization of textiles and polymeric materials aims at imparting them with antifungal properties. This purpose has been achieved by the adherence of the inorganic nano-oxides to different substrates by the ultrasound-assisted method.

9.1 Introduction

Nowadays, there is a growing need for high-quality textiles with antimicrobial properties for hygienic clothing, active wear, and wound healing. It is recognized that neither synthetic nor natural fibers are resistant to bacteria and pathogenic fungi (Purwar and Joshi, 2006). The control of micro-organisms extends from medical institutions to ordinary households. Consumer demands caused a significant growth in the production of antibacterial textiles. An explosive growth is also expected in wound-care production. The wound-care market of the US healthcare system was in excess of \$7 billion in 2007 (Zalesky, 2008). Worldwide industry reports estimated the wound-care market as \$11.8 billion in 2009 and projected a yearly growth for all products (devices for wound closure such as sutures and staples, dressings, and so on) in excess of 7%. European markets accounted for about half of the expenditure (Petrulyte, 2008). The market for textiles using nanotechnologies is predicted to climb dramatically from \$13.6 billion in 2007 to \$115 billion by 2012 (Coyle et al., 2007).

These high demands encourage intensive research and the development of new methods for the antimicrobial treatment of textiles and fibers. The recent achievements in the field of antibacterial textiles were briefly described in a review (Gao and

Cranston, 2008). According to these data, metal and metal salts are one of the major classes of antibacterial and antifungal agents, as are quaternary ammonium compounds, triclosan, chitosan, and chlorine-containing *N*-halamine compounds, and so on.

The fast development of nanotechnology opens up various possibilities for the synthesis of new materials whose properties are influenced very much by their nanosized structure. Nanometals and nanometal oxides have a larger surface than the conventional powders, and can be finely spread on the surface of fibers and fabrics. These unique properties have found wide applications in the textile industry, namely, in the antimicrobial treatment of textiles (Qian and Hinestroza, 2004; Craighead and Leong, 2006). The methods known for stabilizing inorganic nanostructured materials on textile surfaces are based on multistep preparation processes that are time-consuming for up-scale consideration (Wong et al., 2006; Dastjerdi et al., 2009; Rajendran et al., 2010). Moreover, special, and sometimes toxic, binding agents have to be applied to the textiles for the adherence of the nanoparticles (NPs). Binding agents such as isocyanate groups, epoxides, acrylic acid esters, fluoropolymers, and so on (Kathirvelu et al., 2009; Schafer, 2009; Simoncic and Tomsic, 2010; Kim et al., 2011) were used for this purpose. Therefore, the need for effective and environmentally friendly methods for the antimicrobial functionalization of the fabric is highly important.

In previous reviews by our group, we described the development of the sonochemical technique for the fabrication of various kinds of nanomaterials (Gedanken, 2004) and for the doping of NPs into ceramic and polymer bodies (Gedanken, 2007). Other review articles on similar topics have also been published (Suslick and Price, 1999; Vajnhandl et al., 2005; Mason, 2007).

In the current chapter, the unique properties that make ultrasound radiation an excellent technique for the adherence of NPs to a large variety of substrates will be described. We will concentrate on the deposition of inorganic metal nano-oxides (ZnO, CuO, TiO_2, MgO) on textiles and explain the mechanism of their antimicrobial activity. The review will compare the deposition of NPs formed on textiles during the sonochemical process and the NPs purchased from a commercial source. The advantages of sonochemistry as a one-step, environmentally friendly method for the deposition of NPs on different kinds of textiles, such as cotton, wool, nylon, polyester, and so on, will be demonstrated. The review will also scan the work performed by different authors using ultrasound irradiation.

9.2 Deposition of Nanosized Metal Oxides on Solid Surfaces by the Sonochemical Technique

The increasing interest in the synthesis of different kinds of nanomaterials is based on their large specific surface area, and their new, size-dependent physical and chemical properties in comparison with the bulk structures. The potential application of the NPs can be significantly extended by their deposition on different types of substrates. The interest in coatings of different types of substrates with metal oxides lies in the possibility to combine the properties of the two (or more) materials involved in the process, namely, the substrate and the coated layer, with emphasis on the fact that one of the materials will determine the surface properties of the composite, while the other can be responsible for other (optical, catalytic, magnetic, antimicrobial, and so on) properties of the system (Rao and Cheetham, 1991; Schmidt, 2001; Liz-Marzan and Kamat, 2003). The development of scalable methods that effectively bind NPs to surfaces and provide precise patterns is a key step toward the commercial exploitation of the distinctive properties of nanostructured materials (Kilmov et al., 2000; Tricoli et al., 2008).

As mentioned above and discussed previously, many inorganic nanomaterials have been prepared sonochemically. Sonochemistry is the scientific area where chemical reactions occur under ultrasound irradiation. Liquids irradiated with ultrasound produce bubbles. The reaction is dependent on the development of an acoustic bubble in the solution. Ultrasonic waves with the frequency range of 20 kHz to 1 MHz are responsible for the process of acoustic cavitation, that is, the formation, growth, and explosive collapse of the bubbles. There are a number of theories that explain how 20 kHz ultrasonic radiation can break chemical bonds (Suslick et al., 1986; Doktycz and Suslick, 1990; Mason, 1990). We will adopt the hot spot mechanism, one of the theories that explain why, upon the collapse of a bubble, chemical bonds are broken. This theory claims that very high temperatures (5,000–25,000 K) (Suslick et al., 1986) are obtained upon the collapse of the bubble. As this collapse occurs in less than a nanosecond, very high cooling rates in excess of 10^{11} K/s are obtained. These extreme conditions developed when the bubbles' collapse cause the chemical reactions to occur. The high cooling rate prevents the crystallization of the products. This is the reason why amorphous NPs are formed when volatile precursors are used and the gas phase reaction is predominant. However, from this explanation, the reason for the formation of nanostructured material is not clear. Our explanation for the creation of nanoproducts is that the fast kinetics does not permit the growth of the nuclei, and in each collapsing bubble, a few nucleation centers are formed whose growth is limited by the explosive collapse.

The dynamics of cavity growth and collapse during sonication is strictly dependent on the local environment. Cavity collapse in a homogeneous liquid is very different from cavitation near a liquid–solid interface. Suslick and Price (1999) demonstrated that microjets and shock waves produced by acoustic cavitation are able to drive metal particles together at sufficiently high velocities to induce melting upon collision. Metal particles that were irradiated in hydrocarbon liquids with ultrasound underwent collisions at roughly half the speed of sound and generated localized effective temperatures between 2600°C and 3400°C at the particles' point of impact. This approach was developed further in our experiments on the deposition of NPs on different types of solid substrates. Typically, the solid substrate was introduced into the sonication cell containing the precursor solution, leading to the fabrication of NPs under ultrasonic waves. The ultrasonic irradiation passes through the sonication slurry under an inert or oxidizing atmosphere for a specific period of time. This synthetic route is a single-step effective procedure. The microjets formed after the collapse of the bubble throw the just-formed NPs at the surface of the substrate at such a high speed that they strongly adhere to the surface, either via physical or chemical interactions, depending on the nature of the substrate, that is, ceramic, polymer, or textile. The excellent adherence of the NPs to the substrate is reflected, for example, in the lack of leaching of the NPs from the substrate surfaces after many washing cycles. If, instead of forming the NPs sonochemically, we purchase them and use ultrasonic radiation just for "throwing stones" at a solid surface, a good adherence is still obtained, but the amount of the deposited material found on the surface is smaller by a factor of 3–4. The mechanism of NP deposition on the substrate under ultrasound irradiation is illustrated schematically in Fig. 9.1. From this scheme, one can presume that the NPs formed in the precursor solution under ultrasound irradiation are thrown immediately at the solid surface by the microjets. When the suspension of the preliminary-synthesized NPs is irradiated, not all the particles are pushed by shock waves and a large part of them remain in the slurry. Therefore, in the latter case, the coating is less effective.

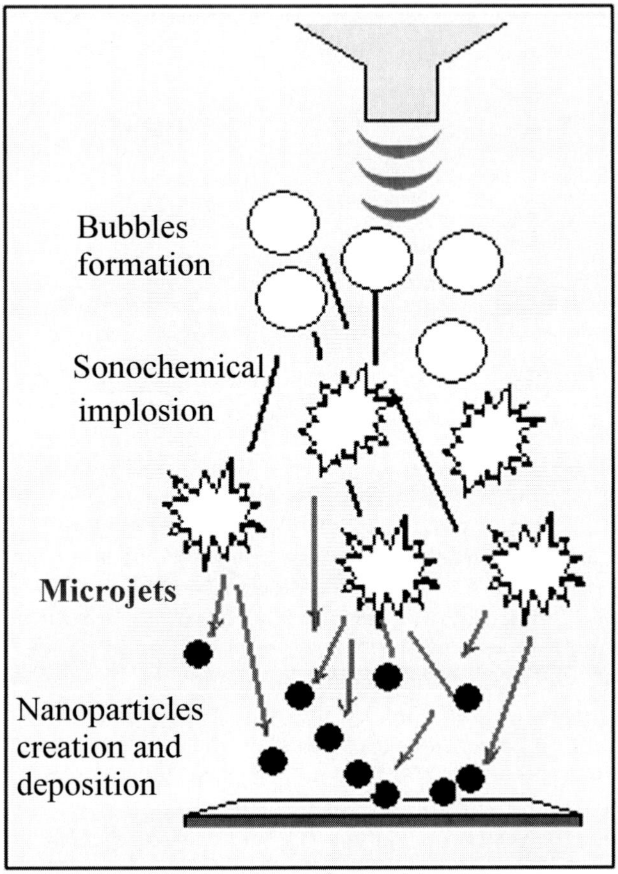

Figure 9.1 Supposed mechanism of NP adhesion to the substrate.

When the NPs hit the solid surface, sintering of the particles and/or interparticle collision between the solid surface and inorganic NPs changes the surface morphology and reactivity, resulting finally in the coating of these particles. Using the sonochemical approach, we studied the coating of the metal oxide NPs on various kinds of surfaces. Ultrasonic irradiation of a decalin solution of iron pentacarbonyl in the presence of alumina submicrospheres resulted in the coating of highly dispersed iron oxides on alumina (Fig. 9.2) (Zhong et al., 1999). The strong interaction between the adhered iron particles and an alumina substrate can hinder the transformation of γ-Fe_2O_3

to α-Fe_2O_3, even at temperatures higher than 700°C. Conversely, the presence of α-Fe_2O_3 can induce the formation of α-Al_2O_3 at high temperatures.

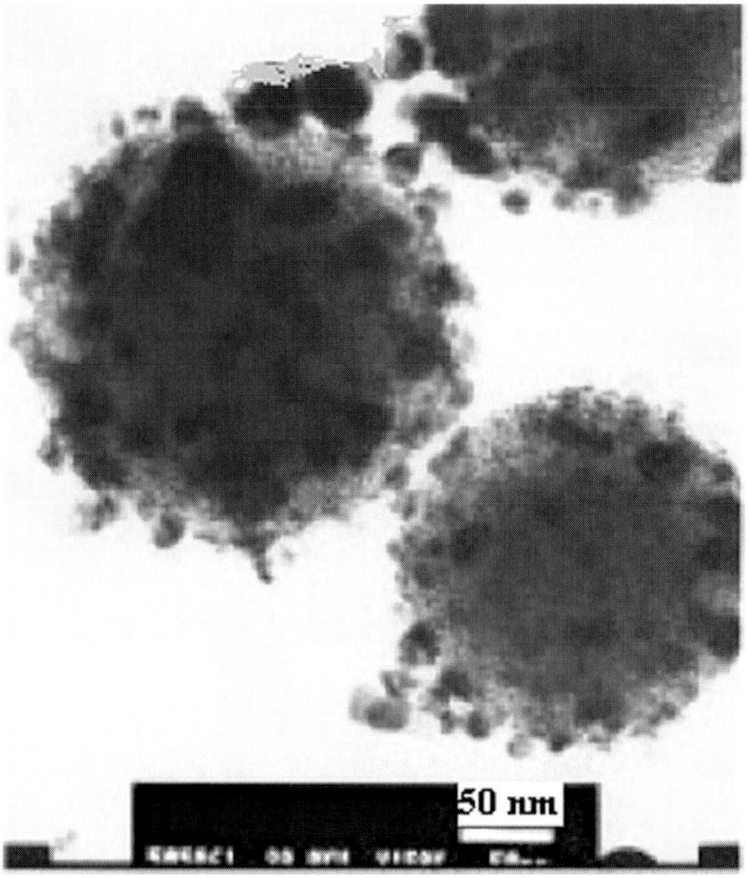

Figure 9.2 TEM micrographs of γ-Fe_2O_3 on alumina microspheres.

The 20–40 nm rare earth oxides, Eu_2O_3, Tb_2O_3, were deposited on a number of ceramic materials such as silica alumina, titania, and yttria-stabilized zirconia (Patra et al., 1999; Gedanken et al., 2000; Pol et al., 2002). These materials demonstrated significant photoluminescence properties and may have wide application in optical and luminescent devices and photocatalysis. A similar sonochemical approach was later used for the synthesis of another photocatalyst, TiO_2/SiO_2 (Guo et al., 2006). TiO_2 clusters in the size range of about 4 nm were deposited on silica particles by the high-intensity ultrasound radiation of a suspension containing $TiCl_4$ as a precursor and silica particles in water. It was demonstrated that the TiO_2/silica photocatalysts exhibited a higher reactivity than bulk TiO_2 (P25, Degussa) in the photo-oxidation of methyl orange. These few examples, and many others that are not related to textiles, show that sonochemistry as a deposition method can prevent the agglomeration of the newly formed NPs. Moreover, the fear that NPs will disperse in air is also inhibited, as their strong adherence to the substrate prevents their scattering to air. The mode of distribution of the NPs on the substrate is always that of nanoclustering, and not of nanolayering. Namely, we coat well-dispersed individual NPs and not a full continuous layer of the deposited material. A special feature of the sonochemical method for the coating of the various substrates is the opportunity to regulate the particle size and the formation of the thickness of the coated layer. The size of the particles formed in the sonochemical reaction depends very much on the reaction conditions. When the sonoreactor is chilled to a low ambient temperature and a low concentration of the precursor solution is used, 5–10 nm inorganic compounds could be obtained and deposited on the surface of the substrate by the sonochemical method. All the above-mentioned properties make the sonochemical method a very perspective technique that provides homogeneous coating and the strong adhesion of the nanostructured materials to the surface of the supporting substrate.

The sonochemical method was developed to deposit NPs on flat and curved surfaces of ceramics (Landau et al., 2001; Perkas et al., 2001; Pol et al., 2002, 2006), polymers (Kotlyar et al., 2007, 2008), metals (Perkas et al., 2009), and paper. It is worth mentioning that the polymer and glass surfaces could realize antibacterial properties by the deposition of nanosilver (Perkas et al., 2007, 2008) or nano ZnO (Applerot et al., 2009). The structure and morphology of the ZnO NPs deposited sonochemically on glass slides were studied as a function of synthesis time. The adjustment of processing time allowed the attainment of ZnO films with various thicknesses. The ZnO nanocrystals were obtained with a mean diameter of 300 nm. The high temperature and the speed of the NPs thrown at the solid surface by sonochemical microjets cause their strong and stable attachment to the glass, with a unique sphere structure of the ZnO. The antibacterial activities of the ZnO-glass composites were tested against *Escherichia coli* (gram-negative) and *Staphylococcus aureus* (gram-positive) cultures. A significant bactericidal effect, even in a 0.13% wt ZnO-coated glass, was demonstrated.

This chapter will cover the research carried out on textiles that were made antimicrobial by the deposition of NPs on their surface and will report on the ultrasound-assisted deposition of antimicrobial NPs on textiles.

9.3 Functionalization of Textiles with Nanoparticles

Nanosilver is one of the most widely used antibacterial agents, generally in textiles and in wound dressing (Duran et al., 2007; Gorenšek and Recelj, 2007; Wang et al., 2007). The antibacterial properties of silver have been known and used for centuries (Searle, 1919). Mineral salts have long been known as a unique and available source of silver. A new method for the delivery of silver into the bacteria-killing medium is by the formation of organic–inorganic nanocomposites combining the properties of textile substrates with antibacterial activity (Shukla et al., 2001; Yeo et al., 2003; Dubas et al., 2006). To achieve the optimum antibacterial effect of nanocomposite fibers, a high concentration of silver ions must be available in the solution. Despite the small number of silver ions released from metallic silver nanocrystals, about 30 times less than that from silver complexes (e.g., silver sulfadiazine), stronger antimicrobial properties have been observed with nanocrystals (Richard et al., 1994).

Different methods have been used for the deposition of silver NPs on fabrics. For example, a poly(ethylene terephthalate) fabric (meadox double velour) was coated with metallic silver using a patented ion beam assisted deposition process developed by the Spire Corporation (Bedford, MA) (Klueh et al., 2001). Antimicrobial fibers were produced by the implementation of nanoscaled silver particles into a solution of cellulose and *N*-methylmorpholine-*N*-oxide (Wendler et al., 2002). Other methods were constant pressure padding (Lee et al., 2003), impregnation in the colloid silver solution (Duran et al., 2007), immersion of the fabric in the silver precursor solution in ethanol or propanol following the boiling procedure for the reduction of silver ions (Yuranova et al., 2006), the magnetron sputter technique (Scholz et al., 2005), and so on. Some of the methods are based on reactions in the liquid medium and require surfactants, reducing agents, or templates for the synthesis of silver NPs, resulting in the presence of toxic impurities in the final products. This method has some disadvantages with regard to the environment. Recently, we reported on the simple and effective ultrasound-assisted deposition of silver NPs on wool fibers (~5–10 nm in size) (Hadad et al., 2007) and on different kinds of fabrics (nylon, polyester and cotton) of about 80 nm in size (Perelshtein et al., 2008). The excellent antibacterial activity of these Ag–fabric composites against *E. coli* (gram-negative) and *S. aureus* (gram-positive) cultures was demonstrated.

Among the heavy metals, silver is considered as nontoxic, in spite of claims that it kills many different disease causing organisms. Its low toxicity to mammalian cells was demonstrated (Joeger et al., 2001). Detailed investigations showed that nanosilver is skin friendly and does not cause irritation (Lee et al., 2005), although some work warned of the danger of nanosilver if its particles are smaller than 55 nm (Carlson, 2006). Thus, once reaching nanoscale, these are the questions that need to be imperatively answered before people rush to participate in the nanosilver boom (Chen and Schluesener, 2008). Lately, more attention is being paid to the application of inorganic metal oxides for the antibacterial finishing of textiles. The present report will not cover the biocidal properties of silver-coated fabrics because (i) this subject has already been exhausted by many papers and reviews, (ii) the FDA has recently started to limit the use of silver, and (iii) in contrast to silver, research on metal oxides as antimicrobial agents is novel. This is the reason why the current chapter will concentrate on ZnO, MgO, and CuO NPs coated on various textiles.

Some metal oxides such as TiO_2, ZnO, MgO, and CuO, are recognized by the FDA as non-toxic for the human body. Nanosized particles of TiO_2, ZnO, and MgO possess photocatalytic ability, UV absorption, and a photo-oxidizing capacity against chemical and biological substrates. During the last decade research involving metal oxide, NPs was intensified, focusing on the production of textiles with antibacterial, self-decontaminating, and UV-blocking functions (Daoud and Xin, 2004; Shi et al., 2008; Sojka-Ledakowicz et al., 2008; Uddin, 2008).

Nanostructured metals and inorganic oxides can be incorporated into textiles by various methods, for example, high energy γ-radiation and thermal treatment assisted impregnation (El-Naggar et al., 2003; Zohdy et al., 2003). In these papers, cotton and cotton/polyester fabrics were immersed in an antimicrobial formulation based on zinc oxide (ZnO), Impron MTP (binder), and Setamol WS (dispersing agent) and subjected to fixation by γ-radiation techniques. The effect of this treatment on the growth of bacteria (*Bacillus subtilis*) was estimated. On the basis of microbial detection, it was found that the ZnO formulation causes a net reduction in the bacterial cells, amounting to 78% and 62% in the case of treated cotton and cotton/polyester fabrics, respectively. However, it was found that treatment with the ZnO formulation caused a reduction in the thermal stability of the fabrics, as indicated by thermogravimetric analysis.

One of the widely used techniques for coating textile substrates is the combination of the sol–gel synthetic procedure with the "pad-dry-cure" method (Schollmeyer, 2007; Wang et al., 2008; Xue et al., 2008). The synthesis process usually involves two main steps, for example, the hexagonally ordered ZnO nanorod arrays might be grown on fiber substrates in the same way as zerogel ZnO (Wang et al., 2004). The growing seeds were formed by coating ZnO nanosol using dip-coating, dip pad-curing, or spraying methods by natural solvent evaporation. In order to stabilize the precursor solution, triethenamine, with the same molar ratio as zinc acetate, was added to form a transparent homogeneous solution. The TiO_2 and TiO_2/SiO_2 nanocomposites prepared by the low-temperature sol–gel synthesis were coated on cotton fabrics by a dip-pad–dry-cure process (Daoud et al., 2005; Qi et al., 2007). The sol–gel immobilization and controlled release of various bioactive liquids from modified silica coatings were investigated by Haufe et al. (2008). The deposition of nano-ZnO onto cotton fabric was performed by padding the textiles in the colloid formulation of the zinc oxide–soluble starch nanocomposite to impart the material with the antibacterial and UV-protection functions (Vigneshwaran et al., 2006).

Very recently, there have been some new publications on the deposition of in situ formed metal oxide NPs on fabrics. A superhydrophobic ZnO nanorod array film on a cotton substrate was fabricated via a wet chemical route of subsequent modification with a layer of *n*-dodecyltrimethoxysilane (Xu and Cai, 2008). ZnO NPs were grown in situ on a SiO_2-coated cotton fabric through the hydrothermal method. After water treatment at 100°C or higher, the cotton fabric was covered with approximately 24 nm diameter needle-shaped ZnO nanorods, which had an excellent UV-blocking property (Mao et al., 2006). ZnO particles were prepared by the wet chemical method using zinc nitrate and sodium hydroxide as precursors and solubilized starch as a stabilizing agent (Kathirvelu et al., 2009). These NPs were impregnated onto cotton fabrics by the "pad-dry-cure" method using an acrylic binder.

Copper is one of the relatively small group of metallic elements that is essential to human health. This element, along with amino and fatty acids and vitamins, is required for normal metabolic processes (Toxicological Profile of Copper 2004, U.S. Department of Health and Human Services). Copper is considered safe to humans, as demonstrated by the widespread and prolonged use of copper intrauterine devices (IUDs) (Bilian, 2002). However, except for the work of Gabbay and coworkers (Borkow and Gabbay, 2005; Gabbay et al., 2006), there are not many publications on the production and application of CuO-textile composites. The copper-containing fibers of cotton and polyester prepared by these authors demonstrated significant antifungal and antimicrobial properties. They inserted the preliminarily synthesized copper oxide powder into polymer fibers during the master-batch stage and impregnated them into the cotton by a multi-phase soaking procedure, including the treatment in formaldehyde.

In summary, most of the methods for the deposition of nanostructured materials on textiles are based on a multistage procedure, including the preliminary synthesis of NPs and the application of some templating agents for anchoring them to the substrates. This approach is rather complicated and can result in the release of some toxic compounds into the waste. The sonochemical method enables avoiding the use of the toxic binders and makes the coating procedure shorter, effective, and environmentally friendly.

9.4 Deposition of Metal Oxide Nanoparticles on Textiles by Ultrasound Irradiation

9.4.1 Synthesis and Deposition of ZnO

The antimicrobial activity of ZnO depends on the particle size: decreasing the particle size leads to an increase in the antimicrobial activity (Yamamoto, 2001). We have developed a simple new method for the preparation of cotton bandages with antimicrobial properties by immobilizing ZnO NPs on the fabric's surface via ultrasound irradiation (Perelshtein et al., 2009). The coating reaction was carried out with a high-intensity ultrasonic horn (Ti horn, 20 kHz, 750 W at 70% efficiency) under a flow of Ar. The sonication flask was placed in a cooling bath maintained at a constant temperature of 30°C during the reaction. Our studies indicated that the yield of the product and the particle size are strongly dependent on the rate of particle collision and on the concentration of the reagents during sonochemical synthesis. Due to this, the experimental parameters such as time and concentration of the precursor were selected as important factors for the optimization of the sonochemical coating reaction. The aim of this work was to obtain a homogeneous coating of small ZnO NPs on the fabrics with a narrow size distribution and to reach a minimal effective ZnO concentration that will still demonstrate antimicrobial activity. This process involves the in situ generation of ZnO under ultrasound irradiation and its deposition on fabrics in a one-step reaction. The sonication was performed in a water–ethanol solution in the presence of a cotton bandage. Zinc acetate was used as a precursor and the pH was adjusted to 8–9 with the addition of $NH_3 \cdot H_2O$.

Previous studies indicated that the product yield and particle size are strongly dependent on the rate of inter-particle collision and on the concentration of the reagents during the sonochemical synthesis (Gedanken, 2004). For this reason, experimental parameters such as time and concentration of the precursor were selected as important factors for the optimization of the sonochemical reaction. The XRD demonstrated that the ZnO NPs on the coated bandage are crystalline and the diffraction patterns matched the hexagonal phase of the wurtzite ZnO structure. No peak characteristics of any impurities were detected. The particle size estimated by the Debye–Scherrer equation is 30 nm. The morphology of the coated bandage before and after the deposition of ZnO NPs studied by HR SEM is presented in Fig. 9.3. Figure 9.3(a) demonstrates the smooth texture of the pristine cotton bandage. After sonication, the bandage fibers were homogeneously coated with NPs (Fig. 9.3b). The inset image in Fig. 9.3(b) was taken at a higher magnification in order to obtain the particles' size distribution. The distribution of the particles is quite narrow, and primary particles are in a very low nanometric range (about 30 nm) that matches well with the XRD results. The selected-area HR SEM image studied with the elemental dot-mapping technique is shown in Fig. 9.3c. The distribution of zinc and oxygen in the mapped area is presented in parts (d) and (e) of Fig. 9.3, respectively. These images verify a homogeneous coating of the fibers with ZnO NPs.

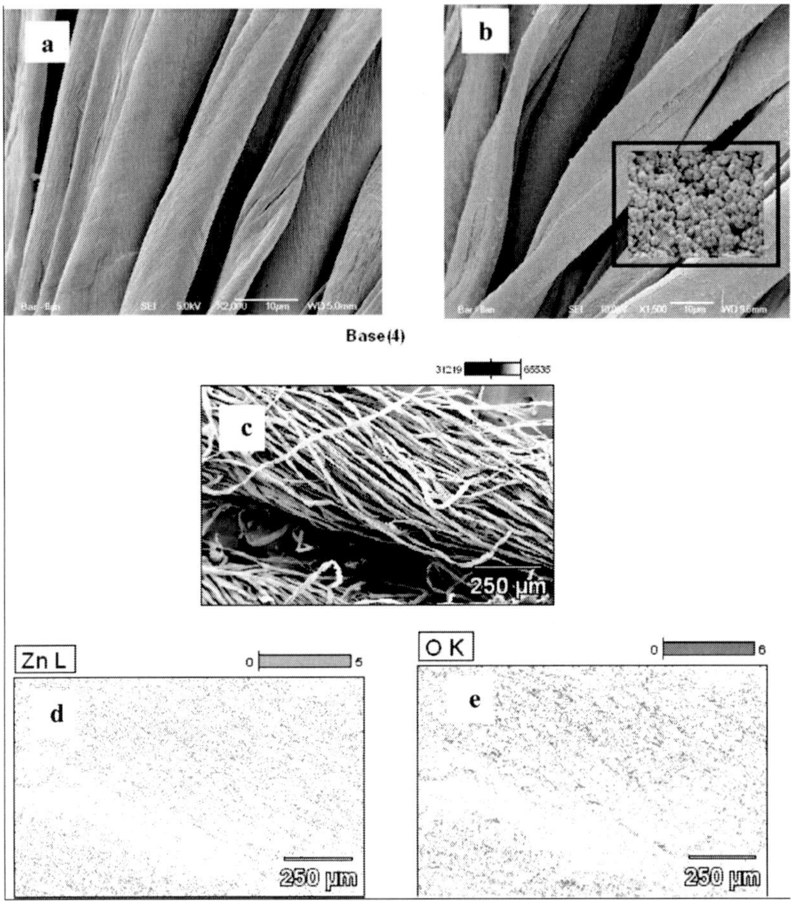

Figure 9.3 HRSEM images of (a) pristine bandage fibers (magnification ×2000), (b) bandage coated with ZnO NPs (magnification ×1500; the inset shows a magnified image (×50000) of the ZnO NPs on the fabric, (c) selected image for X-ray dot mapping, (d) X-ray dot mapping for zinc, (e) X-ray dot mapping for oxygen.

We considered the coating mechanism as follows: it involves the in situ generation of ZnO NPs and their subsequent deposition on fabrics in a one-step reaction via ultrasound irradiation. Zinc oxide is formed during the irradiation according to the following reactions:

$$Zn^{2+}_{[aq]} + 4NH_3{\cdot}H_2O_{[aq]} \longrightarrow [Zn[NH_3]_4]^{2+}_{[aq]} + 4H_2O \qquad \text{(Eq. 9.1)}$$

$$[Zn[NH_3]_4]^{2+}_{[aq]} + 2OH^-_{[aq]} + 3H_2O \longrightarrow ZnO_{[s]} + 4NH_3{\cdot}H_2O_{[aq]} \text{(Eq. 9.2)}$$

Ammonia works as the catalyst of the hydrolysis process, and the formation of zinc oxide takes place through the ammonium complex $[Zn[NH_3]_4]^{2+}$. The ZnO NPs produced by this reaction are thrown at the surface of the bandages by the sonochemical microjets resulting from the collapse of the sonochemical bubble.

As already mentioned above, the sonochemical irradiation of a liquid causes two primary effects, namely, cavitation (bubble formation, growth, and collapse) and heating. When the microscopic cavitation bubbles collapse near the surface of the solid substrate, they generate powerful shock waves and microjets that cause the effective stirring/mixing of the adjusted liquid layer. The after effects of the cavitation are several hundred times greater in heterogeneous systems than in homogeneous systems (Suslick, 1989). In our case, the ultrasonic waves promote the fast migration of the newly formed zinc oxide NPs to the fabric's surface. This fact might cause a local melting of the fibers at the contact sites, which may be the reason why the particles strongly adhere to the fabric.

Here the question rises as to whether sonication does not damage the fabric's substrate? Thus, the tensile mechanical properties of a cotton-impregnated fabric were studied on a universal testing machine, Zwick 1445. A four-fold fabric sample with a gauge length of 60 mm and a width of 40 mm was placed in special grips. The tensile force for the zinc oxide coated sample was observed to be nearly 11% less than that of the pristine bandage (Fig. 9.4). The observed changes in the mechanical behavior of the yarn are in a range that is acceptable for standard cotton fabrics. According to this result, we conclude that the sonochemical treatment of the bandage does not cause any significant change in the structure of the yarn.

One of the factors influencing the commercial exploitation of the antimicrobial bandages is the release of NPs into the surrounding environment. In light of a recent paper (Benn and Westernoff, 2008) in which it has been found that silver NPs of 10–500 nm in diameter leach from sock fabric, we attempted to find the leached ZnO NPs in the washing solution. In the control experiments, we treated the coated bandages with an aqueous solution of 0.9 wt% NaCl overnight at 37°C. The leaching experiment indicated that only 16% of the deposited zinc was removed under these conditions in an ionic form that is dictated by the low K_{sp} of zinc oxide in water.

The dynamic light scattering (DLS) and TEM studies did not reveal the presence of any NPs in the washing solution. This indicates that the sonochemically deposited ZnO NPs are strongly anchored to the textile substrate.

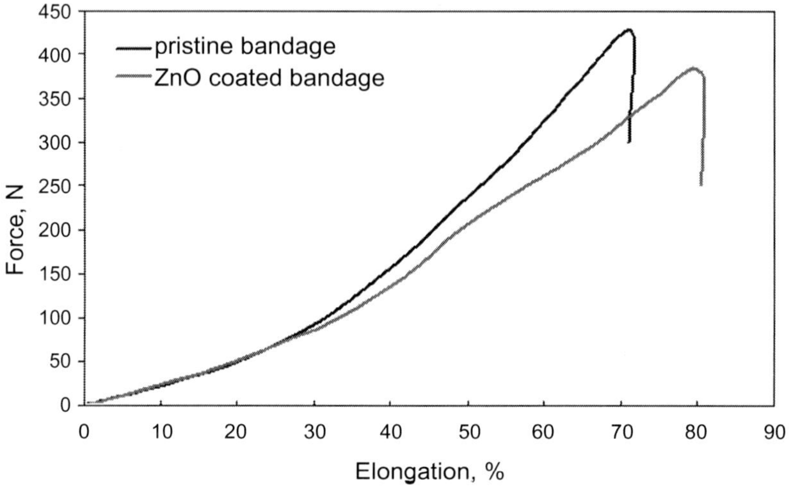

Figure 9.4 Mechanical properties of the cotton bandage before and after the deposition of ZnO NPs.

9.4.2 Synthesis and Deposition of CuO

We extended the sonochemical approach for the deposition of CuO NPs on the textile fabrics, and, as in the case with ZnO NPs, the formation of copper oxide takes place through the ammonium complex. Copper ions react with a solution of ammonia to form a deep blue solution containing $[Cu[NH_3]_4]^{+2}$ complex ions. This complex is hydrolyzed and crystalline CuO NPs are obtained. The CuO NPs produced by these reactions are thrown at the surface of the fabric by the above-described mechanism of the sonochemical microjets and are deposited on the surface of the substrate.

The morphology of the fiber surface before and after the deposition of copper oxide was studied by XRD and HR SEM methods. The XRD revealed the monoclinic structure of CuO nanocrystals. The difference between pristine and coated cotton fabric is clearly demonstrated in Fig. 9.5. The inset image in Fig. 9.5(b) at higher magnification shows that the primary particles are in a very low nanometric range (~10–20 nm).

Figure 9.5 HR-SEM images of cotton fibers: a–pristine cotton (magnification ×20000), inset image–magnification ×100000); b–cotton coated with CuO NPs (magnification ×20000), inset image–magnification ×100000).

Although Cu^{2+} is considered an environmentally safe ion, a much more important and serious issue is the leaching of CuO NPs. DLS and TEM studies of the washing solution after treatment of the CuO-coated fabrics in 0.9 wt% NaCl did not reveal the presence of any NPs. This means that the sonochemically deposited CuO NPs are strongly anchored to the textile substrate, probably due to a local melting of the fibers at the contact sites. Similar results were obtained for the coating of various types of textiles, such as nylon, polyester, and composite types of textiles with ZnO and CuO NPs.

9.4.3 Deposition of MgO

MgO is well known to have a strong antibacterial activity (Huang et al., 2005; Ohira et al., 2008). Different methods have been reported on the synthesis of magnesium oxide NPs, such as the controlled speed of formation following the heating procedure (Huang et al., 2005), microwave-assisted synthesis (Makhluf et al., 2005), the formation of MgO from aqueous droplets in a flame spray pyrolysis reactor (Seo et al., 2003), sonochemically enhanced hydrolysis followed by supercritical drying (Stengl et al., 2003), and so on. However, there is no reported literature related to the deposition of magnesium oxide on textiles.

We have developed a method for depositing MgO NPs on fabrics by ultrasound irradiation. In this case, ultrasound is used just for "throwing stones" at the fabric. Namely, we took commercial MgO nanopowder (Aldrich, <25 nm) and sonicated it in the presence of a cotton fabric. The microjets formed after the collapse of the bubble throw the NPs at high speed at the cotton yarn. This process is different from the deposition of ZnO and CuO where the NPs were in situ generated and deposited on fabrics in a one-step reaction.

The morphology of the fiber surface after the deposition of magnesium oxide NPs studied by HR SEM is presented in Fig. 9.6. Fibers after sonication are homogeneously coated with NPs. The particle size was in the low nanometric range (~10–15 nm) and no aggregation was observed.

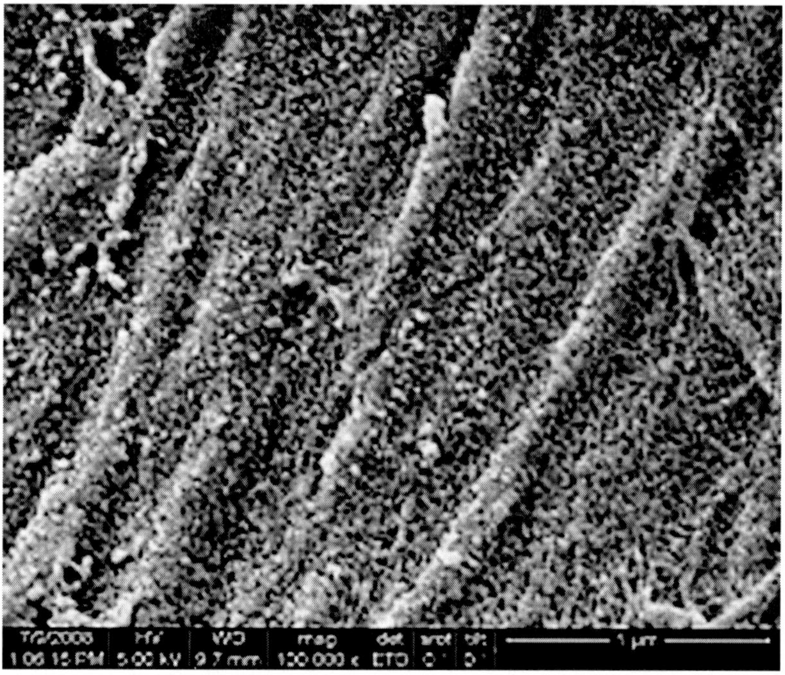

Figure 9.6 HR-SEM images of 0.8 wt% MgO coated cotton fabric.

9.4.4 Up-scale of the Sonochemical Process

We have recently demonstrated the up-scaling of the sonochemical process for coating textiles (Abramov et al., 2009). A small pilot machine was developed that allows the roll-to-roll coating of textiles in a continuous mode. Compared with laboratory experiments, this pilot scale system significantly increased productivity and at the same time decreased the energy consumption by one order of magnitude.

9.5 Mechanism of the Antimicrobial Activity of Metal Nano-Oxides

In this section, we will elucidate the mechanisms and factors that govern antibacterial activity in metal oxide, NP-engineered textile systems. It has been well established that some metal oxide NPs can cause damage to bacterial membranes and, as a consequence, exhibit clear antibacterial properties. In this respect, the key advantages of metal oxides as inorganic antimicrobial agents are the improved safety and stability, as compared with classic organic antimicrobial agents. Moreover, the spreading of antibiotic-resistant pathogens is still a growing concern globally. Therefore, the development of new bactericidal preparations is called for (Anagnostakos et al., 2008).

The mechanism of the antimicrobial activity of these metal oxides is poorly understood and is still controversial. Suggested mechanisms in the literature include the role of reactive oxygen species (ROS) generated on the surface of the particles, ion release, membrane dysfunction, and NP internalization (Li et al., 2008; Neal, 2008; Hu et al., 2009). However, both CuO and TiO_2 are scarcely soluble in water, whereas the Zn^{2+} and Mg^{2+} ions that might be released from oxide NPs are not considered to be harmful to biological systems (Vallee and Falchuk, 1993; Sawai 2000). Moreover, our studies aimed at checking the leaching of NPs from the sonochemically coated textiles did not reveal any escape of the NPs to the washing solution (Perelshtein et al., 2009). These findings make the mechanism of ROS-mediated toxicity the key event in the antibacterial activity of the NP-engineered textile systems, and worthy of further detailed evaluation. In an additional study performed by our group, we have demonstrated that a specific antioxidant, histidine, inhibits completely the antifungal activity of ZnO NPs, strongly indicating that the antimicrobial activity is ROS dependent (Lipovsky et al., 2011). The ROS are probably best known in biology for their ability to cause oxidative stress. They can damage DNA, cell membranes, and cellular proteins, and may lead to cell death (Shen et al., 2009). Hydroxyl radicals (•OH) are highly reactive for the oxidation of organic substances and are able to attack and decompose polyunsaturated phospholipids in bacteria. Other ROS, such as superoxide ions ($\bullet O_2^-$) or hydroperoxyl radicals ($HO_2\bullet$), are less effective against bacteria (Imlay, 2008). The negative charge of superoxide ions prevents them from penetrating bacteria cell membranes. Therefore, these species must be in direct contact with the outer surface of bacteria for inducing the killing effect. Hydrogen peroxide (H_2O_2), as well as hydroperoxyl radicals ($HO_2\bullet$), can enter the cell despite their being less harmful compared with hydroxyl radicals. Furthermore, hydrogen peroxide can cause cell damage via hydroxyl radicals produced by the Fenton reaction (Wright, 2007). Furthermore, hydrogen peroxide in the presence of $\bullet O_2^-$ can generate singlet oxygen, which is very toxic (Rosen and Klebanoff, 1979):

$$\bullet O_2^- + H_2O_2 \longrightarrow OH^- + \bullet OH + {}^1O_2 \quad \text{(Eq. 9.3)}$$

The mechanism underlying the generation of ROS from the metal oxide surface is not yet well understood, but it has been hypothesized that it derives from the highly reactive nature of defect sites (such as oxygen vacancies) on a wet metal oxide surface. In this respect, it was also suggested that the intimate contact possible between the NPs' surface and the bacteria might cause physical damage to the bacterial cell wall, and thus enhance an antibacterial effect. In the following part, the antimicrobial effect of common metal oxides (ZnO, CuO, TiO_2, MgO) will be described in detail.

9.5.1 Antibacterial Activity of ZnO

The excellent studies by Sawai and coworkers (1996, 1998) clearly showed that ROS concentrations increased with the ZnO content of slurries. Following the same paradigm, Applerot et al. (2009) in an innovative study using electron spin resonance (ESR) coupled with the spin trapping probe technique monitored ROS, namely, hydroxyl radicals, and production in water suspensions of ZnO NPs. Their findings showed that hydroxyl radicals were present in water suspensions of ZnO, and their amount was closely related to the size of the ZnO particles, with smaller sizes having greater amounts of $\bullet$OH on the basis of equivalent ZnO mass content. These results were correlated with the increase in the antibacterial effect of the small NPs. Thus, the small size and large specific surface area endow them with high chemical reactivity and intrinsic toxicity. Interestingly, combining gram-negative bacterium *E. coli* and ZnO NP suspensions immediately enhanced the generation of $\bullet$OH for up to an average of 142% (Fig. 9.7). It appears that sub-lethal amounts of ROS may induce excessive intracellular oxidative stress, probably by disturbing the balance between oxidant and anti-oxidant processes.

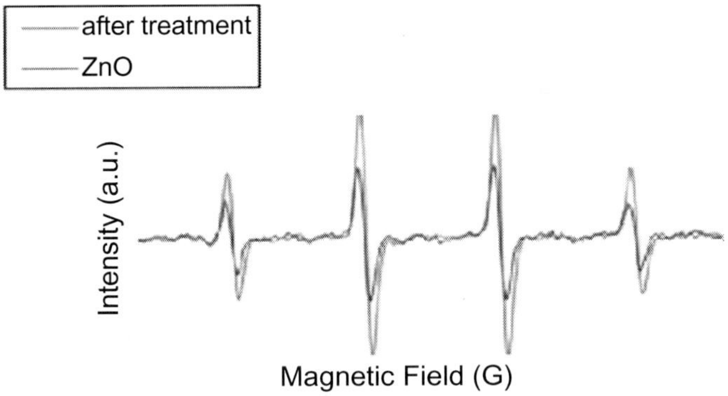

Figure 9.7 ESR spectra demonstrating changes in hydroxyl radical concentration upon antibacterial treatment of *E. coli* with a water suspension of ZnO.

We determined the antibacterial activity of the sonochemically coated cotton fabrics with 0.75 wt% ZnO (Perelshtein et al., 2008) using the gram-negative bacterium *E. coli* and the gram-positive bacterium *S. aureus*. As shown in Table 9.1, treatment for 1 h with the coated cotton leads to the complete inhibition of *E. coli* growth. Regarding *S. aureus*, a 100% reduction in viability was reached after 3 h, while after 1 h of treatment, a reduction of 60% could be seen.

Zinc is an essential micronutrient for prokaryotic organisms. However, at super physiological levels, zinc inhibits the growth of many bacteria (Soderberg et al., 1989). According to a leaching experiment with 0.9 wt% NaCl, the concentration of Zn^{2+} in solution corresponds to 36.7 μM/L (Perelshtein et al., 2009). Compared with the minimum inhibitory concentration reported in the literature of 4–8 mM/L (Lansdown et al., 2007), the amount of zinc released from fabrics in our work is lower at least by a factor of 2. Therefore, we assume that the Zn ions have a minor influence on antibacterial activity. The major components responsible for the bactericidal effect are the ZnO NPs. Although ZnO NPs were not found in the solution, they can generate some species of oxyradicals (Sengupta et al., 1979). In light of these results, the difference in the antibacterial action of ZnO-coated bandages on two strains of bacteria can be explained by the difference in their sensitivity toward oxyradicals. In this respect, it has been found recently that *S. aureus* contains a large amount of carotenoid pigment, which promotes a higher resistance to oxidative stress (Liu et al., 2005).

Table 9.1 Antibacterial activity test of ZnO-coated cotton*

E. coli						
Sample	Duration of Treatment					
	1 h			2 h		
	CFU [mL^{-1}]	N/N_0	% reduction in viability	CFU [mL^{-1}]	N/N_0	% reduction in viability
Pristine fabric	$1.02 \cdot 10^{-7}$	0.98	0.98	$1.34 \cdot 10^{-7}$	1.28	−28.23
No fabric	$1.17 \cdot 10^{-7}$	1.14	−28.57	$1.23 \cdot 10^{-7}$	1.35	−35.16
0.75 wt% ZnO on fabric	$1.71 \cdot 10^{-7}$	$1.58 \cdot 10^{-3}$	99.84	0	$0.9 \cdot 10^{-8}$	100
S. aureus						
Sample	Duration of treatment					
	1 h			3 h		
	CFU [mL^{-1}]	N/N_0	% reduction in viability	CFU [mL^{-1}]	N/N_0	% reduction in viability
Pristine fabric	$0.7 \cdot 10^{-7}$	0.71	20.46	$0.99 \cdot 10^{-7}$	1.125	−12.5
No fabric	$0.98 \cdot 10^{-7}$	1.10	−10.11	$0.67 \cdot 10^{-7}$	0.75	24.72
0.75 wt% ZnO on fabric	$3.9 \cdot 10^{-6}$	0.34	66.4	$7.6 \cdot 10^{-3}$	$6.55 \cdot 10^{-4}$	99.93

*The viable bacteria were monitored by counting the number of colony-forming units [CFU]; N/No: survival fraction.

9.5.2 Antibacterial Activity of CuO

We tested the antimicrobial activity of the cotton bandages coated with CuO via ultrasonic irradiation against *E. coli* and *S. aureus.*

Detailed investigations showed that after 1 h, the growth of both the strains was completely inhibited. One of the factors influencing the antibacterial activity of the developed coating is the release of the active phase into the surrounding medium, namely copper ions or/and copper oxide NPs. The examination of the leaching of the copper ions indicated that only a very low amount (~1.3%) of the deposited copper was removed by washing with a 0.9 wt% NaCl solution that corresponds to a concentration of Cu^{2+} 0.15 ppm. The slight solubility of copper oxide can be explained by the very low K_{sp} of CuO that dictates the very low concentration of Cu^{2+} in the solution. In order to examine the influence of copper ions on the antibacterial effect, a control antibacterial test was performed using a supernatant with the same concentration of Cu^{2+} instead of the coated cotton. After incubation for 24 h at 37°C, no reduction in *E. coli* after 2 h was observed (Fig. 9.8).

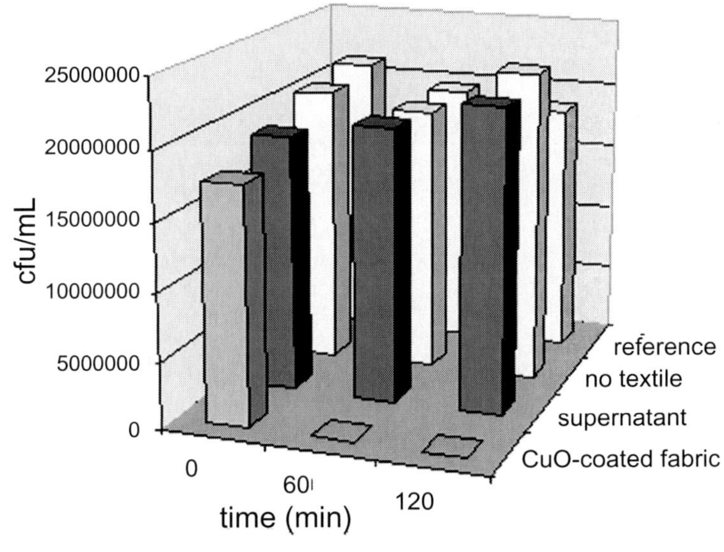

Figure 9.8 Antibacterial test of CuO-coated cotton against *E.coli*.

This result indicates that the Cu^{2+} ions have no influence on the antibacterial activity. Thus, the antibacterial effect can be attributed to the copper oxide NPs. It should be emphasized that no leaching of CuO NPs to the environment was detected in these experiments. Although CuO NPs were not found in the solution, they can generate some active species that are responsible for damaging the bacterial cells. These active species were detected in ESR studies conducted with and without the bacteria present in the ESR tube.

9.5.3 Antibacterial Activity of TiO_2

There is some controversy in the literature regarding the effective inactivation of micro-oganisms by TiO_2, primarily due to the different experimental conditions, such as (UV/vis) irradiance, length of exposure, photocatalyst present in the suspension or powder, range of concentration, and the different TiO_2 photocatalysts and micro-organisms employed, although some effect is generally acknowledged (Caballero et al., 2009). However, a bactericidal effect was reported on food-pathogenic bacteria such as *Salmonella choleraesuis, Vibrio parahaemolyticus*, and *Listeria monocytogenes* (Kim et al., 2003), as well as *Pseudomonas aeruginosa* (Madrid et al., 2002). The formation of a well-adhering bactericidal surface of TiO_2 on organic cellulose fibers was reported by Daoud et al. (2005). In the presence of UV-radiation and O_2, ROS are produced by TiO_2 NPs according to the following reactions (Li et al., 2008):

$$TiO_2 + hv \longrightarrow TiO_2\,[h^+ + e^-] \quad \text{(Eq. 9.4)}$$

$$e^- + O_2 \longrightarrow \bullet O_2^- \quad \text{(Eq. 9.5)}$$

$$\bullet O_2^- + 2H^+ + e^- \longrightarrow H_2O_2 \quad \text{(Eq. 9.6)}$$

$$H_2O_2 + O_2^- \longrightarrow \bullet OH + OH^- + O_2 \quad \text{(Eq. 9.7)}$$

$$h^+ + H_2O \longrightarrow \bullet OH + H^+ \quad \text{(Eq. 9.8)}$$

At the same time, upon the irradiation of a TiO_2 surface with photons of wavelength 385 nm or less, an electron is promoted from the valence band to the conduction band, thus forming an electron–hole pair. The photogenerated holes and electrons react with water molecules attached to TiO_2 surfaces in the presence of oxygen to form hydroxyl radicals and other ROS such as superoxide ions. It is still a subject of investigation as to which of these ROS are directly involved in the damage to bacteria cells. Moreover, it remains unclear what conditions of irradiated light necessitate the activation of ROS generation.

Kangwansupamonkon et al. (2009) have reported on the antibacterial performance of apatite-coated TiO_2 that was fixed on cotton textiles by a dip-coating technique. Their study indicated that the photocatalytic activity of an apatite-coated TiO_2 suspension can help in microbial decomposition in textile applications. Its effectiveness was clearly confirmed against *S. aureus, E. coli* ATCC 25922, MRSA DMST 20627, and *M. luteus* strains of bacteria. The effect of an irradiation source on the antimicrobial activity of cotton fabrics coated with apatite-coated TiO_2 was examined. The highest antibacterial activity was found from black-light irradiation compared with visible light and dark conditions. This could be explained by the presence of more ROS on the surface of TiO_2 particles after irradiation with black light (at wavelengths <385 nm), whereas the amount of ROS with visible light and dark condition would be less pronounced.

9.5.4 Antibacterial Activity of MgO

During the application of MgO NPs as a household paint, it was reported that it exhibited good bactericidal activity against *Bacillus subtilis var. niger* and *S. aureus* strains of bacteria and is effective even in the absence of light irradiation (Huang et al., 2005). Makhluf et al. (2005) investigated the nature of MgO nanocrystalline as a bacteriocide. They demonstrated a clear nanoscale effect, where the amount of eradicated bacteria increased with a decrease in the NP size. No publications were found out in the literature concerning an antibacterial textile coated with MgO.

We determined the antibacterial activity of cotton fabrics sonochemically coated with 0.8 wt% MgO against *E. coli* and *S. aureus.* As shown in Fig. 9.9(a), treatment for 1 h with the coated cotton leads to the complete inhibition of *E. coli* growth. Regarding *S. aureus* (Fig. 9.9b), a 99% reduction in viability was reached after 3 h, while after 1 h of treatment, a reduction of 60% could be seen.

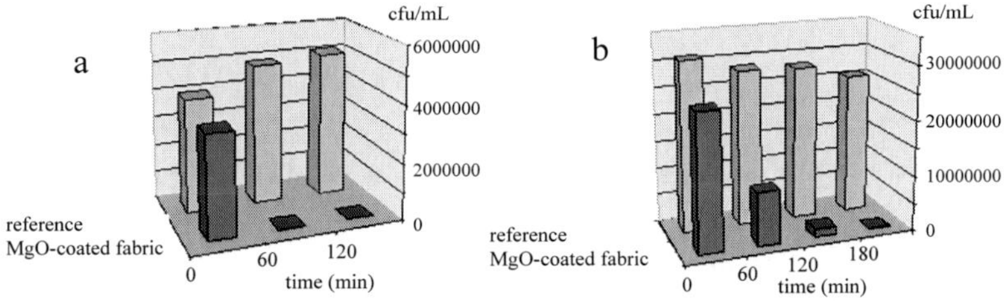

Figure 9.9 Antibacterial test of MgO-coated cotton against: (a) *E. coli* and (b) *S. aureus.*

These results confirmed that the sonochemcial method is effective for the production of antibacterial fabrics irrespective of NP deposition, that is, the creation of NPs in the sonochemcial reaction and the simultaneous deposition or sonochemical-assisted deposition of the preliminary-synthesized NPs.

By using the chemiluminescence method, Sawai et al. (2000) showed that the dominant ROS generated on MgO surfaces were superoxide anions. As mentioned above, these ROS are not very reactive themselves to bacterial cells; however, these species are in equilibrium, as shown in the following reaction:

$$\bullet O_2^- + H^+ \longrightarrow HO_2\bullet \qquad \text{(Eq. 9.9)}$$

When $\bullet O_2^-$ capture hydrogen ions are generated near the bacterial membrane during the respiratory process, hydroperoxyl radicals ($HO_2\bullet$) are produced. The $HO_2\bullet$ is much more reactive than $\bullet O_2^-$ and is able to penetrate the cell membrane. This concept will support the fact that contact between bacterial cells and MgO powders is important for MgO-induced bacterial death. Moreover, as a consequence of this contact, a local alkaline effect due to basic sites on MgO surfaces may also enhance the antibacterial activity of MgO (Ardizzone et al., 1998).

9.5.5 Antifungal Activity of ZnO

Although metal oxide NPs have been widely studied for their antibacterial properties, there are almost no studies regarding their antifungal activities. For example, Sawai and Yoshikawa (2004) found that ZnO powder (at the micro scale) exhibited a very weak antifungal activity against *C. albicans*. Though a growth inhibition was observed, it was revealed only at a concentration above 100 mg/mL. Consequently, ZnO powder was found to be unsuitable for the control of fungi. Later, Gomes et al. (2009) used ZnO (at the micro scale) with calcium hydroxide + 2% chlorhexidine to treat microbial infections at the external root surface, but without success. More promising results were shown by Fang et al. (2006) where ZnO in a "whisker" form was found to inhibit the growth of *C. albicans.* There are relatively few agents that can be used to treat fungal infections. Although the major antibiotic groups available for treating *C. albicans* infections are polyenes or the azoles drugs, resistance is documented for all of the antifungal drugs that have been widely deployed therapeutically.

In a recent study (the paper has been submitted but not yet accepted), we have tested the ability of ZnO NPs to affect the viability of the pathogenic yeast *Candida albicans* (*C. albicans*). A concentration-dependent effect of ZnO on the viability of *C. albicans* was observed. The minimal fungicidal concentration of ZnO was found to be 0.1 mg/mL. ZnO in the above concentration caused an inhibition of over 95% in the growth of *C. albicans*. ZnO NPs also inhibited the growth of *C. albicans* when it was added at the logarithmic phase of growth. In addition, the involvement of ROS, including a superoxide anion, hydroxyl radicals, and singlet oxygen in cell death, is also discussed. The addition of histidine (a quencher of hydroxyl radicals and singlet oxygen) caused a reduction in the effect of ZnO on *C. albicans* corresponding to its concentration. Almost a complete reduction of the antimytotic effect was achieved following the addition of 5 mM of histidine. Excitation of ZnO by visible light increased the death of the yeast cells.

9.5.6 Antifungal Activity Test of ZnO-Coated Cotton

The antimicrobial activity of the coated bandages was very recently tested against the yeast, *C. albicans*. Cultures of the yeast were grown in plates containing nutrient broth with 2% glucose (overnight at 37°C) and transferred the next day into a fresh medium (NB + 4% glucose) at an initial optical density (OD) of 0.1 at 660 nm. When the culture reached an OD nm of 0.3 at 660 nm, the cells were harvested by centrifugation and washed twice with a 0.9% solution of NaCl. A total of 3 cm^2 of the coated fabric and/or the uncoated fabric, which served as the control, was placed in a vial containing 4 mL of a 0.9% NaCl solution with a yeast cell. The initial concentration of *C. albicans* in the vial was approximately 10^6 CFU/mL. An additional control in the experiment was the yeast in a saline solution alone. The suspensions were incubated for up to 4 h at 37°C and at 170 rpm. Samples of 100 μl each were taken at the beginning and after 1 h and 3 h. Serial dilutions were then transferred onto nutrient agar plates (supplemented with 2% glucose). The plates were allowed to grow overnight at 37°C. The viable yeasts were monitored by counting the number of colony-forming units from the appropriate dilution. The reduction percentage in the *C. albicans* viability is shown in Table 9.2.

Table 9.2 Antifungal activity test of ZnO-coated cotton

	Duration of treatment	
	1 h	3 h
	% reduction in viability	% reduction in viability
0.75 wt% ZnO on fabric	73.1 %	84.5 %

9.6 Conclusion

Metal oxide NPs were uniformly deposited onto the surface of different kinds of textiles by the sonochemical method. The coating was performed by a simple, efficient, one-step procedure using environmentally friendly reagents. The physical and chemical analyses demonstrated that nanocrystals nearly 20–30 nm in size are finely dispersed onto the fabric surfaces without significant damage to the structure of the yarn. The mechanism of nano-oxide formation and adhesion to the textile was discussed. It is based on the local melting of the substrate due to the high rate and temperature of NPs thrown at the solid surface by sonochemical microjets. The strong adhesion of metal nano-oxides to the substrate was confirmed in terms of the absence of the leaching of the NPs into the washing solution. The performance of fabrics coated with a low content of nano-oxides (<1 wt%) as an antimicrobial agent was investigated, and their excellent bactericidal and antifungal effect was demonstrated.

Acknowledgment

This research was carried out as part of the activities of the LIDWINE Consortium, Contract No.NMP2-CT-2006-026741. LIDWINE is an IP Project of the 6th EC Program.

References

Abramov, O. V., Gedanken, A., Koltypin, Y., Perkas, N., Perelshtein, I., Joyce, E. and Mason, T. J. (2009). Pilot scale sonochemical coating of nanoparticles onto textiles to produce biocidal fabrics. *Surf. Techn.*, **204**, pp. 718–722.

Anagnostakos, K., Hitzler, P., Pape, D., Kohn, D. and Kelm, J. (2008). Persistence of bacteria growth on antibiotic-loaded beads: is it actually a problem? *Acta. Orthop.*, **79**, pp. 302–307.

Applerot, G., Lipovsky, A., Dror, R., Perkas, N., Nitzan, Y., Lubart, R. and Gedanken, A. (2009). Enhanced antibacterial activity of nanocrystalline ZnO due to increased ROS-mediated cell injury. *Adv. Funct. Mater.*, **19**, pp. 842–852.

Applerot, G., Perkas, N., Amirian, G., Girshevitz, O. and Gedanken, A. (2009). Coating of glass with ZnO via ultrasonic irradiation and a study of its antibacterial properties. *Appl. Sur. Sci.*, **256S**, pp. S3–S8.

Ardizzone, S., Bianchi, C. L. and Vercelli, B. (1998). MgO powders: interplay between adsorbed species and localization of basic sites. *Appl. Surf. Sci.*, **126**, pp. 169–175.

Benn, T. and Westernoff, P. (2008). Nanoparticle silver released into water from commercially available sock fabrics. *Environ. Sci. Technol.*, **42**, pp. 4133–4139.

Bilian, X. (2002). Intrauterine devices. *Best. Pract. Res. Clin. Obstet. Gynaecol.*, **16**, pp. 155–168.

Bolard, J. (1986). How do the polyene macrolide antibiotics affect the cellular membrane properties? *Biochim. Biophys. Acta.*, **864**, pp. 257–304.

Borkow, G. and Gabbay, J. (2005). Copper as a biocidal tool. *Current Med. Chem.*, **12**, pp. 2163–2175.

Caballero, L., Whitehead, K. A., Allen, N. S. and Verran, J. (2009). Inactivation of *Escherichia coli* on immobilized TiO_2 using fluorescent light. *J. Photochem. Photobiol. A*, **202**, pp. 92–98.

Carlson, K. (2006). In vitro toxicity assessment of silver nanoparticles in ratalveolar macrophages. MS Thesis, USA, Wright State University.

Chen, X. and Schluesener, H. J. (2008). Nanosilver: A nanoproduct in medical application. *Toxicology Let.*, **176**, pp. 1–12.

Coyle, S., Wu, Y., Lau, K.-T., De Rossi, D., Wallace, G. and Diamond D. (2007). Smart nanotextiles: a review of materials and applications. *MRS Bulletin*, **32**, pp. 434–442.

Craighead, H., and Leong, K. 2000. Nanotechnology Research Direction: Biotechnology, Medicine, and Healthcare, Ed. M. C. Roco, R. S. Williams, and P. Alivisatos, 153–172. Netherlands: Kluwer Acad. Publishers.

Daoud, W. A. and Xin, J. H. (2004). Low temperature sol-gel processed photocatalytic titania coating. *J. Sol-Gel Sci. Tech.*, **29**, pp. 25–29.

Daoud, W. A., Xin, J. H. and Zhang, Y. H. (2005). Surface functionalization of cellulose fibers with titanium dioxide nanoparticles and their combined bactericidal activities. *Surf. Sci.*, **599**, pp. 69–75.

Dastjerdi, R., Montazera, M. and Shahsavan, S. (2009). A new method to stabilize nanoparticles on textile surfaces. *Coll. Surf. A: Physicochem. Eng. Aspects,* **345**, pp. 202–210.

Doktycz, S. J. and Suslick, K. S. (1990). Interparticles collisions driving by ultrasound. *Science,* **247**, pp. 1067–1069.

Dubas, S. T., Kumlangdudsana, P. and Potiyaraj P. (2006). Layer-by-layer deposition of antimicrobial silver nanoparticles on textile fibers. *Coll. Sur. A: Physicochem. Eng. Aspects,* **289**, pp. 105–109.

Duran, N., Marcato, P. D., De Souza, G. I. H, Alves, O. L. and Esposito, E. (2007). Antibacterial effect of silver nanoparticles produced by fungal process on textile fabrics and their effluent treatment. *J. Biomed. Nanotechnol.,* **13**, pp. 203–208.

El-Naggar, A. M., Zohdy, M. H., Hassan, M. S. and Khalil, E. M. (2003). Antimicrobial protection of cotton and cotton/polyester fabrics by radiation and thermal treatments. I. Effect of ZnO formulation on the mechanical and dyeing properties. *J. Appl. Polym. Sci.,* **88**, pp. 1129–1137.

Fang, M., Chen, J. H., Xu, X. L., Yang, P. H. and Hildebrand, H. F. (2006). Antibacterial activities of inorganic agents on six bacteria associated with oral infections by two susceptibility tests. *Int. J. Antimicrob. Agents,* **27**, pp. 513–517.

Gabbay, J., Borkow, G., Mishal, J., Magen, E., Zatcoff, R. and Shemer-Avni, Y. (2006). Copper oxide impregnated textiles with potent biocidal activities. *J. Ind. J. Ind. Textiles,* **35**, pp. 323–335.

Gao, Y. and Cranston, R. (2008). Recent advances in antimicrobial treatments of Textiles. *Textile Res. J.,* **78**, pp. 60–72.

Gedanken, A., Reisfeld, R., Sominski, E., Palchik, O., Koltypin, Yu., Panczer, G., Gaft, M. and Minti, H. (2000). Sonochemical preparation and characterization of europium oxide doped in and coated on ZrO_2 and yttrium-stabilized zirconium (YSZ). *J. Phys. Chem. B,* **104**, pp. 7057–7065.

Gedanken, A. (2004). Using sonochemistry for the fabrication of nanomaterials. *Ultrason. Sonochem.,* **11**, pp. 47–55.

Gedanken, A. (2007). Doping nanoparticles into polymers and ceramics using ultrasound radiation. *Ultrason. Sonochem.,* **14**, pp. 418–430.

Gomes, B. P., Montagner, F., Berber, V. B., Zaia, A. A., Ferraz, C. C., de Almeida, J. F. and Souza-Filho, F. J. (2009). Antimicrobial action of intracanal medicaments on the external root surface. *J. Dent.,* **37**, pp. 76–81.

Gorenšek, M. and Recelj, P. (2007). Nanosilver functionalized cotton fabric. *Textile Res. J.,* **77**, pp. 138–141.

Guo, W.-L., Yang, Z.-X. and Wang, X.-K. (2006). Sonochemical deposition and characterization of nanophasic TiO_2 on silica particles *Mater. Res. Innovations,* **10**, pp. 11–20.

Hadad, L., Perkas, N., Gofer, Y., Calderon-Moreno, J., Ghule, A. and Gedanken, A. (2007). Sonochemical deposition of silver nanoparticles on wool fibers. *J. Appl. Polym. Sci.,* **104**, pp. 1732–1737.

Haufe, H., Muschter, K., Siegert, J. and Bottcher, H. (2008). Bioactive textiles by sol-gel immobilised natural active agents. *J. Sol-Gel Sci. Technol.,* **45**, pp. 97–101.

Hu, X, Cook, S. Wang, P. and Hwang, H. (2009). In vitro evaluation of cytotoxicity of engineered metal oxide nanoparticles. *Sci. Total Environ.,* **407**, pp. 3070–3072.

Huang, L, Li, D. Q, Lin, Y. J., Wei, M, Evans, D.G. and Duan X. (2005). Controllable preparation of nano-MgO and investigation of its bactericidal properties. *J. Inorg. Biochem.,* **99**, pp. 986–993.

Imlay, J. A. (2008). Cellular defenses against superoxide and hydrogen peroxide. *Annu. Rev. Biochem.,* **77**, pp. 755–776.

Joeger, T. K., Joeger, R., Olsson, E. and Granqvist, C. G. (2001). Bacteria as workers in the living factory: metal-accumulating bacteria and their potential for materials science. *Trends in Biotechnol.,* **19**, pp. 15–20.

Kangwansupamonkon, W., Lauruengtana, V., Surassmo, S. and Ruktanonchai, U. (2009). Antibacterial effect of apatite-coated titanium dioxide for textiles applications. *Nanomed. Nanotechnol.,* **5**, pp. 240–249.

Kathirvelu, S., D'Souza, L. and Dhurai, B. (2009). A study on functional finishing of cotton fabrics using nano-particles of zinc oxide. *Mater. Sci. [Medziagotyra],* **15**, pp. 75–79.

Kim, B., Kim, D., Cho, D., and Cho, S. (2003). Bactericidal effect of TiO_2 photocatalyst on selected food-borne pathogenic bacteria. *Chemosphere,* **52**, pp. 277–281.

Kim, S. S., Park, J. E. and Lee, J. (2011). Properties and antimicrobial efficacy of cellulose fiber coated with silver nanoparticles and 3-mercaptopropyltrimethoxysilane (3-MPTMS). *J. Appl. Polym. Sci.,* **119**, pp. 2261–2267.

Klimov, V. I., Mikhailovsky, A. A., Xu, S., Malko, A., Hollingsworth, J. A., Leatherdale, C. A., Eisler, H. J. and Bawendi, M. G. (2000). Optical gain and stimulated emission in nanocrystal quantum dots. *Science,* **290**, pp. 314–317.

Klueh, U., Wagner V., Kelly, S., Johnson, A. and Bryers, J. D. (2000). Efficacy of silver-coated fabric to prevent bacterial colonization and subsequent device-based biofilm formation. *J. Biomed. Mater. Res.*, **53**, pp. 621–631.

Kotlyar, A., Perkas, N., Amiryan, G., Meyer, M., Zimmermann, W. and Gedanken, A. (2007). Coating silver nanoparticles on poly(methyl methacrylate) chips and spheres via ultrasound irradiation. *J. Appl. Polym. Sci.*, **104**, pp. 2868–2876.

Kotlyar, A., Perkas, N., Koltypin, Y., Meyer, M., Zimmermann, W. and Gedanken, A. (2008). Coating zinc oxide submicron crystals on poly[methyl methacrylate] chips and spheres via ultrasound irradiation. *Ultrason. Sonochem.*, **15**, pp. 839–845.

Landau, M. V., Vradman, L., Herskowitz, M., Koltypin Y. and Gedanken, A. (2001). Ultrasonically controlled deposition–precipitation Co–Mo HDS catalysts deposited on wide-pore MCM material. *J. Catal.*, **201**, pp. 22–36.

Lansdown, A. B. G., Mirastschijski, U., Stubbs, N., Scanlon, E. and Agren, M. S. (2007). Zinc in wound healing: Theoretical, experimental, and clinical aspects. *Wound Rep. Reg.*, **15**, pp. 2–16.

Lee, H. J., Yeo, S. Y. and Jeong S. H. (2003). Antibacterial effect of nanosized silver colloidal solution on textile fabrics. *J. Mat. Sci.*, **38**, pp. 2199–2204.

Lee, H. J. and Jeong, S. H. (2005). Bacteriostasis and skin innoxiousness of nanosize silver colloids on textile fabrics. *Textile Res. J.*, **75**, pp. 551–556.

Li, N., Xia, T. and E. Nel, A. (2008). The role of oxidative stress in ambient particulate matter-induced lung diseases and its implications in the toxicity of engineered nanoparticles. *Free Radic. Biol. Med.*, **44**, pp. 1689–1699.

Lipovsky, A., Nitzan, N., Gedanken, A. and Lubart, R. (2011). Antifungal activity of ZnO nanoparticles—the role of ROS mediated cell injury. *Nanotechnology*, **22**, p. 105101.

Liu, G. Y., Essex, A., Buchanan, J. T., Datta, V., Hoffman, H. M., Bastian, J. F., Fierer, J. and Nizet, V. (2005). *Staphylococcus aureus* golden pigment impairs neutrophil killing and promotes virulence through its antioxidant activity. *J. Exp. Med.*, **202**, pp. 209–215.

Liz-Marzan, L. M. and Kamat, P. V. (2003). *Nanoscale materials.* Norwell, MA: Kluwer Academic Publishers.

Madrid, P. A., Moorillon, G. V. N., Borunda, E. O. and Yoshida, M. M. (2002). Photoinduced bactericidal activity against *Pseudomonas aeruginosa* by TiO_2 based thin films. *FEMS Microbiol Lett.*, **211**, pp. 183–188.

Makhluf, S., Dror, R., Nitzan, Y., Abramovich, Y., Jelinek, R. and Gedanken, A. (2005). Microwave-assisted synthesis of nanocrystalline MgO and its use as a bacteriocide. *Adv. Funct. Mater.*, **15**, pp. 1708–1715.

Mao, Z., Shi, Q., Zhang, L. and Cao, H. (2006). The formation and UV-blocking property of needle-shaped ZnO nanorod on cotton fabric. *Thin Solid Films,* **517**, pp. 2681–2686.

Mason, T. J. (1990). *Sonochemistry.* London: Royal Society of Chemistry Press.

Mason, T. J. (2007). Development in ultrasound non-medical. *Prog. Biophys. Molec. Biol.,* **93**, pp. 166–175.

Neal, A. L. (2008). What can be inferred from bacterium–nanoparticle interactions about the potential consequences of environmental exposure to nanoparticles? *Ecotoxicology,* **17**, pp. 362–371.

Ohira, T., Kawamura, M., Iida, Y., Fukuda, M. and Yamamoto, O. (2008). Influence of the mixing ratio on antibacterial characteristics of MgO-ZnO solid solution in two phase coexistence region. *J. Ceram. Soc. Japan,* **116**, pp. 1234–1237.

Patra, A., Sominska, E., Ramesh, S., Koltypin, Yu., Zhong, H., Minti, H., Reisfeld, R. and Gedanken, A. (1999). Sonochemical preparation and characterization of Eu_2O_3 and Tb_2O_3 doped in and coated on silica and alumina nanoparticles. *J. Phys. Chem. B,* **103**, pp. 3361–3365.

Perelshtein, I., Applerot, G., Perkas, N., Guibert, G., Mikhailov, S. and Gedanken, A. (2008). Sonochemical coating of silver nanoparticles on textile fabrics (nylon, polyester and cotton) and their antibacterial activity. *Nanotechnology*, **19**, art. 245705, p. 6.

Perelshtein I., Applerot G., Perkas N., Wehrschetz-Sigl, E., Hasmann, A., Guebitz, G. M. and Gedanken, A. (2009). Antibacterial properties of an in situ generated and simultaneously deposited nanocrystalline ZnO on fabrics. *ACS Appl. Mater. Interfaces,* **1**, pp. 361–366.

Perkas, N., Wang, Y., Koltypin, Y., Gedanken, A. and Chandrasekaran, S. (2001). Mesoporous iron-titania catalysts for cyclohexane oxidation. *Chem. Comm.*, pp. 988–989.

Perkas, N, Amirian, G., Dubinsky, S., Gazit, S. and Gedanken, A. (2007). Ultrasound—assisted coating of nylon 6,6 with silver nanoparticles, and its antibacterial activity. *J. Appl. Polym. Sci.,* **104**, pp. 1423–1430.

Perkas, N., Amirian, G., Applerot, G., Efendiev, E., Kaganovskii, Y., Ghule, A. V., Chen, B-J., Ling, Y-C. and Gedanken, A. (2008). Depositing silver nanoparticles on/in a glass slide by the sonochemical method. *Nanotechnology,* **19**, art. 435604, p. 9.

Perkas, N., Shuster, M., Amirian, G., Koptypin, Y. and Gedanken, A. (2008). Sonochemical immobilization of silver nanoparticles on porous polymers. *J. Polym. Sci. Part A: Polym. Chem.*, **46**, pp. 1719–1729.

Perkas, N., Amirian, G., Rottman, C., De La Vega, F. and Gedanken, A. (2009). Sonochemical deposition of magnetite on silver nanocrystals. *Ultrason. Sonochem.*, **16**, pp. 132–135.

Petrulyte, S. (2008). Advanced textile materials and biopolymers in wound management. *Dan. Med. Bull.*, **55**, pp. 72–77.

Pol, V. G., Reisfeld, R. and Gedanken, A. (2002). Sonochemical synthesis and optical properties of europium oxide nanolayer coated on titania. *Chem. Mater.*, **14**, pp. 3920–3924.

Pol, V. G., Srivastava, D. N., Palchik, O., Palchik, V., Slifkin, A., Weiss, A. M. and Gedanken, A. (2002). Sonochemical deposition of silver nanoparticles on silica spheres. *Langmuir,* **18**, pp. 3352–3357.

Pol, V. G., Wildermuth, G., Felsche, J., Gedanken, A. and Calderon-Moreno, J. (2006).

Sonochemical deposition of Au nanoparticles on titania and the significant decrease in the melting point of gold. *J. Nanosci. Nanotech.,* **5**, pp. 975–979.

Purwar, R. and Joshi, R. M. (2004). Recent developments in antimicrobial finishing of textiles. *AATCC Rev.,* **4**, pp. 22–25.

Qi, K. H., Chen, X. Q., Liu, Y. Y., Xin, J. H., Mak, C. L. and Daoud, W. A. (2007). Facile preparation of anatase/SiO_2 spherical nanocomposites and their application in self-cleaning textiles. *J. Mater. Chem.*, **17**, pp. 3504–3508.

Qian, L. and Hinestroza, J. P. (2004). Application of nanotechnology for high performance textile. *J. Textile and Apparel. Technol. Management.,* **4**, pp. 1–7.

Rajendran, R., Balakumar, C, Ahammed H. A. M., Jayakumar, S., Vaideki, K. and Rajesh, E. M. (2010). Use of zinc oxide nano particles for production of antimicrobial textiles. *Int. J. Eng. Sci. Technol.*, **2**, pp. 202–208.

Rao, C. N. R. and Cheetham, A. K. (2001). Science and technology of nanomaterials: current status and future prospects. *J. Mater. Chem.,* **11**, pp. 2887–2894.

Richard, P., LeFloch, R., Chamoux, C., Pannier, M., Espaze, E. and Richet, H. (1994). Pseudomonas aeruginosa outbreak in a burn unit: role of antimicrobials in the emergence of multiply resistant strains. *J. Inf. Dis.*, **170**, pp. 377–383.

Rincon, A. G., Pulgarin., C., Laub, D., Xantopoulos, M., Mathieu, H.J. and Kiwi, J. Yuranova, T. (2006). Performance and characterization of Ag-cotton and Ag/TiO_2 loaded textiles during the abatement of *E. coli*. *J. Photochem. Photobiol.* A, **181**, pp. 363–369.

Rosen, H. and Klebanoff, S. J. (1979). Bactericidal activity of a superoxide anion-generating system. *J. Exp. Med.*, **149**, pp. 27–39.

Sawai, J., Igarashi, H., Hashimoto, A., Kokugan, T. and Shimizu, M. (1996). Effect of particle size and heating temperature of ceramic powders on antibacterial activity of their slurries. *J. Chem. Eng. Jpn.*, **29**, pp. 251–256.

Sawai, J., Kawada, E., Kanou F., Igarashi, H., Hashimoto, A., Kokugan, T. and Shimizu, M. (1996). Detection of active oxygen generated from ceramic powders having antibacterial activity. *J. Chem. Eng. Jpn.*, **29**, pp. 627–633.

Sawai, J., Shoji, S., Igarashi, H., Hashimoto, A., Kokugan, T., Shimizu, M. and Kojima, H. (1998). Hydrogen peroxide as an antibacterial factor in zinc oxide powder slurry. *J. Ferment. Bioeng.*, **86**, pp. 521–522.

Sawai, J., Kojima, H., Igarashi, H., Hashimoto, A., Shoji, S., Sawaki, T., Hakoda, A., Kawada, E., Kokugan, T. and Shimizu, M. (2000). Antibacterial characteristics of magnesium oxide powder. *World J. Microb. Biot.*, **16**, pp. 187–194.

Sawai, J. and Yoshikawa, T. (2004). Quantitative evaluation of antifungal activity of metallic oxide powders (MgO, CaO and ZnO) by an indirect conductimetric assay. *J. Appl. Microb.*, **96**, pp. 803–809.

Schäfer, K. (2009). Nanotechnology in Textile Finishing. www.exzellenz.nrw.de/highlights.

Schmidt, H. (2001). Nanoparticles by chemical synthesis, processing to materials and innovative applications *Appl. Organometal. Chem.*, **15**, pp. 331–343.

Schollmeyer, E. (2007). Nanotechnology for functionalization of textile materials. 3rd International Textile Clothing and Design Conference, 2006, Dubrovnik, Croatia. *Tekstil* **56**, pp. 75–80.

Scholz, J., Nocke, G., Hollstein, F. and Weissbach, A. (2005). Investigations on fabrics coated with precious metals using the magnetron sputter technique with regard to their anti-microbial properties. *Surf. Coat. Technol.*, **192**, pp. 252–256.

Searle, A. (1919). *The Use of Metal Colloids in Health and Disease*. Sutton: New York.

Sengupta, G., Ahluwalia, H. S., Banerjee, S. and Sen, S. P. (1979). Chemisorption of water-vapor on zinc oxide. *J. Colloids Interface Sci.*, **69**, pp. 217–224.

Seo, D. J., Park, S. B., Kang, Y. C. and Choy, K. L. (2003). Formation of ZnO, MgO and NiO nanoparticles from aqueous droplets in flame reactor. *J. Nanopart.Res.*, **5**, pp. 199–210.

Shen, Q., Nie, Z., Guo, M., Zhong, C. J., Lin, B., Li, W. and Yao, S. (2009). Simple and rapid colorimetric sensing of enzymatic cleavage and oxidative damage of single-stranded DNA with unmodified gold nanoparticles as indicator. *Chem. Commun.*, pp. 929–931.

Shi, L., Zhao, Y., Zhang, X., Su, H. and Tan, H. (2008). Antibacterial and anti-mildew behavior of chitosan/nano-TiO_2 composite emulsion. *Korean J. Chem. Eng.*, **25**, pp. 1434–1438.

Shukla, S., Seal, S., Schwartz, S. and Zhou, D. (2001). Synthesis and characterization of nanocrystalline silver coating of fly ash cenosphere particles by electroless process. *J. Nanosci. Nanotechnol.*, pp. 417–424.

Simoncic, B. and Tomsic, T. (2010). Structures of novel antimicrobial agents for textiles. A review. *Textile Res. J.*, **80**, pp. 1721–1737.

Soderberg, T., Agren, M., Tengrup, I., Hallmans, G. and Banck, G. (1989). The effects of an occlusive zinc medicated dressing on the bacterial flora in excised wounds in the rat. *Infection*, **17**, pp. 81–85.

Sojka-Ledakowicz, J., Lewartowska, J., Kudzin, M., Jesionowski, T., Siwińska-Stefańska, K. and Krysztafkiewicz, A. (2008). Modification of textile materials with micro- and nano-structural metal oxides. *Fibers and Textiles in Eastern Europe A*, **16**, pp. 112–116.

Stengl, V., Bakardjieva, S., Marikova, M., Bezdicka, P. and Subrt, J. (2003). Magnesium oxide nanoparticles prepared by ultrasound enhanced hydrolysis of Mg-alkoxides. *Mater. Lett.*, **57**, pp. 3998–4003.

Suslick, K. S., Hammerton, D. A. and. Cline R. E. (1986). Sonochemical hot spot. *J. Am. Chem. Soc.*, **108**, pp. 5641–5642.

Suslick, K. S. (1989). The chemical effects of ultrasound. *Sci. Am.*, **260**, pp. 80–86.

Suslick, K. S. and Price, G. J. (1999). Application of ultrasound to materials chemistry. *J. Annu. Rev. Mater.*, **29**, pp. 295–326.

Toxicological profile of copper. (2004). *U.S. Department of Health and Human Services, Public Health Service Agency for Toxic Substances and Disease Registry. Textiles*, **35**, pp. 323–335.

Tricoli, A., Graf, M., Mayer, F., Kuhne, S., Hierlemann, A. and Pratsinis, S. E. (2008).

Micropatterning layers by flame aerosol deposition-annealing. *Adv. Mater.*, **20**, pp. 3005–3010.

Uddin, F. (2008). Some aspects in advancement of textile nanofinishing. 86th Textile-Institute World Conf., Conf. Proceed. **2**, U291–U307. Hong Kong: Polytechnic Univ., Inst. Textiles and Clothing Press.

Vajnhandl, S. and Le Marechal, A. M. (2005). Ultrasound in textile dyeing and the decolouration/mineralization of textile dyes. *Dyes and Pigm.*, **65**, pp. 89–101.

Vallee, B. L. and Falchuk, K. H. (1993). The biochemical basis of zinc physiology. *Physiol Rev.*, **73**, pp. 79–118.

Vigneshwaran, N., Kumar. S., Kathe, A. A., Varadarajan, P. V. and Prasad, V. (2006). Functional finishing of cotton fabrics using zinc oxide-soluble starch nanocomposites. *Nanotechnology,* **17**, pp. 5087–5095.

Wang, C., Li, M., Wu, M. and Chen L. (2008). Cotton fabric properties with water-repellent finishing via sol-gel process. *Sur. Rev. Let.*, **15**, pp. 833–839.

Wang, H., Wang, J., Hong, J., Wei, Q., Gao W. and Zhu, Z. (2007). Preparation and characterization of silver nanocomposite textile. *J. Coat. Technol. Res.*, **4**, pp. 101–106.

Wang, R., Xin, J. H., Tao, X. M. and Daoud, W. A. (2004). ZnO nanorods grown on cotton fabrics at low temperature. *Chem. Phys. Lett.,* **398**, pp. 250–255.

Wendler, F., Meister, F., Montigny, R. and Wagener, M. (2007). A new antimicrobial ALCERU fibre with silver nanoparticles. *Fibers and Tex. in East. Europe,* **15**, pp. 64–65.

White, T. C., Marr, K. A. and Bowden, R. A. (1998). Clinical, cellular, and molecular factors that contribute to antifungal drug resistance. *Clin. Microbiol. Rev.,* **11**, pp. 382–402.

Wong, Y. W. H., Yuen, C. W. M., Leung, M. Y. S., Ku, S. K. A. and Lam, H. L. I. (2006). Selected applications of nanotechnology in textiles. *AUTEX Research J.,* **6**, pp. 1–8.

Wright, G. D. (2007). On the road to bacterial cell death. *Cell,* **130**, pp. 781–783.

Xu, B. and Cai, Z. S. (2008). Fabrication of a superhydrophobic ZnO nanorod array film on cotton fabrics via a wet chemical route and hydrophobic modification. *Appl. Sur. Sci.*, **254**, pp. 5899–5904.

Xue, C. H., Jia, S. T., Chen, H. Z. and Wang, M. (2008). Superhydrofobic cotton fabrics prepared by sol-gel coating of TiO_2 and surface hydrofobization. *Sci. Technol. Adv. Mater.,* **9**, art. 035001, p. 5.

Yamamoto, O. (2001). Influence of particle size on the antibacterial activity of zinc oxide. *Int. J. Inorg. Mater.*, **3**, pp. 643–646.

Yeo, S. Y., Lee H. J. and Jeong, S. H. (2003). Preparation of nanocomposite fibers for permanent antibacterial effect. *J. Mater. Sci.*, **38**, pp. 2143–2147.

Zalesky, T. (2008). "Active" products drive wound-care market. *GEN Biobusiness Biomarket Trends*, **9** (May 1), 1–2. http://genengnewa.com/articles/chitem.

Zhong, Z. Y., Prozorov, T., Felner, I. and Gedanken A. (1999). Sonochemical synthesis and characterization of iron oxide coated on submicrospherical alumina: a direct observation of interaction between iron oxide and alumina. *J. Phys. Chem. B*, 103, pp. 947–956.

Zohdy, M. H., Kareem, H. A., El-Naggar, A. M. and Hassan, M. S. (2003). Microbial detection, surface morphology, and thermal stability of cotton and cotton/polyester fabrics treated with antimicrobial formulations by a radiation method. *J. Appl. Polym. Sci.*, **89**, pp. 2604–2610.

Chapter 10

Ultrasonic Processing for Synthesis of Nanocomposite via in situ Emulsion Polymerization and Their Applications

Shirish H. Sonawane,[d] Bharat A. Bhanvase,[b] Ravindra D. Kulkarni,[a] and Pawan K. Khanna[c]

[a]*Department of Chemical Technology, North Maharashtra University, P.O. Box 80, UMAVI Nagar, Jalgaon 425001, Maharashtra, India*

[b]*Department of Chemical Engineering, Laxminarayan Institute of Technology, Amaravati Raod, Bharat Nagar, Nagpur 440033, Maharashtra, India*

[c]*Department of Applied Chemistry, Defense Institute of Advanced Technology (DIAT) Deemed University, Girinagar, Pune 411025, Maharashtra, India*

[d]*Department of Chemical Engineering, National Institute of Technology, Warangal 506004, Andhra Pradesh, India*

shirishsonawane@rediffmail.com

10.1 Introduction

In chemical process industries, a number of unit operations and processes are being used such as emulsification, dispersion, homogenization, grinding, milling, sieving, extraction, crystallization, and so on. These operations and processes are being used with the

Cavitation: A Novel Energy-Efficient Technique for the Generation of Nanomaterials
Edited by Sivakumar Manickam and Muthupandian Ashokkumar

ISBN 978-981-4411-54-7 (Hardcover), 978-981-4411-55-4 (eBook)
www.panstanford.com

aim of maximization of recovery, to improve purity of product, to generate stable colloidal particles and to reduce the particle size, and so on. Major mechanical operations such as sand/bead mill are being used for dispersion, while solvents extraction process is being used for the extraction of valuable products; mechanical agitation is used to enhance the rate of transport of chemical species from one phase to other phase.

These chemical and physical effects can be effectively achieved by the assistance of ultrasound technique. By application of ultrasound, these chemical and physical processes can be enhanced. Improvement in the mass transfer and heat transfer may occur due to micro streaming of the fluid motion around a (weakly) pulsating bubble (Elder, 1959). Penetration of solute and reacting molecules reaches to active sites of catalyst surface because of microjets (Mason and Lorimer, 1988, Mason, 1997, Mason and Cordemans de Meulenaer, 1998). Some applications of ultrasound which are used in industrial chemical processes are as follows: formation of fine emulsion droplets can be achieved when cavitation bubbles collapse at the interface of liquid, which is immiscible in continuous phase (Behrend and Schubert, 2000). Due to high shearing of ultrasound, formation of aggregates is prevented. Ultrasound has been used for the preparation of inorganic nanomaterials by reducing the micron size particles into nanosize, for example, silica and many ceramic/oxide particles (Hielscher Ultrasonics). Further ultrasound cavitation facilitates the wet grinding processing of organic materials by reduction in the particle size and making stable dispersion of pigments. For the preparation of pharmaceutical sol and lubricants, sonolysis process has been attracted by many chemical processing industries, especially paint and surface-coating industries. The improper encapsulation of these inorganic particles affects percentage of filler loading, pigment to volume ratio, and preparation of homogeneous dispersion, which finally affects the performance of final coating (Lorimer et al., 1991).

Microemulsion is a clear, thermodynamically stable liquid mixture of oil, water, and surfactant, usually along with a cosurfactant. In comparison to ordinary emulsion, microemulsion is formed upon simple mixing of the components and do not require the high shear conditions. In case of miniemulsion, stable droplets are obtained in the size range of 50–500 nm. Although 10–100 nm droplets are obtained in case of microemulsion that is thermodynamically stable, 1–10 μm size droplets are obtained in conventional emulsion. Polymer synthesis and polymer encapsulation are the important areas in the materials science. Ultrasound-assisted miniemulsion polymerization has several advantages over the conventional emulsion polymerization. In this chapter, an attempt has been made to describe the role of ultrasound in technically complex systems such as miniemulsion polymerization. In commercial processes, the semi-batch miniemulsion polymerization is being used for controlling the molecular weight and rate of polymerization. In this chapter, role of ultrasound and initiator has been described in detail for semi-batch miniemulsion polymerization. Specific examples of nanocomposite synthesized via ultrasound assisted in situ emulsion polymerization process, and applications of ultrasound in encapsulation, in situ polymerization, anticorrosion properties of such formulated coating, and latex applications of photo anode and sensor have been described in detail.

10.1.1 Role of Ultrasound in Emulsion Polymerization

Synthesis of polymer nanoparticles is one growing research area in material science and engineering due to number of technological applications of polymer nanoparticles in nano medicine formulations, drug delivery (Allen and Cullis, 2004), (Bisht et al., 2007) surface coatings, hydrophilic paints, adhesives (Wang et al., 2010), optoelectronics, medical diagnostics, and so on. Polymeric nanoparticles required for various noted applications are predominantly prepared by wet chemical synthesis routes such as suspension polymerization, micro, mini, and reverse emulsion processes. Compared with miniemulsion, microemulsion is thermodynamically stable; this non-equilbirium state of miniemulsion polymerization contains all the monomer droplets with intiator that leads to an increase in the particle size as compared with microemulsion (Klein and Daniels, 1997). Polymers such as acrylonitrile butadiene styrene (ABS), polystyrene (PS), polymethyl methacrylate (PMMA), and so on can be easily synthesized via emulsion polymerization processes. It is well known that for batch emulsion polymerization, hydrophobic monomer and hydrophilic initiator are the main constituents. In addition, the use of a suitable ionic or nonionic surfactant could be used in the water phase (continuous phase). In the initial stage, an oil in water emulsion is formed. Subsequently, the dissociation of initiator generates number of radicals to facilitate the polymerization and the reaction goes to completion over time. Such conventional emulsion polymerization processes have various disadvantages, for example, high polydispersity of particles and instability of colloidal particles (Bradley and Grieser, 2002). Similarly, the decomposed constitutes of the initiator may also be trapped, hence affecting the polymer purity. Some of the by-products formed after termination process may undergo series of reaction leading to discoloration of the final polymer product. These problems can be overcome by the elimination of chemical initiator and if the mode of reaction is altered. This is possible by avoiding thermal polymerization and by employing the sonolysis technique that does not require the use of a catalyst for polymerization (Bradley and Grieser, 2002). Emulsion polymerization has its benefits, as it has control over the reactor conditions, which leads to a narrow particle size distribution, smaller range of particle size (typically between 1 μm and 10 μm), less exothermic reactions, and so on. In addition, the solventless reactions make the process environmentally friendlier than other suspension and solution polymerization (Klein and Daniels, 1997; Landfester et al., 2003; Asua, 2004; Odian, 2004). In order to achieve smaller size of nanoparticles with higher surface area for higher loading of functional attachments of molecules, miniemulsion polymerization is the most preferred method compared with conventional polymerization processes (Liu and Yang, 2003). Recently, ultrasound is being used for emulsion polymerization (Price, 2003). Ultrasound has been found to be one of the most effective methods for the preparation of stable monomer droplets and to avoid aggregation and adhesion of polymer onto reactor wall during suspension and emulsion polymerization process (Mason and Lorimer, 1988; Kobayashi et al., 2008). Acoustic cavitation generates small cavities that grow and collapse in a microsecond, thereby allowing an adiabatic process that generates very high local temperatures (>1000°C) and pressures (>1000 atm) (Mason and Lorimer, 1988). Because of this extreme environment, decomposition of organic molecules such as monomers and oligomers takes place that gives radical generation (Price, 2003; Cao et al., 2008; Kobayashi et al., 2008).

Use of ultrasound during emulsion polymerization has the following advantages. Degassing mechanism of ultrasound helps to remove oxygen from the reaction mass that acts as an inhibiting agent. Small quantity of polymer chains may be degraded during polymerization, which helps in providing the radicals by autocatalysis. Degradation of polymer chain of degree *"n"* takes place due to acoustic cavitation results into the formation of radicals, which are responsible for enhancement of the polymerization rate.

$$P_{\mathrm{n}} + [\mathrm{CS}] \xrightarrow{\text{Ultrasound}} P_{\mathrm{m}}^{\bullet} + P_{\mathrm{n-m}}^{\bullet} \tag{R.10.1}$$

where [CS] is cavity density.

Rapid dissociation of initiator therefore increases the production rate of radicals leading to faster polymerization rate. The chemical initiator, "*I*," is thermally decomposed into two radical entities:

$$I \rightarrow 2R_{\mathrm{aq}}^{\bullet} \tag{R.10.2}$$

Reduction in the induction time of polymerization reaction helps in the formation of stable emulsion. Initial monomer droplet size is reduced and remains almost the same as the final latex particle size (Bradley and Grieser, 2002). The chemical initiators or costabilizers are not required during polymerization; hence, ultrasound-assisted polymerization is an eco-friendly process (Bradley and Grieser, 2002; Yin et al., 2006; Nie et al., 2008; Teo et al., 2008). Bradley and Grieser (2002) reported the overall mechanism and reaction processes involved in the preparation of polymer latex particles by ultrasound-assisted miniemulsion polymerization. There are two kinds of effects during sonochemical reactions: (i) physical effects and (ii) chemical effects. In physical effects, due to acoustic cavitation, very efficient mixing and strong dispersion of the constituents of reaction mixture takes place. The high shear forces around the cavitation bubbles also control the growth of polymer particle due to reduced Ostwald ripening process because of constant fragmentation of larger oil droplets into smaller ones. In chemical effects, the acoustic cavitation helps in the generation of radicals that initiates the polymerization process. The micro bubbles are produced in solution due to number of cavitation events when ultrasound waves pass through a liquid medium. On collapse of these micro bubbles in an aqueous solution, primary radical species •OH and •H are most likely formed. Further, volatile monomers such as methyl methacrylate

(MMA) and butyl acrylate (BA) could enter a cavitation bubble and are thus thermally decomposed to produce a variety of primary organic radicals. In ultrasound-assisted miniemulsion process, very fine and stable uniform monomer droplets are normally formed at or near the interface of immiscible oil phase. These so-called cavitation microjets help radicals get transported toward the surface of micelle (Price, 2003; Teo et al., 2008, 2009, 2011). It is also possible that sometimes radicals may recombine to give water and H_2O_2, which will adversely affect the reaction and in fact this can lead to reduction in radical activity (Bhanvase et al., 2012). It is therefore advisable to add a small quantity of initiator during the emulsion polymerization. Kobayashi et al. (2008) used ultrasonic generators for the intensification of emulsion polymerization process as well as alternative initiators in polymerization processes. Styrene was polymerized under indirect ultrasonic irradiation. With change in irradiation conditions at different stages of polymerization process, intensification of PS process was observed. Several researchers have employed ultrasound technique for emulsion polymerization of various monomers such as MMA, BA, styrene, aniline, and so on (Chau and Stoffer, 1999; Cheung and Gaddam, 2000; Kojima et al., 2001; Xia and Wang, 2001; Xia et al., 2002; Yin and Chen, 2004; Wang and Cheung, 2005). All physical and chemical effects of ultrasound during emulsion polymerization are schematically represented in Fig. 10.1. As shown in Fig. 10.1, initially cavities or bubbles are formed due to the presence of ultrasound waves; these bubbles collapse and generate •OH and •H radicals. Further, if the initiator is present, then dissociation of initiator occurs very fast.

10.1.2 Synergistic Effect of Ultrasound and Initiator in Semibatch Polymerization: A Case Study

In this section, combined influence of ultrasound and initiator has been investigated for the MMA emulsion polymerization in the semi-batch process (Bhanvase et al., 2012). It has been investigated that ultrasound and initiator concentration on the monomer helps in effective conversion, particle size, and on polymerization rate. Some research has described the use of 13 mm ultrasound probe for synthesis of PMMA, which was operated at a frequency of 22 kHz with the power output of 750 W. Typically, the synthesis can be carried

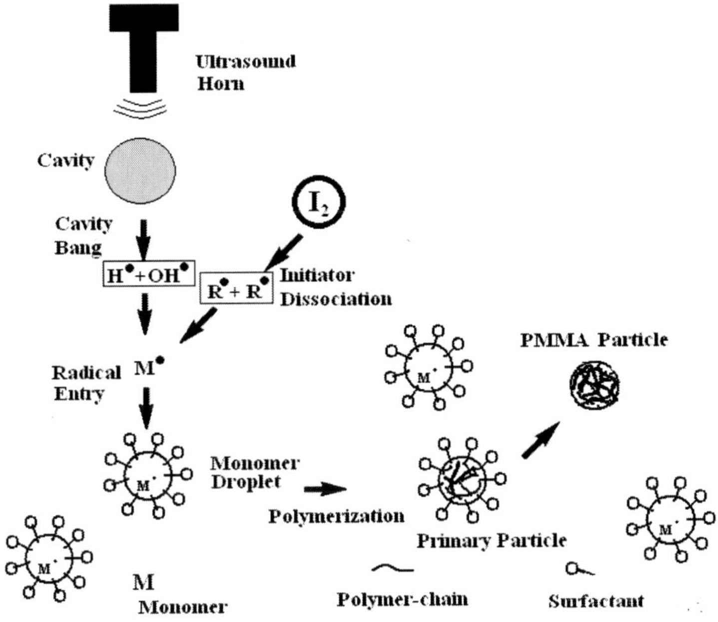

Figure 10.1 Mechanism of the formation of polymer nanoparticles in the presence of ultrasound and initiator during miniemulsion polymerization.

out by using an appropriate amount of the required reagents, for example, a little amount of the monomer can be initially added to the reactor followed by the continuous addition of the rest of the amount into a reactor over a period of time. This may be done at reasonable mild reaction temperature, suitable for the dissociation of initiator to generate more radicals to initiate the polymerization. The effect of monomer, surfactant, and initiator concentration on the emulsion polymerization process can be thus studied using the conversion and the polymerization rate. To understand the effect of ultrasound, emulsion polymerization can be carried out with ultrasound alone (without initiator) (Bhanvase et al., 2012) (Fig. 10.2 defined by symbol ◊). Radical generation due to ultrasound alone can be very small, typically in the range of 7–8%. The generation of radicals in such case could be considered either due to monomer degradation or due to surfactant and water dissociation because of cavitation (Cao et al., 2008). Such experiment shows that fast dissociation of initiator and formation of uniform monomer droplets are present in the reactor due to violent shearing action of cavitation. Small radical generation could be good enough for polymerization of about 8%. Figure 10.3 shows a comparative study of conversion of MMA for 4% KPS with and without ultrasonic irradiation (Bhanvase et al., 2012). As shown in Fig. 10.3, conversion of MMA is 87% and 71% with and without ultrasound, respectively. Polymerization of MMA is possible at room temperature using ultrasound as reported by Bradley and Grieser (2002) with the use of external initiator in the presence of ultrasound. It has been described that in the initial stage, rapid increase in the conversion is observed due to high rate of radical generation (Fig. 10.2), which enhances the initial rate of polymerization because of sonication. The effect of 4% initiator was found to generate particle size of 106 nm. Figure 10.4 shows the rate of polymerization (g/L/min) as a function of sonication time (min) at different initiator concentrations. Low polymerization rate up to 0.5% initiator concentration and high polymerization rate (high conversion) at higher initiator concentration (1–5% initiator) was observed (Bhanvase et al., 2012). Figure 10.4 shows that the polymerization rate attains maxima and then reduces at a later time. Due to ultrasonic irradiation, the rates are altered, because the cavitation jets facilitate the transport of generated free radicals rapidly toward the monomer droplet, which results in an initial increase in the rates reaching maxima. The polymerization rate decreases in the later stage possibly due to limited monomer availability. TEM micrograph of PMMA nanoparticles produced for the optimized condition (based on conversion) is shown in Fig. 10.5 (Bhanvase et al., 2012). It is found that the size of PMMA particles is around 50 nm, with spherical morphology, and compared with conventional emulsion polymerization, the particle size obtained by the current described process is smaller. It is observed that the ratio of generated radical to monomer is lower in conventional batch emulsion polymerization than in ultrasonically assisted polymerization. Hence, the conversion decreases with an increase in the monomer concentration in batch polymerization, whereas it increases with an increase in the monomer concentration (Bhanvase et al., 2012).

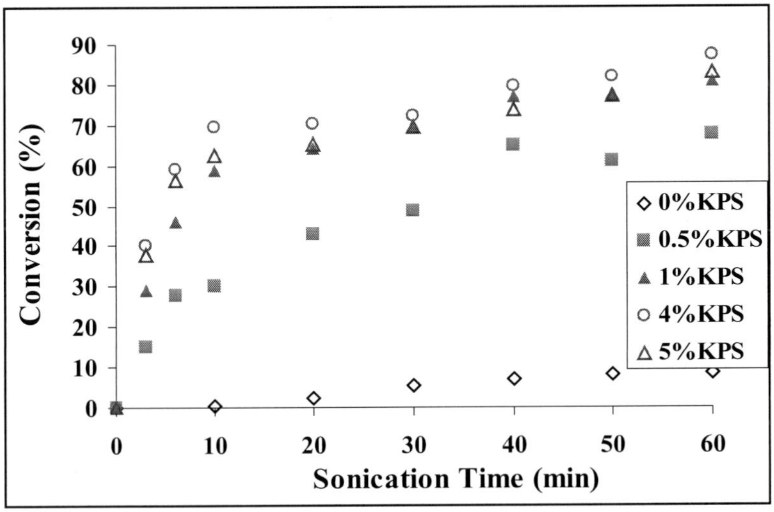

Figure 10.2 Effect of percentage of initiator concentration (of MMA) on the conversion of MMA (MMA concentration = 7.2 wt%, surfactant concentration = 2% of MMA, temperature = 65°C) (Bhanvase et al., 2012).

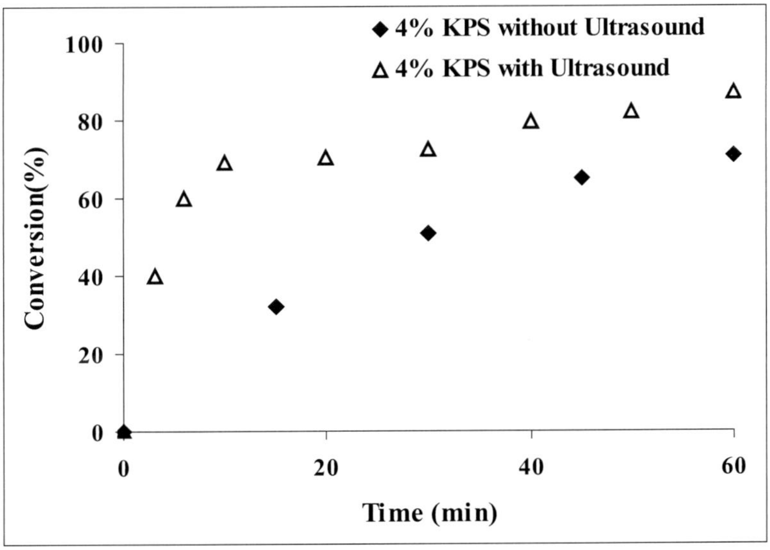

Figure 10.3 Comparative study of conversion of MMA for 4% KPS with and without ultrasound at MMA concentration = 7.2 wt%, surfactant concentration = 2% of MMA, temperature = 65°C (Bhanvase et al., 2012).

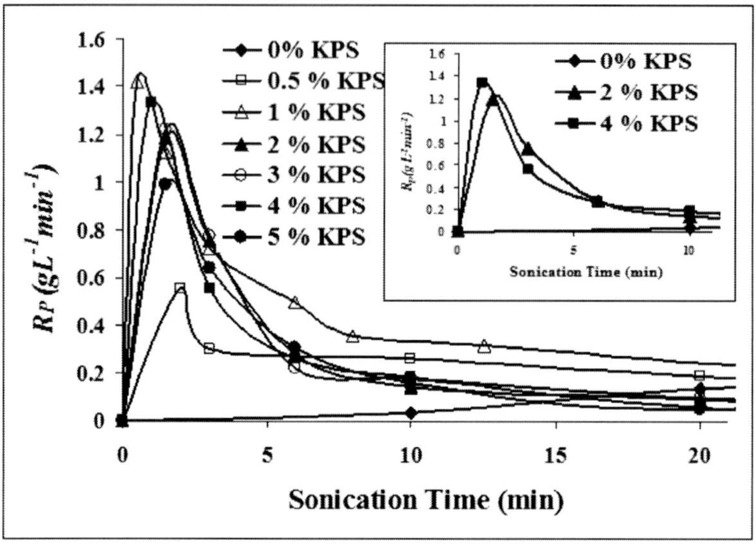

Figure 10.4 *R*p (g/L min) vs. sonication time (min) at different conditions of initiator concentration and only ultrasound-assisted initiation (0 % initiator) (MMA concentration = 7.2 wt%, surfactant concentration = 2% of MMA, temperature = 65°C) (Bhanvase et al., 2012).

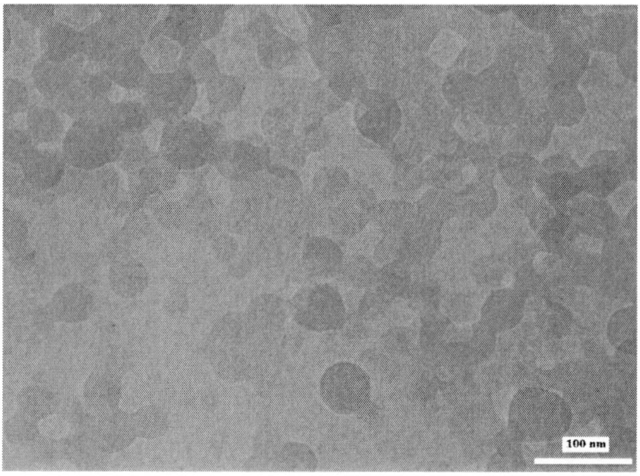

Figure 10.5 TEM analysis of PMMA nanoparticles (initiator concentration = 4% of MMA, temperature = 65°C, surfactant concentration = 2% of MMA, MMA concentration = 7.2 wt %) (Bhanvase et al., 2012).

10.2 Ultrasound-Assisted in situ Emulsion Polymerization

Inorganic-organic nanocomposite materials with well-defined morphology and structure at the nanometer scale have gained enormous interest by offering prominent changes in various properties, for example, optical, thermal, electrical, magnetic, and mechanical and barrier properties in comparison to pure polymers. These materials correspond to an innovative class of polymeric materials that provide the properties of the inorganic particles (mechanical strength, modulus, thermal stability) and the flexibility of the organic polymer matrix. For the preparation of polymer nanocomposites, either organic molecules or inorganic particles can be encapsulated into the polymer. Several methods have been adopted to achieve a fine dispersion of the inorganic particles in polymers and to improve the bonding between the polymer and inorganic particles. Nanocomposites have attracted researchers because of the potential applications in many areas such as paints, adhesives, cosmetics, optics, capacitors, controlled release, and electronics (Siegel, 1995; Erdem et al., 2000; Sinha and Okamoto, 2003; Huang et al., 2004; Pothukuchi et al., 2004; Patel et al., 2006; Paul and Obeson, 2008).

In situ polymerization is an effective way for the synthesis of polymer nanocomposite materials. In this method, the nanoparticles are impregnated with a monomer and then polymerization is carried out by different methods such as emulsion polymerization (Tiarks et al., 2001), intercalative polymerization (Kong et al., 1999), solution method (Liu et al., 2004), hybrid latex polymerization (Tissot et al., 2001), and so on. Out of these methods, in situ emulsion polymerization is the most convenient way for the encapsulation and preparation of polymer/inorganic composite particles on a large scale (Ma et al., 2006; Qi et al., 2006; Soares et al., 2006). In emulsion polymerization, dispersed monomer droplet acts as a nanoreactor for the growth of polymer, while polymer nanocomposite prepared by conventional in situ emulsion polymerization may find improper dispersion of inorganic nanoparticles, which in turn leads to inferior properties of polymer nanocomposite. In order to obtain fine and uniform dispersion of inorganic nanoparticles in polymer, ultrasound-assisted in situ emulsion polymerization can be highly effective. Ultrasound-assisted polymerization is the well-accepted method for the synthesis of finely dispersed inorganic particles in polymer for the preparation of nanocomposites. Fillers such as titanium dioxide, calcium carbonate, clays, other metal oxides, and metal particles have been encapsulated via in situ emulsion polymerization (Xu et al., 2003; Ryu et al., 2004; Ni et al., 2006; Diaconu et al., 2008; Matusinovic et al., 2009). Furthermore, preparation of different types of polymer nanocomposites containing various inorganic oxides, clay, and magnetic particles can be prepared by using ultrasound-assisted method, which is attractive because of shorter reaction times and easy handling. Liu et al. (2009) have developed core–shell magnesium hydroxide/PS hybrid nanoparticles by the use of ultrasound copolymerization. It has been reported that ultrasound-assisted in situ emulsion polymerization is an excellent approach for the uniform dispersion of vinylated magnesium hydroxide into PS, which eventually may lead to core–shell structure of magnesium hydroxide/PS hybrid nanoparticles. Jiang (2007) has synthesized PMMA/$Ni_{0.5}Zn_{0.5}Fe_2O_4$ nanocomposite by in situ emulsion polymerization of MMA monomer in the presence of $Ni_{0.5}Zn_{0.5}Fe_2O_4$ colloidal suspension assisted by ultrasonic irradiation. It was found that this procedure is effective in reducing the agglomeration of $Ni_{0.5}Zn_{0.5}Fe_2O_4$ nanoparticles during the preparation of PMMA/$Ni_{0.5}Zn_{0.5}Fe_2O_4$ nanocomposite. It has also been reported that $Ni_{0.5}Zn_{0.5}Fe_2O_4$ nanoparticles with the particle sizes of about 12 nm are well dispersed in the PMMA matrix. Haldorai et al. (2009) have prepared electrically conducting copolymer poly(aniline-co-p-phenylenediamine) and silica (SiO_2) nanocomposites by ultrasonically assisted in situ emulsion polymerization. They have claimed that this approach can solve problems in the dispersion and stabilization of SiO_2 nanoparticles in the copolymer matrix and they also found that the aggregation of SiO_2 nanoparticles is reduced under ultrasonic irradiation resulting into spherical nanocomposite particles. They also reported on the formation of core–shell nanocomposite by adsorbing co-monomer molecules and polymerization on the surface of SiO_2 particles. Zhang et al. (2007), Zhang et al. (2005), and Xia and Wang (2002) have also used ultrasonic irradiation technique for preparation of SiO_2/poly (3-aminophenylboronic acid) composites, SiO_2/PS nanocomposite, and polyaniline/nanocrystalline TiO_2 composite particles, respectively. They have addressed the issue of aggregation of silica and TiO_2, which is reduced in the presence of ultrasonic irradiation and promote finely dispersed core–shell polymer nanocomposites. They have also reported that the thermal and crystalline properties of core–shell SiO_2/poly(3-aminophenylboronic acid) nanocomposites prepared via ultrasonication method shows significant enhancement in comparison with nanocomposite prepared via conventional method. Further, the fine dispersion of TiO_2 in polyaniline in the presence of ultrasonic irradiation leads to an increase in the conductivity of the nanocomposite compared with conventional stirring.

Sonawane et al. (2010) and Bhanvase and Sonawane (2010) have studied the ultrasound-assisted preparation of ZnO-polymer (ZnO-poly(butyl methacrylate) and ZnO-poly(butyl methacrylate)-polyaniline) and polyaniline (PANI)/$CaCO_3$ nanocomposite for the improvement in anticorrosion properties. It has been reported that the anticorrosion performance of the pigments used in the study increases, when ultrasound-assisted in situ emulsion polymerization technique is used for the synthesis of polymer nanocomposites. It is attributed to the fine dispersion of nanosized pigments in polymer matrix due to micro-mixing caused by ultrasound, which produces fine emulsion droplets, each containing nanosized pigment. Sonawane et al. (2010) have also prepared metal oxide-loaded poly(butylmethacrylate) films by using ultrasound irradiation and prepared composites used as photoanodes in photoelectrochemical cells. It is reported that photocurrents generated were directly proportional to the amount of metal oxide loading, whereas a decrease in photocurrent was observed with an increase in the thickness of the latex film. Similar attempts have also been made for the preparation of polymer-metal nanocomposites by several researchers using ultrasound polymerization (Sivakumar and Gedanken, 2005; Feng et al., 2006; Xu et al., 2007; Barkade et al., 2011). It has been found that an innovative and novel sonochemical method facilitates the reaction by the generation of free radicals during the reaction and also produces finely dispersed polymer-metal nanocomposites compared with conventional methods. Ultrasonic irradiation facilitates compatibility of metal/inorganic particle attachments to conducting polymers when compared with conventional stirring methods during the preparation of nanocomposites

(Wang et al., 2003; Park et al., 2005; Xu et al., 2005; Lu et al., 2006; Mosqueda et al., 2006; Kowsari and Faraghi, 2010; Hosseini et al., 2011). It is also true that ultrasound treatment emerges to be more effective than conventional methods, such as magnetic stirring, to produce more homogeneous materials containing a large amount of the metal oxide particles of smaller size. Further, studies on the preparation of polymer-clay nanocomposites (Borthakur et al., 2010) and polymer-carbon nanotubes (CNTs) (Xia et al., 2003; Ginic-Markovic et al., 2006; Xia et al., 2006; Li et al., 2009; Cheng et al., 2010; Lahelin et al., 2011) have also been carried out by several researchers. It has been reported that sonication contributes to the improvement of the encapsulation rate of CNTs in the polymer matrix. The reported reasons are as follows: (i) ultrasonic cavitation can break the aggregates of CNTs and effectively disperse the nanotubes in solution, (ii) the use of ultrasound for polymer synthesis means that no chemical initiators are required, hence the reaction can be conducted at ambient temperatures. At this lower temperature, the adsorption capability of monomer is higher than a high temperature in conventional emulsion polymerization, (iii) ultrasonic cavitation creates fine emulsion of nanometer-sized monomer droplets of size in nanometers, which increases the chance of collision between monomer droplet and CNTs, and (iv) the radicals are preferably formed on the surface of CNTs, which leads to improved encapsulation of CNTs in polymer latex.

10.3 Specific Examples of Nanocomposite Synthesis via Ultrasound-Assisted Polymerization and Their Applications

10.3.1 Anticorrosive ZnO-Encapsulated Functional Nanolatex

Sonawane et al. (2010) have carried out the synthesis of PBMA, PBMA-ZnO, and PBMA-ZnO-PANI nanocomposites by a hydrothermal-sonochemical emulsion polymerization of butyl methacrylate using sodium dodecyl sulfate (SDS) as a surfactant.

Argon gas was passed over the surface of the liquid mixture to avoid contact of air with solution during the polymerization reaction. The sonochemical polymerization reaction was completed in less than 40 min at a reaction temperature of 20°C. Samples were tested at regular intervals to monitor the percentage conversion by gravimetric analysis. Myristic acid capped ZnO particles prepared in the presence of ethanol with ultrasound irradiation were used for such encapsulation experiment. PANI was modified using dodecylbenzenesulfonic acid (DBSA) and HCl in ethanol and then dried. A 0.005 g sample of treated PANI was mixed with 1 mL of BMA and then added to the ultrasound reactor. Figure 10.6 shows the conversion percentage of the monomer into polymer for PBMA, PBMA-ZnO, and PBMA-ZnO-PANI (Sonawane et al., 2010). For neat PBMA, the reported conversion was 77%, whereas in the presence of ZnO, the conversion was reduced slightly to 66% (Sonawane et al., 2010). The presence of inorganic particles and organic solvents was found to decrease the rate of sonochemical polymerization. Polymer/ZnO nanocomposites were analyzed by using transmission electron microscope and thermogravimetric analysis. Figure 10.7(a) depicts the cryo-TEM image of latex particles embedded with ZnO with size less than 50 nm (Sonawane et al., 2010). It has been predicted that the latex particles are densely packed with ZnO particles. The reported average particle size of ZnO nanoparticles is about 15 nm (Sonawane et al., 2010) (Fig. 10.7b). The thermogravimetric analysis demonstrates that water-based ZnO-PBMA latex is thermally stable at high temperatures and hence is suitable for use in high-temperature applications. The electrochemical measurements were carried out by using a standard three-electrode assembly consisting of a working electrode (ZnO-PBMA latex), an Ag/AgCl reference electrode, and a platinum counter electrode. They have used steel strips for corrosion testing. The Tafel plots were recorded at a scan rate of 2 mV/s at 25°C. Figure 10.8 reports a Tafel plot for bare steel sample and the samples coated with ZnO-PBMA and PBMA latex in 3.5 wt% HCl solution. In corrosion processes, as with any redox system, cathodic and anodic reactions occur simultaneously. The Tafel plot [E as a function of $\log(|I|)$] can isolate these two processes. Estimated values of corrosion current, I_{corr}, and corrosion potential, E_{corr}, are listed in Table 10.1 for different coatings used (Sonawane et al., 2010). On the basis of this analysis, it was reported that in HCl solution, the sample without coating (bare sample) shows the maximum negative corrosion potential (Sonawane et al., 2010). The bare sample shows a corrosion potential of about −0.68 V, while PBMA and ZnO-PBMA coated samples have potentials of −0.49 V and −0.48 V, respectively, as shown in Table 10.1 (Sonawane et al., 2010). No difference in the corrosion potential was reported for the bare steel sample, PBMA coating, and ZnO-PBMA coating in NaCl solution. This indicates that the ZnO-PBMA coating does not show any anticorrosion effect in salt solution. The coating containing PBMA-ZnO gives a self-healing effect and shows a slight shift toward the cathodic region compared with the PBMA coating. The lowest corrosion current density of 3.2 μA/cm^2 was reported for the sample coated with ZnO-PBMA, whereas the highest corrosion current (31 μA) was reported for the PBMA-PANI coating. Also, ZnO-PBMA-PANI shows a corrosion current density of 30 μA/cm^2, and the PANI coating shows a 12 μA/cm^2 current density in HCl solution. This is an indication of poor performance for corrosion inhibition of PANI coatings. Form these results, it can be predicted that the PBMA-PANI and PANI coatings show relatively higher currents in both HCl and NaOH. The possible reason may be the reaction of PANI with the electrolyte solution and thereby generating a higher current (Sonawane et al., 2010).

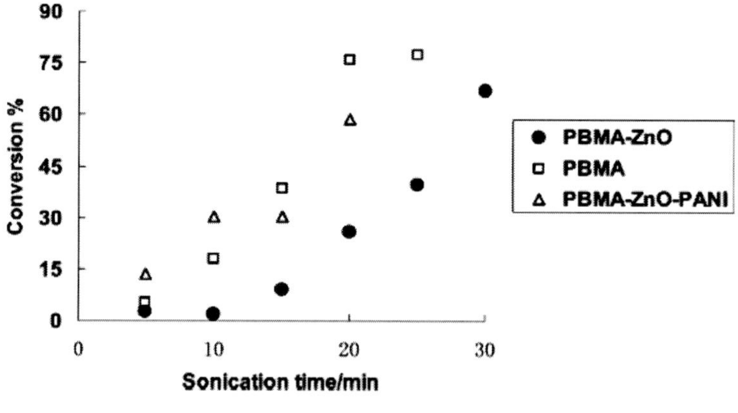

Figure 10.6 Extent of monomer conversion as a function of sonication time for PBMA alone, PBMA-ZnO, and PBMA-ZnO-PANI (Sonawane et al., 2010).

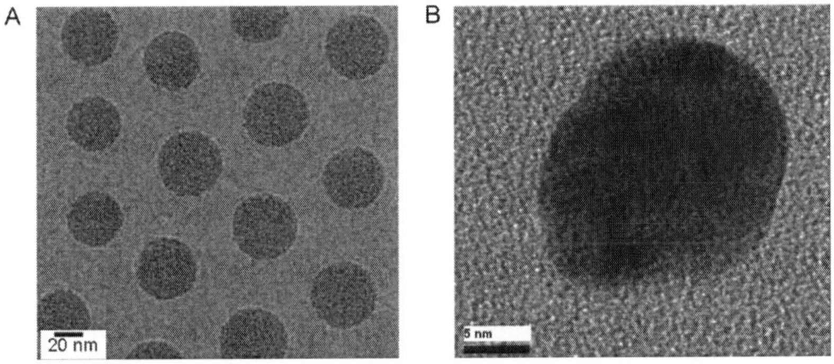

Figure 10.7 (a) Cryo-TEM image of nano PBMA-ZnO latex particles synthesized using sonochemical emulsion polymerization (100000× magnification). (b) TEM image of ZnO nanoparticle synthesized using ultrasound-assisted hydrothermal technique (Sonawane et al., 2010).

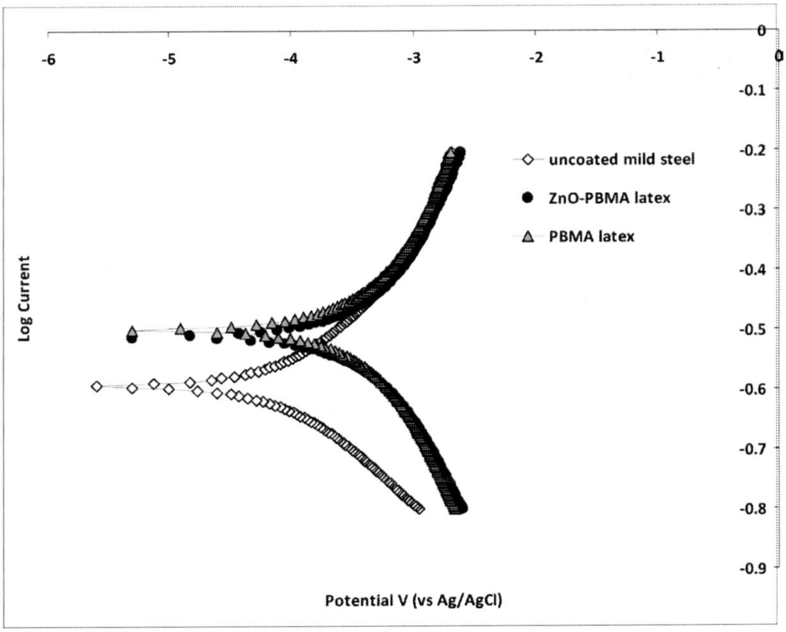

Figure 10.8 Tafel plots of mild steel samples coated with different coatings in 3.5 wt% HCl solution (Sonawane et al., 2010).

Table 10.1 Corrosion parameters of steel samples coated with different coatings in 3.5 wt% HCl solution

Sample	I_{corr} (μA/cm^2)	$-E_{corr}$ (V)
Uncoated mild steel	4.0	0.68
PBMA latex	3.7	0.49
ZnO-PBMA latex	3.2	0.48
ZnO-PBMA latex-PANI	30	0.60
PBMA-PANI	31	0.58
PANI	27	0.41

Source: Sonawane et al., 2010.

Corrosion rate (R_c) of the samples was estimated by using the following equation:

$$R_c = \frac{\Delta W_1}{Atd} \tag{10.1}$$

where ΔW_1 is the weight loss in grams, A is the exposed area of the sample in cm^2, t is the time of exposure in years, and d is the density of the metallic species in g/cm^3. It has been reported that the PBMA-ZnO coating showed the lowest corrosion rate (0.20 mm/year) in comparison to the bare sample (0.62 mm/year). It was also reported that in HCl solution, PBMA-ZnO showed 67% efficiency, which is the highest in comparison to other coatings. In NaCl solution, the sample coated with PBMA-ZnO-PANI latex showed the lowest corrosion rate (0.03 mm/year) and the highest inhibition efficiency (83%). In the case of NaOH solution, the PBMA coating showed the lowest corrosion rate (0.18 mm/year), while the inhibition efficiency for the PBMA-ZnO coating was found to be 61%. The performance of the ZnO-PBMA coating was reported to be better in HCl solution due to the better dispersion of ZnO in the latex facilitating the formation of a well-adhered, dense, and continuous network-like structure, which retards the penetration of the corrosive ions through the metal substrate and inhibits the metal from attack by corrosive species. Hence, the PBMA-ZnO coating acts as an excellent inhibitor and protects metals from corrosion. It has been reported that the iron passivation takes place due to the presence of ZnO layer and the O_2 barrier increases due to the thick polymer matrix. Finally, Sonawane et al. (2010) have reported the possible mechanism for corrosion inhibition as follows:

Generation of electrons from the substrate:

$$Fe \rightarrow Fe^{2+} + 2e^- \quad \text{(R.10.3)}$$

Generation of electrons from the iron substrate:

$$Fe^{2+} \rightarrow Fe^{3+} + e^- \quad \text{(R.10.4)}$$

Neutralization at ZnO-PBMA coating:

$$O_2(g) + 2H_2O + 4e^- \rightarrow 4OH^- \quad \text{(R.10.5)}$$

Combination of electrons in pinholes of PBMA-ZnO coating:

$$2Fe^{2+}(aq) + O_2(g) + 2H_2O \rightarrow 2FeOOH + 2H^+ \quad \text{(R.10.6)}$$

10.3.2 Anticorrosive PANI-$CaCO_3$ Nanocomposites Using Indirect Ultrasound Technique

Bhanvase and Sonawane (2010) have reported the synthesis of PANI/$CaCO_3$ nanocomposites by semi-batch in situ emulsion polymerization by indirect ultrasound technique (ultrasonication in bath) by using distilled aniline as a monomer, ammonium persulphate as an initiator, and sodium lauryl sulphate as a surfactant. Xylene and alkyd resin (soya alkyd semi-drying type) was used during the preparation of PANI/$CaCO_3$/alkyd coatings. Reaction was carried out in semi-batch mode (continuous addition of aniline during the course of reaction) in 90 min at 4°C. All the experiments were carried out in an ultrasonic bath (Sonics and Materials, 20 kHz, 600 W) so as to enhance the reaction rate and micromixing of the reaction mass. Myristic acid coated nano $CaCO_3$ percentage was varied from 2% to 8% of monomer quantity. XRD shows that the PANI/$CaCO_3$ nanocomposites exhibit a semi-crystalline nature. It has also been reported that most of the $CaCO_3$ particles are finely dispersed in the PANI matrix. The possible reasons are due to the hydrophobic nature and micro-mixing caused by ultrasound produce fine $CaCO_3$ embedded emulsion droplets, which leads to the formation of finely dispersed PANI/$CaCO_3$ nanocomposites. Fine dispersion of myristic acid treated $CaCO_3$ in PANI/$CaCO_3$ nanocomposites has also been confirmed by TEM and EDAX analysis (Bhanvase and Sonawane,

2010) (Fig. 10.9). The reported anticorrosion properties of PANI/$CaCO_3$ nanocomposites have been tested by using electrochemical corrosion tests at room temperature (25°C). The Tafel plot (log |*I*| vs. *E*) was carried out in 5% NaCl solution as an electrolyte. The corrosion rates of pure alkyd, PANI/alkyd, and PANI/$CaCO_3$/alkyd coatings were monitored for a period of 200 h by using Eq. 10.1. The thickness of the coating film on mild steel (MS) plate was maintained close to 50 μm. Rate of corrosion is reported in Fig. 10.10. Pure alkyd coatings disappear when placed in different corrosive media, that is, NaOH, HCl, and NaCl, exhibiting a rapid corrosion rate of the pure alkyd organic coating. Corrosion rate of alkyd coating was found to decrease by increasing the addition of PANI and PANI/$CaCO_3$ nanocomposites in alkyd resin. It has been found that for 1% loading of PANI and PANI/$CaCO_3$ nanocomposites, the reported corrosion rate was 0.7 and 0.42 cm/year, respectively. Further, it gets reduced to 0.35 and 0.1 cm/year, when 5% PANI and PANI/$CaCO_3$ nanocomposites were loaded in alkyd resin, respectively (Bhanvase and Sonawane, 2010). It is also reported that the value of corrosion rate is very similar for 5% PANI/$CaCO_3$ nanocomposite and PANI loading, when coated panels were dipped into NaOH and NaCl solutions (0.25 and 0.21 cm/year for PANI and PANI/$CaCO_3$ nanocomposite in NaOH and close to 0.1 for NaCl solution). Both PANI and PANI/$CaCO_3$ nanocomposites for 5% loading in alkyd resin showed the best performance in the case of NaCl solution. It has been predicted that 5% loading of PANI/$CaCO_3$ nanocomposite in alkyd resin is more effective for decreasing the corrosion rate in acid, alkali, and salt solutions. Phenomena such as removal of film and color change observed for alkyd coatings are minimized with an increase in percentage of PANI and PANI/$CaCO_3$ nanocomposites loading. It means that the loading of PANI/$CaCO_3$ nanocomposite or neat PANI can improve the adhesion onto the metal substrate. In case of PANI/alkyd, a compact iron/dopant complex layer formation at the metal-coating interface acts as a passive protective layer. PANI has redox capability to undergo a continuous charge transfer reaction at the metal-coating interface, in which PANI is reduced from emeraldine salt (ES) form to an emeraldine base (EB) (Jose et al., 2005). The protective behavior depends on the size and charge of the dopant, that is, as the size of the dopant increases, the strength of the iron/dopant complex film increases, which improves the protective efficiency (Nakai et al., 2004; Alam et al., 2008; Jiang and Ai, 2008). The corrosion rate of PANI/$CaCO_3$ nanocomposite/alkyd was found to decrease appreciably with an increase in the PANI/$CaCO_3$ nanocomposites loading in alkyd (Fig. 10.10a). In the case of PANI/$CaCO_3$/alkyd coatings, the inhibition effect of the nanocomposite coatings can be attributed to the presence of $CaCO_3$ particles. It has been reported that the small pore size and uniform dispersion of the PANI/$CaCO_3$ nanocomposite in alkyd helps in the formation of a well-adhered, dense, and continuous network-like structure, which slows down the penetration of the corrosive ions through the metal substrate and inhibits the MS from the attack of corrosive species (Zarras et al., 2003). Figure 10.10(b) depicts the Tafel plot for neat alkyd resin, PANI/alkyd, and PANI/$CaCO_3$/alkyd coating. It has been found that electrochemical current (I_{corr}) gets decreased from 0.89 $\mu A/cm^2$ to 0.03 $\mu A/cm^2$, when neat alkyd resin and PANI/$CaCO_3$ nanocomposites were tested in NaCl electrolyte solution. The reported corrosion current for neat alkyd coating was 0.89 $\mu A/cm^2$, while for PANI/alkyd and PANI/$CaCO_3$/alkyd coating, it was 0.50 $\mu A/cm^2$ and 0.03 $\mu A/cm^2$, respectively. Furthermore, E_{corr} value shows shifts in positive direction from −1.74 V to −1.47 V by the addition of PANI-$CaCO_3$ nanocomposites into the alkyd coating. Overall, results indicate that PANI/$CaCO_3$ nanocomposites show an improvement in anticorrosive properties, which supports the corrosion rate data obtained by the dip test method. Furthermore, it has been concluded that the adhesion of coating could be an important factor that controls the corrosion rate. PANI/Alkyd and PANI/$CaCO_3$/Alkyd coating undergo surface crazing phenomenon due to the diffusion of oxygen, acidic, and alkaline ions. Due to the attack of acid, salt, and alkali, metal iron complexes (Fe^{2+} and so on) are formed. Due to alkyd cross-linking and hydrogen bonding of alkyd with PANI/$CaCO_3$ nanocomposites, coating becomes more compact (Chisholm et al., 2005; Yu et al., 2006). Compact cross-linking of alkyd resin and hydrophobic nature of $CaCO_3$ inhibit the moisture infiltration and moisture contact with metal through barrier mechanism, which is inferior in case of neat alkyd coatings. PANI has the ability to store some charges (Fe^{2+} and so on) generated by corrosion mechanism on mild steel panel (passivation of metal). It has been reported that small quantity of PANI acts as an electrochemical inhibitor and electrochemical properties of PANI nanocomposites film provide additional backup for protection (Bhanvase and Sonawane, 2010). Both electrochemical and barrier mechanism communally gives the corrosion protection to mild steel (Abu and Aoki, 2005; Dispenza et al., 2006).

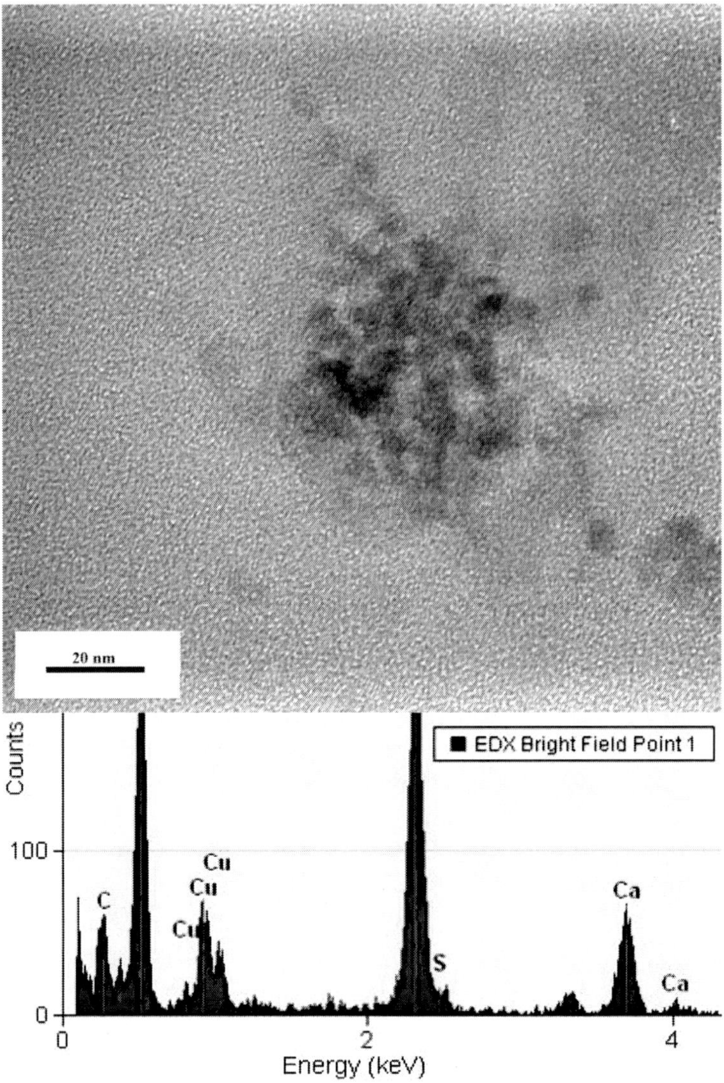

Figure 10.9 EDAX analysis and transmission electron microscopic images of PANI/$CaCO_3$ composite (Bhanvase and Sonawane, 2010).

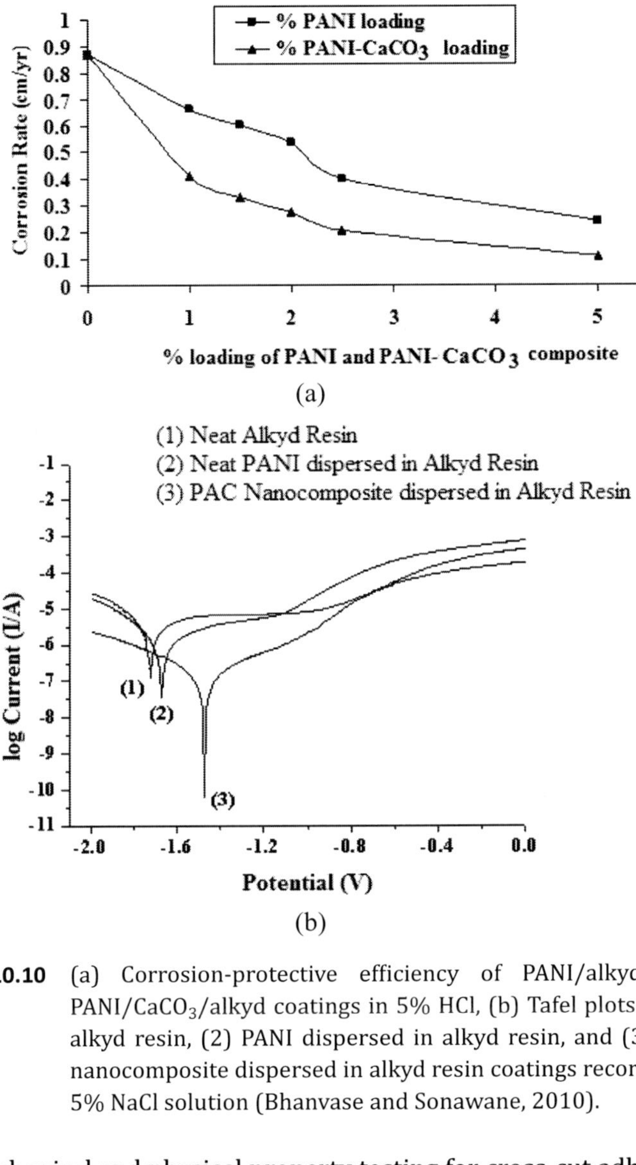

Figure 10.10 (a) Corrosion-protective efficiency of PANI/alkyd and PANI/$CaCO_3$/alkyd coatings in 5% HCl, (b) Tafel plots of (1) alkyd resin, (2) PANI dispersed in alkyd resin, and (3) PAC nanocomposite dispersed in alkyd resin coatings recorded in 5% NaCl solution (Bhanvase and Sonawane, 2010).

Mechanical and physical property testing for cross-cut adhesion (as per ASTM D3359-87, cross-cut adhesion test is used to determine the adhesive property of coating on which it is applied), impact resistance test (ASTM D2794), and gloss at 45^0 (Fig. 10.11) has been evaluated as per American Society for Testing and Materials (ASTM) standards. The adhesion strength of alkyd coating was found to be an important factor in determining the coating performance to control the corrosion. The cross-cut adhesion of pure alkyd resin shows poor adhesion. It has been reported that the cross-cut adhesion increases as the loading of PANI increased from 1% to 5%. The possible reason reported is the enhanced adhesion between the PANI/alkyd coatings with mild steel plate. For aniline, the amino group and the aromatic ring are in the same plane. This coplanar orientation with respect to the metallic surface renders PANI a greater capacity to form more homogeneous films, which, in turn, translates into a better adhesion to the metal substrate. Furthermore, the presence of a long pair of electrons in polyaniline structure enhances the electrostatic interaction between the coatings and the metal substrate, resulting in superior cross-cut adhesion. The cross-cut adhesions of PANI/ $CaCO_3$/alkyd coatings were found to be higher than the PANI/alkyd coatings. The cross-cut adhesion at identical loading was found to be different because of the variation in the morphology of the PANI/ alkyd and PANI/$CaCO_3$/alkyd coatings (Bhanvase and Sonawane, 2010). The presence of myristic acid coated $CaCO_3$ in PANI/ $CaCO_3$ nanocomposites promotes the adhesive property as well as toughness of the coatings. The impact strength of pure alkyd is found to be 60 kg/cm^2 and it is increased to 70 kg/cm^2 by addition of 1% PANI in alkyd resin. The maximum impact strength is found to be 78 kg/cm^2 at 5% PANI loading in alkyd resin. A drastic improvement is reported when 4% of nano $CaCO_3$ is added along with varying percentage of PANI (92–102 kg/cm^2, Table 10.2). The improvement of impact strength from 92 kg/cm^2 to 102 kg/cm^2 indicates that $CaCO_3$ nanoparticles also contribute to impact strength of alkyd resin. An enhancement in impact strength is due to the nanosize of both polymer matrix and $CaCO_3$ particles, which impart positive effect on the alkyd coating. Further, the gloss values increases with % loading of PANI and PANI/$CaCO_3$ nanocomposites, attains optimum at 1.5% loading of PANI or PANI/$CaCO_3$ nanocomposites, and then decreases with the increase in % loading of PANI and PANI/$CaCO_3$ nanocomposite in alkyd resin. As compared with neat alkyd resin, the gloss values are found to increase with % loading of PANI and PANI/$CaCO_3$ nanocomposites in alkyd resin. It has been concluded that the physico-mechanical properties of the PANI/alkyd coatings are found to be significantly enhanced with the loading of the PANI/ $CaCO_3$ nanocomposite in alkyd (Bhanvase and Sonawane, 2010).

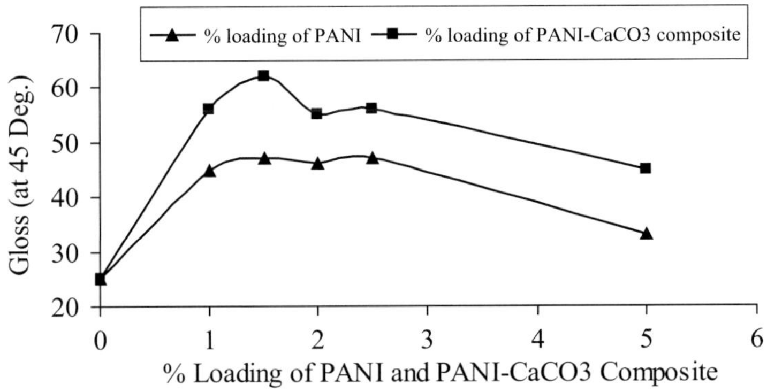

Figure 10.11 Variation of gloss with loading of PANI and PANI/$CaCO_3$ in alkyd (Bhanvase and Sonawane, 2010).

Table 10.2 Physico-mechanical characteristics of PANI/alkyd and PANI/ $CaCO_3$/alkyd coatings

Resin	Cross-cut adhesion ASTM D 3359-87	Impact resistance (kg/cm^2) ASTM D2794
Alkyd	1 mm Fail	60
1.0%PANI/Alkyd	1 mm Pass	70
1.5%PANI/Alkyd	1 mm Pass	71
2.0%PANI/Alkyd	1 mm Pass	73
2.5%PANI/Alkyd	1 mm Pass	74
5.0%PANI/Alkyd	1 mm Pass	78
1.0% PANI/$CaCO_3$/Alkyd	1 mm Pass	92
1.5% PANI/$CaCO_3$/Alkyd	1 mm Pass	95
2.0% PANI/$CaCO_3$/Alkyd	1 mm Pass	97
2.5% PANI/$CaCO_3$/Alkyd	1 mm Pass	98
5.0% PANI/$CaCO_3$/Alkyd	1 mm Pass	102

Source: Bhanvase and Sonawane, 2010.

10.3.3 Preparation of Semiconductor/Polymer Photoanodes for Photoelectrochemical Properties

Semiconductor materials have been used as photoanodes for the conversion of solar energy into electrical energy in photoeletrochemical cells (Sonawane et al., 2010). Devices based on polymer-semiconductor metal oxide materials offer cost-effectiveness for the energy conversion process. Two approaches are used in the preparation of thin film based photoanodes: (i) metal or semiconductor metal oxides are deposited onto a conductive glass at higher temperature using chemical vapor deposition or electrochemical deposition and (ii) semiconductors incorporating conducting polymers, such as polyaniline (PANI)/polypyrrole/porphyrin, are coated onto the nonconducting transparent matrix such as poly(methylmethacrylate) and polyethylene terphthalate. Sonawane et al. (2010) have prepared metal oxide encapsulated polybutyl methacrylate (PBMA) latex by ultrasound-assisted in situ miniemulsion polymerization to make stable photoelectrodes, which may find applications in solar cells. $Bi_2O_3/ZrTiO_4$ oxide particles that were initially encapsulated were treated with myristic acid in the presence of ethanol at 60°C in an ultrasound bath (Sonawane et al., 2010]. It has been reported that the particle sizes, determined by TEM analysis (Fig. 10.12a–c), of TiO_2, ZnO, and $Bi_2O_3/ZrTiO_4$ embedded into PBMA latex were found to be 21 nm, 15 nm, and 7 nm, respectively (Sonawane et al., 2010). Fringes are also observed on the particles in the case of $Bi_2O_3/ZrTiO_4$ (Fig. 10.12d), which indicates that the semiconductor nanocomposite oxides synthesized using ultrasound-assisted methods are crystalline in nature. The reported cryo-TEM images in Fig. 10.13a,b show that the nanosized TiO_2 and ZnO particles are incorporated into the polymer latex, respectively (Sonawane et al., 2010).

Electrochemical measurements have been carried out using a standard three-electrode assembly, consisting of a working electrode (semiconductor oxide-PBMA latex photoanode), a standard calomel electrode, and a platinum counter electrode under UV irradiation. For the photo-anode preparation, the oxide-loaded latex was cast as a film on ITO. Aqueous solutions containing 0.01 M NaOH, KI, and formic acid or oxalic acid have been used as electrolytes. In the photo-

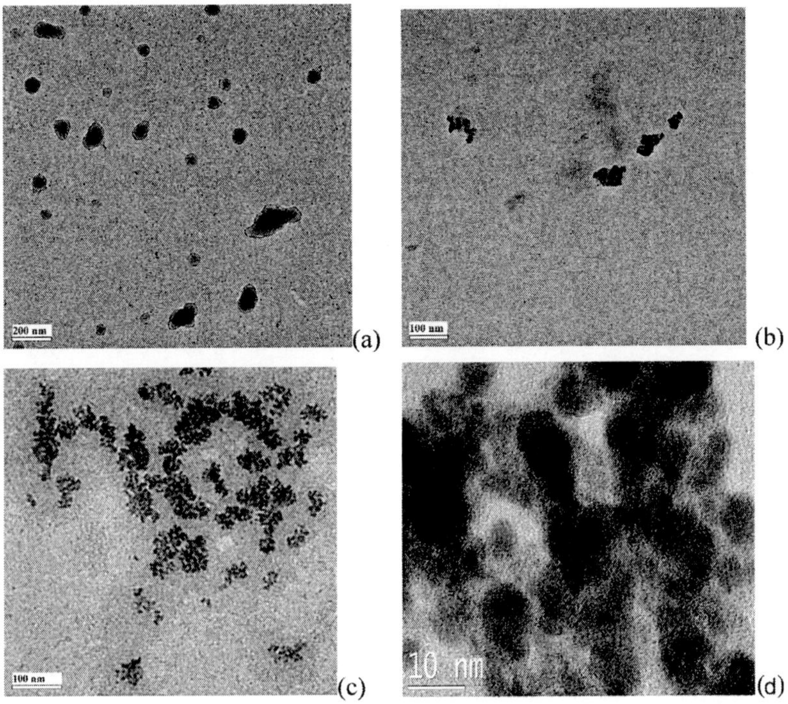

Figure 10.12 TEM image of (a) TiO_2 nanoparticles embedded in the PBMA matrix, (b) ZnO nanoparticles embedded in the PBMA matrix, (c) $Bi_2O_3/ZrTiO_4$ nanoparticles in the PBMA matrix, and (d) $Bi_2O_3/ZrTiO_4$ nanoparticles (Sonawane et al., 2010).

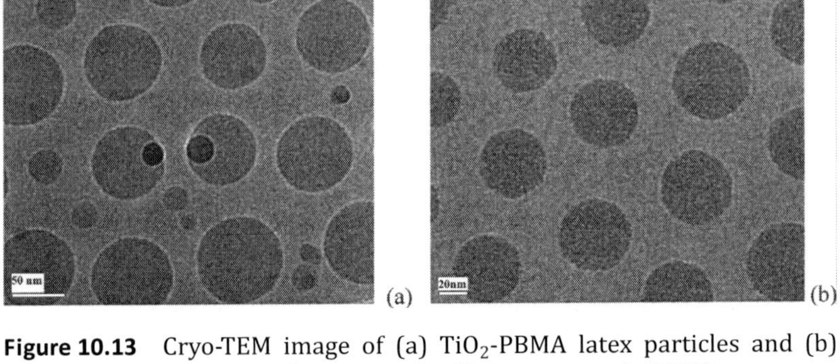

Figure 10.13 Cryo-TEM image of (a) TiO_2-PBMA latex particles and (b) ZnO-PBMA latex particles (Sonawane et al., 2010).

anodes containing the semiconductor-loaded latex particles, the generation of photo-currents can be expected to occur via a similar mechanism as that in a conventional photoelectochemical cell. The advantage of using the polymer latex is that the semiconductor nanoparticles are protected from photo-dissolution, which increases the stability of the photo-anodes against photo-corrosion (Heng et al., 2006). Figure 10.14 shows photocurrent as a function of voltage curves for the ZnO/PBMA system (Sonawane et al., 2010). It can be clearly seen from the photogenerated currents that the electrode is highly photoresponsive. The device exhibited an open-circuit voltage (V_{oc}) of −0.03 V, a short-circuit current (J_{sc}) of 0.8×10^{-4} mA/cm^2, and a fill factor (FF) of 0.20 (FF is defined as the ratio of the maximum power to the product of open-circuit voltage and short-circuit current). It has been reported that the addition of a small amount of PANI increased the conductivity. Photocurrents of 4×10^{-5} and 1×10^{-5} mA/cm^2 have been reported in the presence and absence of polyaniline, respectively, at 0.78 V applied potential (Sonawane et al., 2010). Overall, the photoelectrochemical results of the polymer latex for three different systems (ZnO, TiO_2, and Bi_2O_3/$ZrTiO_4$) in various electrolyte solutions have been summarized in Table 10.3. Sonawane et al. (2010) have tested three latex materials

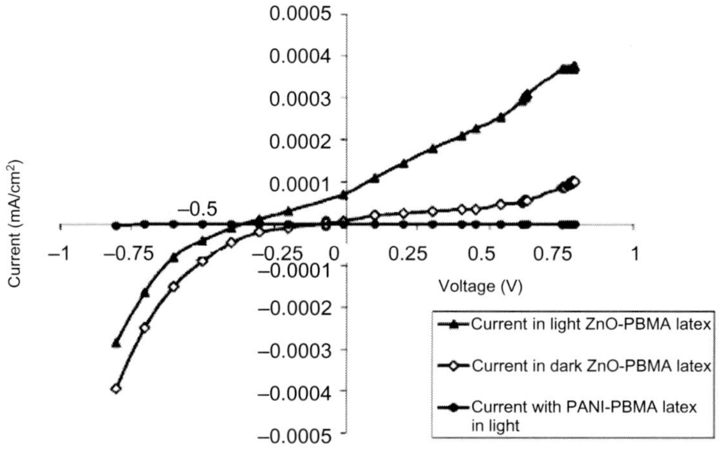

Figure 10.14 Representative example of the photocurrent of ZnO-PBMA latex nanocomposites (0.1 g loading) (Sonawane et al., 2010).

Table 10.3 Summary of the photoelectrochemical properties of the photoelectrodes in different electrolyte solutions

Composition	Electrolyte 0.01 mol/dm^3	I_{Sc} Short circuit photocurrent	I_{max} photocurrent @ 0.798 V (μA)	Open circuit potential (V_{oc}) (Volts)
TiO_2-PBMA 0.1 g	Oxalic acid	0.05	0.06	0.9
	Formic acid	0.06	0.08	0.1
	NaOH	0.09	0.10	−0.05
	KI	0.04	0.05	−0.03
ZnO–PBMA 0.1 g	Oxalic acid	0.005	0.10	−0.26
	Formic acid	0.02	0.3	−0.18
	NaOH	0.1	0.4	−0.31
	KI	0.1	0.4	−0.33
$Bi_2O_3/ZrTiO_4$ PBMA 0.1 g	Oxalic acid	0.1	0.3	−0.18
	Formic acid	0.03	0.6	0.018
	NaOH	0.1	0.4	−0.020
	KI	0.1	0.3	−0.008
TiO_2–PBMA 0.4g	Formic acid	0.04	0.50	−0.34
	Oxalic acid	0.70	0.30	−0.68
	NaOH	0.11	0.11	−0.78
	KI	0.04	0.04	−0.62
ZnO–PBMA 0.4 g	Oxalic acid	0.07	0.14	−0.34
	Formic acid	0.07	0.70	−0.23
	NaOH	0.17	0.27	−0.61
	KI	0.10	0.28	−0.41
$Bi_2O_3/ZrTiO_4$ PBMA 0.4 g	Oxalic acid	0.10	0.26	−0.14
	Formic acid	0.10	0.90	0.02
	NaOH	0.22	0.42	−0.50
	KI	0.11	0.46	−0.20

Source: Sonawane et al., 2010.

with two different loadings (0.1 g and 0.4 g) and four different electrolyte solutions (NaOH, KI, formic acid, and oxalic acid). It has been reported that, at 0.1 g loading, TiO_2 showed a 0.1 μA/cm^2 maximum photocurrent at 0.798 V in NaOH electrolyte and that ZnO latex showed a photocurrent of 0.4 μA/cm^2 at 0.798 V in both KI and NaOH solutions. $Bi_2O_3/ZrTiO_4$ has shown a maximum photocurrent of 0.6 μA/cm^2 at 0.798 V in the presence of formic acid under UV light. Sonawane et al. (2010) have also studied the effect of catalyst loading on the photocurrent. It was found that, in the case of TiO_2-PBMA latex, the short circuit photocurrent was 0.7 μA/cm^2 in the presence of oxalic acid at a 0.4 g loading. At the same level of loading (0.4 g), 0.5 and 0.9μA/cm^2 photocurrents were observed with ZnO-PBMA and $Bi_2O_3/ZrTiO_4$ in the presence of formic acid. Overall, it has been reported that the $Bi_2O_3/ZrTiO_4$ generates the maximum current. It has also been reported that the generation of photocurrent is highly dependent on the encapsulation, particle size of the semiconductor, and their morphology. Further, there has been an effect of electrolyte solution on the performance of the semiconductor oxide and $Bi_2O_3/ZrTiO_4$ shows a better performance in the presence of acid electrolytes. Finally, it has been concluded that the generation of photocurrent increases with an increase in the semiconductor oxide loading, whereas a reverse effect was observed with an increase in the thickness of the latex. $Bi_2O_3/ZrTiO_4$ showed photocurrents of 0.6 μA/cm^2 and 0.9 μA/cm^2 at 0.1 g and 0.4 g loadings, respectively, in formic acid electrolyte. ZnO showed photocurrents of 0.4 μA/cm^2 and 0.6 μA/cm^2 in NaOH and KI, respectively, at a 0.1 g loading. The order of photocurrent generation efficiency with the different catalysts was $Bi_2O_3/ZrTiO_4$>ZnO> TiO_2 (Sonawane et al., 2010).

10.3.4 Synthesis of Polyaniline/Ag Nanocomposite for Ethanol Vapor Sensing Application

Barkade et al. (2011) have attempted ultrasound-assisted in situ miniemulsion polymerization of aniline along with different loading of silver nanoparticles and have studied its application for ethanol vapor sensing. Reaction of ultrasound-assisted synthesis of polyaniline/Ag nanocomposite synthesis by miniemulsion polymerization is reported in Fig. 10.15 (Barkade et al., 2011).

Ultrasound-assisted in situ miniemulsion polymerization of aniline in the presence of different loading of silver nanoparticles has been carried out by using the following procedure. Separately prepared surfactant solution by adding sodium dedocyl sulfate in water and initiator solution by adding ammonium persulfate in water was transferred to the semi-batch sonochemical reactor (sonochemical reactor in which continuous addition of monomer has been made). Around 1 M aqueous solution of hydrochloric acid (dopant) has been added into the reactor for protonation. Monomer (aniline) was added into the reactor in a semibatch manner within 60 min to control the reaction rate at 4°C. Schematic representation of ultrasound-assisted in situ emulsion polymer nanocomposite formation process is shown in Fig. 10.15. When ultrasound waves pass through the emulsion polymerization reaction medium, the formation of cavities occur, which contains vapor of the liquid reaction medium. During the collapse, these vapors will be subjected to extreme conditions of high temperatures and pressures, causing molecules to fragment and generate highly reactive radical species. Because of this rigorous environment decomposition of water, surfactant, monomers, and oligomers take place to generate radicals that help in propagating polymerization. Because of ultrasonic irradiation, the size of synthesized PANI/Ag nanocomposite is 50 nm obtained by transmission electron microscopic analysis (Fig. 10.16). It has been reported that Ag nanoparticles (5–10 nm) are finely dispersed into polyaniline matrix without any aggregation (Barkade et al., 2011). This uniform dispersion also attributed to strong affinity of silver for nitrogen. The electrical conductivity of nanocomposite is affected by the adsorption of vapor molecules on its surface. Change in resistance of polyaniline/Ag and polyaniline film sensors was characterized at room temperature in air. The sensor response (S) is defined as,

$$S = \frac{(R_g - R_a)}{R_a} \tag{10.2}$$

where R_g and R_a are the resistances of the sensitive film in measuring ethanol vapor and in clean air, respectively. The initial resistance of pure polyaniline and that of nanocomposite before exposure to ethanol vapors were 240 and 150 Ω, respectively (Barkade et al., 2011).

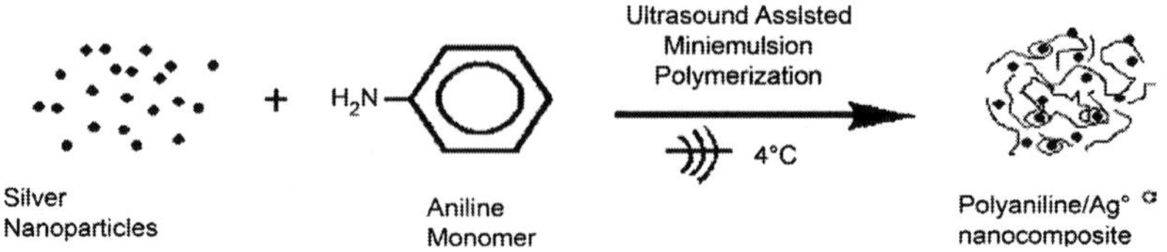

Figure 10.15 Ultrasound-assisted synthesis of polyaniline/Ag nanocomposite (Barkade et al., 2011).

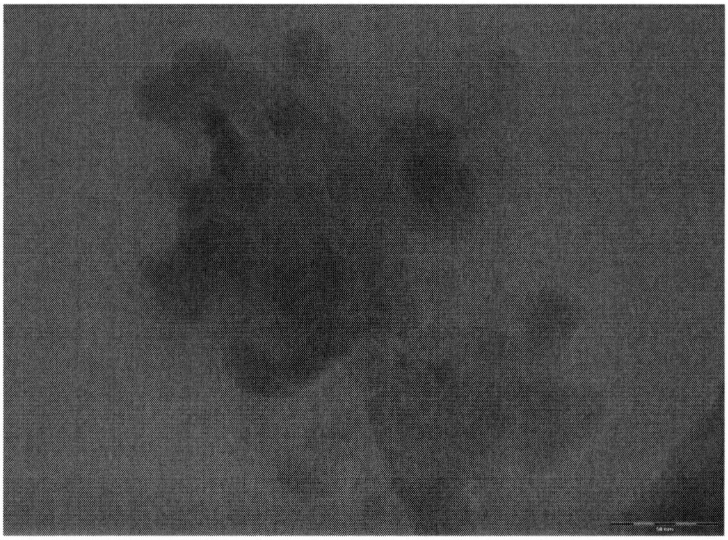

Figure 10.16 TEM image of PANI-Ag nanocomposite synthesized by ultrasound-assisted emulsion polymerization (Barkade et al., 2011).

From Fig. 10.17(a), it has been found that the nanocomposite exhibits a linear response up to 100 ppm. Further, the change in resistance is found to be independent of concentration (Barkade et al., 2011). An increase in resistance of sensor on exposure to ethanol may originate due to the interaction of –OH groups of ethanol molecules and nitrogen of polyaniline, leading to electron delocalization and charge transport through the polymer chain. In comparison to pure polyaniline, sensor response of polyaniline/Ag nanocomposite shows more stability as well as good reproducibility to ethanol vapors under the same conditions. It has been observed that ethanol vapor sensing ability of the nanocomposite sensor increases with the addition of silver nanoparticles in the polymer matrix. These results show that polyaniline/Ag nanocomposite serves as an efficient sensor compared with pure polyaniline. The improvement in sensitivity of polyaniline/Ag is due to the enhanced degree of interactions between the nanocomposite and the ethanol vapors. Figure 10.17(b) shows steady linear response up to 2100 s at 100 ppm, which on further increase in time leads to saturation of the nanocomposite film. This can be attributed to the decrease in available free volume for vapor permeability into the nanocomposite (Barkade et al., 2011).

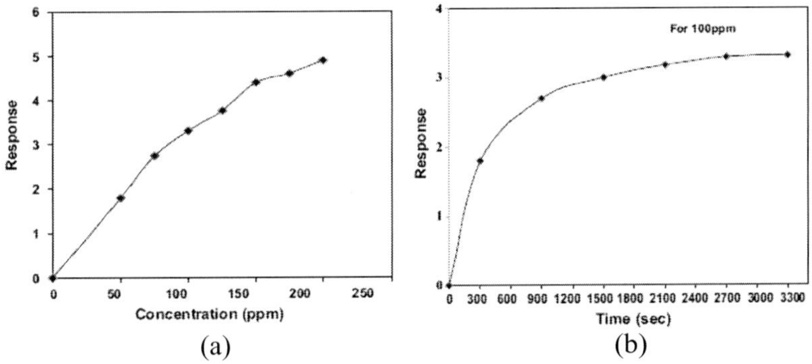

Figure 10.17 Response of polyaniline/Ag film sensor (a) at different ethanol concentrations (50–200 ppm) and (b) to ethanol (100 ppm) (Barkade et al., 2011).

Similarly, the synthesis and sensor application of Ag/polypyrrole nanocomposites prepared by sonochemical method have been studied in which we have postulated that nanoparticles of silver can be first generated via in situ reduction process followed by the use of silver nanoparticles as oxidant for the polymerization of pyrrole. Sensor studies on such thick films showed that it is best suited for ammonia gas sensing (Khanna et al., unpublished work).

10.4 Conclusions

PMMA nanoparticles were successfully synthesized in a reactor operated in a semibatch manner by emulsion polymerization of MMA using high-intensity ultrasound. PMMA particles formed are of nanosize and spherical in nature, which has been confirmed with the help of TEM and particle size analysis. It was found that the particle size decreases with an increase in the initiator concentration (up to 1.5%) and then increases again for higher initiator concentration. ZnO-PBMA and PANI-PBMA nanocomposite latexes were synthesized using a sonochemical emulsion polymerization technique.

The corrosion performance of the PBMA/ZnO coatings synthesized using ultrasound-assisted emulsion polymerization technique was evaluated. The corrosion rate of PBMA/ZnO was found to be the lowest in HCl solution, with an inhibition efficiency of 67%. The functionalized latex coatings were found to be superior in performance to the normal latex coatings. The sonochemical synthesis of semiconductor particles loaded polymer latex is achieved. The photo-electrochemical activities of different semiconductors-loaded latex particles have also been evaluated. It is observed that the generation of photocurrent increases with an increase in the semiconductor oxide loading, whereas reverse effect was observed with an increase in the thickness of the latex. Bi_2O_4/$ZrTiO_4$ showed photocurrents of 0.6 μA and 0.9 μA at 0.1 g and 0.4 g loading, respectively, in the formic acid electrolyte.

PANI/Ag nanocomposite was synthesized by in situ miniemulsion polymerization. This method has been found to be effective for the controlled polymerization. Presence of ultrasound effectively enhances the dispersion ability of silver nanoparticles of size nearly 10 nm in the polymer matrix. Sensor device was fabricated using PANI/Ag films and its sensitivity toward ethanol vapor was recorded. The linear response was obtained for 100 ppm upto 2100 s after which the response is independent of concentration. Sensitivity of fabricated device shows good reproducibility and stability of response. Response time of sensor decreases with an increase in loading of Ag nanoparticles. Sensor studies on Ag/polypyrrole thick films showed that it is best suited for ammonia gas sensing.

References

Abu, Y. M. and Aoki, K. (2005). Corrosion protection by polyaniline-coated latex microspheres, *J. Elec. Chem.*, **583,** pp. 133–139.

Alam, J., Riaz, U., Ashraf, S. M. and Ahmad, S. (2008). Corrosion-protective performance of nano polyaniline/ferrite dispersed alkyd coatings, *J. Coat. Technol. Res.*, **5,** pp. 123–128.

Allen, T. M., and Cullis, P. R. (2004). Drug delivery systems: entering the mainstream, *Science.,* **303**, pp. 1818–1822.

Asua, J. M. (2004). Emulsion polymerization: From fundamental mechanisms to process developments, *J Polym Sci: Part A: Polym. Chem.,* **42**, pp. 1025–1041.

Barkade, S. S., Naik, J. B. and Sonawane, S. H. (2011). Ultrasound assisted miniemulsion synthesis of polyaniline/Ag nanocomposite and its application for ethanol vapor sensing, *Colloid. Surface A,* **378,** pp. 94–98.

Behrend, O. and Schubert, H. (2000). Influence of continuous phase viscosity on emulsification by ultrasound, *Ultrason. Sonochem.,* **7**, pp. 77–85.

Bhanvase, B. A., Pinjari, D. V., Sonawane, S. H., Gogate, P.R. and Pandit A.B. (2012). Analysis of semibatch emulsion polymerization: Role of ultrasound and initiator, *Ultrason. Sonochem.*, **19,** pp. 97–103.

Bhanvase, B. A. and Sonawane, S. H. (2010). New approach for simultaneous enhancement of anticorrosive and mechanical properties of coatings: Application of water repellent nano $CaCO_3$–PANI emulsion nanocomposite in alkyd resin, *Chem. Eng. J.,* **156,** pp. 177–183.

Bisht, A., Feldmann, G., Soni, S., Ravi, R., Karikar, C., Maitra, A. and Maitra, A. (2007). Polymeric nanoparticle-encapsulated curcumin ("nanocurcumin"): a novel strategy for human cancer therapy, *J. Nanobiotechnology*, **5,** pp. 1–18.

Borthakur, L. J., Das, D. and Dolui, S. K. (2010). Development of core–shell nano composite of poly (styrene-co-methyl acrylate) and bentonite clay by ultrasonic assisted mini-emulsion polymerization, *Mater. Chem. Phys.,* **124,** pp. 1182–1187.

Bradley, M. A. and Grieser, F. (2002). Emulsion polymerization synthesis of cationic polymer latex in an ultrasonic field, *J. Coll. Interface Sci.,* **251,** pp. 78-84.

Cao, Y., Zheng, Y. and Pan, G. (2008). Radical generation process studies of the cationic surfactants in ultrasonically irradiated emulsion polymerization. *Ultrason. Sonochem.*, **15,** pp. 320–325.

Chau, H. C. J. and Stoffer, J. O. (1999). Ultrasonically initiated free radical-catalyzed emulsion polymerization of methyl methacrylate (I), *J. Appl. Polym. Sci.,* **72,** pp. 797–825.

Cheng, Z., Pan, Q. and Rempel, G. L. (2010). Modification of multiwall carbon nanotubes via soap-free emulsion polymerization of acrylonitrile, *J. Polym. Sci. Pol. Chem.,* **48,** pp. 2057–2062.

Cheung, H. M. and Gaddam, K. (2000). Ultrasound-assisted emulsion polymerization of methyl methacrylate and styrene, *J. Appl. Polym. Sci.*, **76,** pp. 101–104.

Chisholm, N., Mahfuz, H., Rangari, V. K., Ashfaq, A. and Jeelani, S. (2005). Fabrication and mechanical characterization of carbon/SiC-epoxy nanocompositesm, *Comp. Struct.*, **67,** pp. 115–124.

Diaconu, G., Paulis, M. and Leiza, J. R. (2008). Towards the synthesis of high solids content waterborne poly(methyl methacrylate-co-butyl acrylate)/montmorillonite nanocomposites, *Polym.*, **49,** pp. 2444–2454.

Dispenza, C., Presti, C. L., Belfiore, C., Spadaro, G. and Piazza, S. (2006). Electrically conductive hydrogel composites made of polyaniline nanoparticles and poly(N-vinyl-2-pyrrolidone), *Polym.*, **47,** pp. 961–971.

Elder, S. A. (1959). Cavitation microstreaming, *J. Acoust. Soc. Am.*, **31**, pp. 54–64.

Erdem, B., Sudol, E. D., Dimonie, V. L. and El-Aaser, M. S. (2000). Encapsulation of inorganic particles via miniemulsion polymerization I. Dispersion of titanium dioxide particles in organic media using OLOA 370 as stabilizer, *J. Polym. Sci. Part A Polym. Chem.*, **38**, pp. 4419–4430.

Feng, X., Mao, C., Yang, G., Hou, W. and Zhu, J.-J. (2006). Polyaniline/Au composite hollow spheres: synthesis, characterization, and application to the detection of dopamine, *Langmuir, 22,* pp. 4384–4389.

Ginic-Markovic, M., Matisons, J. G., Cervini, R., Simon, G. P. and Fredericks, P. M. (2006). Synthesis of new polyaniline/nanotube composites using ultrasonically initiated emulsion polymerization, *Chem. Mater.*, **18,** pp. 6258–6265.

Haldorai, Y., Long, P. Q., Noh, S. K., Lyoo, W. S. and Shim, J.-J. (2009). Ultrasonically assisted *in situ* emulsion polymerization: a facile approach to prepare conducting copolymer/silica nanocomposites, *Polym. Adv. Technol.*, **22,** pp. 781–787.

Heng, L., Zhai, J., Zhao, Y., Xu, J., Sheng, X. and Jiang, L. (2006). Enhancement of photocurrent generation by honeycomb structures in organic thin films, *Chem. Phys. Chem.*, **7,** pp. 2520–2525.

Hosseini, S. H., Rahimi, R. and Kerdar, H. (2011). Preparation of a nanocomposite of magnetic, conducting nanoporous polyaniline and hollow manganese ferrite, *Polym. J.*, **43,** pp. 745–750.

Huang, Z., Lin, Z., Cai, Z. and Mai, K. (2004). Physical and mechanical properties of nano-$CaCO_3$/PP composites modified with acrylic acid, *Plast. Rubbers Comp.*, **33,** pp. 343–352.

Jiang, J. (2007). Ultrasonic-assisted synthesis of PMMA/$Ni_{0.5}Zn_{0.5}Fe_2O_4$ nanocomposite in mixed surfactant system, *Eur. Polym. J.*, **43**, pp. 1724–1728

Jiang, J., and Ai, L. H. (2008). Microemulsion-mediated in-situ synthesis and magnetic characterization of polyaniline/$Zn_{0.5}Cu_{0.5}Fe_2O_4$ nanocomposite, *Appl. Phys. A: Mat. Sc. and Eng.*, **92**, pp. 2341–2344.

Jose, E., Pereira, S., Susana, I., Cordoba, T. and Roberto, M. T. (2005). Polyaniline acrylic coatings for corrosion inhibition: the role played by counter-ions, *Corros. Sci.*, **47**, pp. 811–822.

Khanna, P. K., Kate, K. H., Banerjee, S., Jain, G. H. and Sonawane, S. H. Application of Ag nano-particles in ammonia gas sensing via polymerization of pyrrole (unpublished work).

Klein, A. and Daniels, E. S. (1997). *Chapter 6,* eds. Lovell, P. A. and Asserm, E. I. *Emulsion Polymerization and Emulsion Polymers* (Wiley, New York), p. 207.

Kobayashi, D., Matsumoto, H. and Kuroda, C. (2008). Improvement of indirect ultrasonic irradiation method for intensification of emulsion polymerization process, *Chem. Eng. J.*, **135**, pp. 43–48.

Kojima, Y., Koda, S. and Nomura, H. (2001). Effect of ultrasonic frequency on polymerization of styrene under sonication, *Ultrason. Sonochem.*, **8**, pp. 75–79.

Kong, Y. X., Kan, C. and Sun, C. (1999). Encapsulation of calcium carbonate by styrene polymerization, *Polym. Adv. Technol.*, **10**, pp. 54–59.

Kowsari, E. and Faraghi, G. (2010). Ultrasound and ionic-liquid-assisted synthesis and characterization of polyaniline/Y_2O_3 nanocomposite with controlled conductivity, *Ultrason. Sonochem.*, **17**, pp. 718–725.

Lahelin, M., Annala, M., Nykänen, A., Ruokolainen, J. and Seppälä, J. (2011). In situ polymerized nanocomposites: Polystyrene/CNT and Poly(methyl methacrylate)/CNT composites, *Compos. Sci. Technol.*, **71**, pp. 900–907.

Landfester, K., Schork, F. J. and Kusuma, V. A. (2003). Particle size distribution in mini-emulsion polymerization, *C. R. Chimie*, **6**, pp. 1337-1342.

Li, M., Wang, X., Tian, R., Liu, F., Hu, H., Chen, R., Zheng, H. and Wan, L. (2009). Preparation, solubility, and electrorheological properties of carbon nanotubes/poly(methyl methacrylate) nanocomposites by *in situ* functionalization, *Compos. Part A-Appl. S.*, **40**, pp. 413–417.

Liu, P., Liu, W. M. and Xue, Q. J. (2004). *In situ* radical transfer addition polymerization of styrene from silica nanoparticles, *Eur. Polym. J.*, **40**, pp. 267–271.

Liu, S., Ying, J., Zhou, X. and Xie, X. (2009). Core-shell magnesium hydroxide/polystyrene hybrid nanoparticles prepared by ultrasonic wave-assisted in-situ copolymerization, *Mater. Lett.,* **63,** pp. 911–913.

Liu, Y. Y. and Yang, M. J. (2003). Preparation of MMA-BA-DMAEMA nanosized latex, *Chinese Chem. Lett.*, **14,** pp. 863–865.

Lorimer, P., Mason, T. J. and Kershaw, D. (1991). The effect of ultrasound on the encapsulation of titanium dioxide pigment, *Colloid. Polym. Sci.*, **269,** pp. 392–397.

Lu, X. Mao, H., Chao, D., Zhang, W. and Wei, Y. (2006). Ultrasonic synthesis of polyaniline nanotubes containing Fe_3O_4 nanoparticles, *J. Solid. State. Chem.,* **179,** pp. 2609–2615.

Ma, X. F,; Wang, M., Li, G., Chen, H. Z. and Bai, R. (2006). Preparation of polyaniline–TiO_2 composite film with *in situ* polymerization approach and its gas-sensitivity at room temperature, *Mat. Chem. Phys.,* **98,** pp. 241–247.

Mason, T. J. (1997). Ultrasound in synthetic organic chemistry, *Chem. Soc. Rev.,* **26**, pp. 443–451.

Mason, T. J. and Cordemans de Meulenaer, E. (1998) Practical considerations for process optimisation, eds. Luche, J-L., and Bianchi, C., *Synthetic Organic Sonochemistry* (Plenum Press), pp. 301–328.

Mason, T. J. and Lorimer, P. J. (1988) *Sonochemistry: Theory, Applications and Uses of Ultrasound in Chemistry* (Ellis Horwood: Chichester).

Matusinovic, Z., Rogosic, M. and Sipusic J. (2009). Synthesis and characterization of poly(styrene-co-methyl methacrylate)/layered double hydroxide nanocomposites via in situ polymerization, *Polym. Degrad. Stab.,* **94,** pp. 95–101.

Mosqueda, Y., Perez-Cappe, E., Arana, J., Longo, E., Ries, A., Cilense, M., Nascente, P. A. P., Aranda, P. and Ruiz-Hitzky, E. (2006). Preparation and characterization of $LiNi_{0.8}Co_{0.2}O_2$/PANI microcomposite electrode materials under assisted ultrasonic irradiation, *J. Solid. State. Chem.,* **179,** pp. 308–314.

Nakai, G., Fukuda, T. and Hosoi, K. (2004). Surface-coated calcium carbonate particles, method for manufacturing same, and adhesive, U. S. Patent 6,686,044.

Ni, Y., Hao, H., Cao, X., Su, S., Zhang, Y. and Wei, X. (2006). Preparation, characterization, and optical, electrochemical property research of CdS/PAM nanocomposites, *J. Phys. Chem. B.* **110**, pp. 17347–17352.

Nie, M., Wang, Q. and Qiu, G. (2008). Enhancement of ultrasonically initiated emulsion polymerization rate using aliphatic alcohols as hydroxyl radical scavengers, *Ultrason. Sonochem.*, **15,** pp. 222–226.

Odian, G. (2004) *Principles of Polymerization,* 4th ed. (John Wiley & Sons, Inc.: New Jersey).

Park, J.-E., Atobe, M. and Fuchigami, T. (2005). Sonochemical synthesis of conducting polymer–metal nanoparticles nanocomposite, *Electrochim. Acta.,* **51,** pp. 849–854.

Patel, H. A., Somani, R. S., Bajaj, H. C. and Jasra, R. V. (2006). Nanoclays for polymer nanocomposites, paints, inks, greases and cosmetics formulations, drug delivery vehicle and waste water treatment, *Bull. Mater. Sci.,* **29,** pp. 133–145.

Paul, D. R. and Robeson, L. M. P. (2008). Polymer nanotechnology: nanocomposites, *Polym.,* **49,** pp. 3187–3204.

Pothukuchi, S., Li, Y. and Wong, C. P. (2004). Development of a novel polymer–metal nanocomposite obtained through the route of *in situ* reduction for integral capacitor application, *J. Appl. Polym. Sci.,* **93,** pp. 1531–1538.

Price, G. J. (2003). Recent developments in sonochemical polymerization, *Ultrason. Sonochem.,* **10,** pp. 277–283.

Qi, D. M., Bao, Y. Z., Weng, Z. X. and Huang, Z. M. (2006). Preparation of acrylate polymer/silica nanocomposite particles with high silica encapsulation efficiency via miniemulsion polymerization, *Polym.,* **47,** pp. 4622–4629.

Ryu, J. G., Park, S. W., Kim, H. and Lee J. W. (2004). Power ultrasound effects for *in situ* compatibilization of polymer–clay nanocomposites, *Mater. Sci. Eng. C,* **24,** pp. 285–288.

Siegel, R. W. (1995). Nanostructured materials-mind over matter, *Nanost. Mat.,* **3,** pp. 1–18.

Sinha, S. R. and Okamoto, M. (2003). Polymer/layered silicate nanocomposites: a review from preparation to processing, *Prog. Polym. Sci.,* **28,** pp. 1539–1641.

Sivakumar, M., and Gedanken, A. (2005). A sonochemical method for the synthesis of polyaniline and Au–polyaniline composites using H_2O_2 for enhancing rate and yield, *Synthetic Met.,* **148,** pp. 301–306.

Sonawane, S. H., Neppolian, B., Teo, B. M., Grieser, F. and Ashokkumar, M. (2010). Ultrasound-assisted preparation of semiconductor/polymer photoanodes and their photoelectrochemical properties, *J. Phys. Chem. C,* **114,** pp. 5148–5153.

Sonawane, S. H., Teo, B. M., Brotchie, A., Grieser, F. and Ashokkumar, M. (2010). Sonochemical synthesis of ZnO encapsulated functional nanolatex and its anticorrosive performance, *Ind. Eng. Chem. Res.,* **49,** pp. 2200–2205.

Soares, B. G., Amorim, G. S., Souza, F. G., Oliveira, M. G. and Pereira da Silva, J. E. (2006). The *in situ* polymerization of aniline in nitrile rubber, *Synth. Metals.,* **156,** pp. 91–98.

Teo, B. M., Ashokkumar, M. and Grieser, F. (2008). Microemulsion polymerizations via high-frequency ultrasound irradiation, *J. Phys. Chem. B,* **112,** pp. 5265–5267.

Teo, B. M., Ashokkumar, M. and Grieser, F. (2011). Sonochemical polymerization of miniemulsions in organic liquids/water mixtures, *Phys. Chem. Chem. Phys.,* **13,** pp. 4095–4102.

Teo, B. M., Grieser, F. and Ashokkumar, M. (2009). High intensity ultrasound initiated polymerization of butyl methacrylate in mini- and microemulsions, *Macromol.,* **42,** pp. 4479–4483.

Teo, B. M., Prescott, S. W., Ashokkumar, M. and Grieser, F. (2008). Ultrasound initiated miniemulsion polymerization of methacrylate monomers, *Ultrason. Sonochem.,* **15,** pp. 89–94.

Tiarks, F., Landfester, K. and Antonietti, M. (2001). Silica nanoparticles as surfactants and fillers for latexes made by miniemulsion polymerization, *Langmuir.,* **17,** pp. 5775–5780.

Tissot, I., Novat, C., Lefebvre, F. and Bourgeat-Lami, E. (2001). Hybrid latex particles coated with silica, *Macromol.* **34,** pp. 5737–5739.

Wang, J., Hu, Y., Tang, Y. and Chen, Z. (2003). Preparation of nanocomposite of polyaniline and gamma-zirconium phosphate (γ-ZrP) by power ultrasonic irradiation, *Mater. Res. Bull.,* **38,** pp. 1301–1313.

Wang, R. and Cheung, H. M. (2005). Ultrasound assisted polymerization of MMA and styrene in near critical CO_2, *J. Supercrit. Fluids.,* **33,** pp. 269–274.

Wang, R.-M., Lv, W.-H., He, Y.-F., Wang, Y. and Guo J.-F. (2010). An emulsifier-free core–shell polyacrylate/diacetone acrylamide emulsion with nano-SiO_2 for room temperature curable waterborne coatings, *Polym. Adv. Technol.,* pp. 128–134.

www.hielscher.com/ultrasonics/nano_01.htm

Xia, H., Qiu, G. and Wang, Q. (2006). Polymer/carbon nanotube composite emulsion prepared through ultrasonically assisted *in situ* emulsion polymerization, *J. Appl. Polym. Sci.,* **100,** pp. 3123–3130.

Xia, H. and Wang, Q. (2001). Synthesis and characterization of conductive polyaniline nanoparticles through ultrasonic assisted inverse microemulsion polymerization, *J. Nanopart. Res.,* **3,** pp. 399–409.

Xia, H. and Wang, Q. (2002). Ultrasonic irradiation: a novel approach to prepare conductive polyaniline/nanocrystalline titanium oxide composites, *Chem. Mater.,* **14,** pp. 2158–2165.

Xia, H., Wang, Q., Liao, Y., Xu, X., Baxter, S. M., Slone, R. V., Wu, S., Swift, G. and Westmoreland, D. G. (2002). Polymerization rate and mechanism of ultrasonically initiated emulsion polymerization of n-butyl acrylate, *Ultrason. Sonochem.*, **9**, pp. 151–158.

Xia, H., Wang, Q. and Qiu, G. (2003). Polymer-encapsulated carbon nanotubes prepared through ultrasonically initiated in-situ emulsion polymerization, *Chem. Mater.*, **15**, pp. 3879–3886.

Xu, G. C., Xiong, J. Y., Ji, X. L. and Wang, Y. L. (2007). Synthesis of nanosilver/PMMA composites via ultrasonically bi-*in situ* emulsion polymerization, *J. Thermoplast. Compos.*, **20**, pp. 523–533.

Xu, J., Li, X., Liu, J., Wang, X., Peng, Q. and Li, Y. (2005). Solution route to inorganic nanobelt-conducting organic polymer core-shell nanocomposites, *J. Polym. Sci. Part. A: Polym. Chem.*, **43**, pp. 2892–2900.

Xu, M., Choi, Y. S., Kim, Y. K., Wang, K. H. and Chung I. J. (2003). Synthesis and characterization of exfoliated poly(styrene-co-methyl methacrylate)/clay nanocomposites via emulsion polymerization with AMPS, *Polymer*, **44**, pp. 6387–6395.

Yin, N. and Chen, K. (2004). Ultrasonically initiated emulsifier-free emulsion copolymerization of n-butyl acrylate and acrylamide. Part I: Polymerization mechanism, *Polymer*, **45**, pp. 3587–3594.

Yin, N., Chen, K. and Kang, W. (2006). Preparation of BA/ST/AM nano particles by ultrasonic emulsifier-free emulsion polymerization, *Ultrason. Sonochem.*, **13**, pp. 345–351.

Yu, H. J., Wang, L., Shi, Q., Jiang, G. H., Zhao, Z. R. and Dong, X. C. (2006). Study on nano-$CaCO_3$ modified epoxy powder coatings, *Prog. Org. Coat.*, **55**, pp. 296–300.

Zarras, P., Anderson, N., Webber, C., Irvin, D. J., Guenthner, A. and Stengersmith, J. D. (2003). Progress in using conductive polymers as corrosion-inhibiting coatings, *Rad. Phy. Chem.*, **68**, pp. 387–394.

Zhang, Y.-P., Lee, S.-H., Reddy, K. R., Gopalan, A. I. and Lee, K.-P. (2007). Synthesis and characterization of core-shell SiO_2 nanoparticles/poly(3-aminophenylboronic acid) composites, *J. Appl. Polym. Sci.*, **104**, pp. 2743–2750.

Zhang, S.-W., Zhou, S.-X., Weng, Y.-M. and Wu, L.-M. (2005). Synthesis of SiO_2/polystyrene nanocomposite particles via miniemulsion polymerization, *Langmuir*, **21**, pp. 2124–2128.

Chapter 11

Controlled Sonochemical Fabrication of Mesoporous Surfaces and Metal Sponges

Ekaterina V. Skorb and Dmitry G. Shchukin
Stephenson Institute for Renewable Energy, School of Physical Sciences, University of Liverpool, L69 7ZF Liverpool, UK
d.shchukin@liverpool.ac.uk

This review comes timely, as there have been many developments in the recent years to fabricate new types of materials based on using high-intensity ultrasound. Application of ultrasound in physico-chemical processes is environmentally friendly and also allows one to achieve high temperature and high pressure chemistry locally on micro-level with the bulk reactor near the room temperature. This leads to the formation of novel microstructure and nanostructure due to non-equilibrium conditions of cavitation interface. Despite its great potential, research in the area of sonochemical treatment of the solid surfaces has not grown, especially related to the quantitative understanding of the cavitation bubble, solid surface interactions. Solving this fundamental question, the next step will involve the

Cavitation: A Novel Energy-Efficient Technique for the Generation of Nanomaterials
Edited by Sivakumar Manickam and Muthupandian Ashokkumar

ISBN 978-981-4411-54-7 (Hardcover), 978-981-4411-55-4 (eBook)
www.panstanford.com

development of new surfaces in a controlled way. It is shown that there are conceptual solutions for the controlled sonochemical fabrication of mesoporous surfaces and metal sponges; moreover, there are still great prospects in the development of such materials, as it is shown in the Section 11.4: Advanced Functional Materials.

11.1 Introduction

The first report on cavitation concerned the notice that the propeller of submarine was pitted and eroded (Thornycroft and Barnaby, 1895). Since the early work of Lord Rayleigh, it is known that ultrasound may form cavitation bubbles inside liquids, which upon collapse create transiently temperatures around 5000 K and pressures around 1000 atm with cooling rates above 10^8 K/s (Rayleigh, 1917). Structures may also be formed and quenched far from equilibrium. One of the most basic concepts of sonochemistry is that free radicals are formed as a result of cavitation of microbubbles that are created during the rarefaction (or negative pressure) period of sound waves. The growth of microbubbles can be also described by an unequal transfer of mass across the bubble interface during oscillation (Harvey et al., 1994). It was also shown that bubble-induced microstreaming was one of the factors leading to the well-known ultrasonic cleaning effects in heterogeneous systems (Elder et al., 1954; Elder, 1959). Later on, the understanding about the physical effects of ultrasound in liquid systems was also increased. Thus, the hypothesis of microjets formed during asymmetric cavitation was suggested (Naude and Ellis, 1961). Moreover, the sonolysis of water was found to be possible, thus allowing one to discuss about sonochemistry during ultrasound exposure (Makino et al., 1982).

Although ultrasound has been used as a cleaning tool in many chemical and metallurgical labs, the knowledge about influence of ultrasound onto surfaces and metals is poor. Ultrasonic cleaning baths have been introduced in the middle of last century and became popular in many chemical and metallurgical labs for cleaning glassware and forming dispersions. Moreover, during ultrasonic treatment of solid surfaces, the resulting surface morphology is not initiated by purpose but happens in an unwanted way in most of the cases, thus making surface control difficult. So, the rules for the development of certain surface structures have to be found. For example, the oldest findings of cavitation erosion of the propellers

(Silberrad, 1912) initiated studies of bubble dynamics, but even after nearly a century of research, the mechanisms are not yet fully understood. Bubbles are created by high pressure gradients and large flow rates in the liquids near a solid surface. The collapse creates locally extreme pressures and temperatures as well as shock waves and a liquid yet impinging on the surface. This may not only reduce efficiency and lifetimes of turbines or pumps but also may be useful, for example, destroy kidney stones by lithotripsy (Delius et al., 1988).

Ultrasonic frequencies range from 20 kHz to 1 GHz and it is known that the threshold for cavitation increases with frequency. Thus, the higher frequencies are typically used for ultrasonic diagnostics whereas the lower ones are for chemistry purposes. They encompass wavelengths between 10 cm and 1 μm and, hence, do not directly interact on the molecular level. The mechanism of bubble formation may be qualitatively understood as follows (Fig. 11.1). In the zone of negative pressure of the sound wave, the small volumes of lower liquid density or containing gas clusters may expand to form certain bubble nuclei. These do not fully collapse during the high pressure phase and are further expanded in the next low pressure phases. This process continues until reaching a maximum critical diameter depending on ultrasound frequency and solvent. In the following very quick cavitation collapse, the surface energy of the cavitation microbubble is converted into mechanical energy, chemical energy, heat, and, under certain conditions, light.

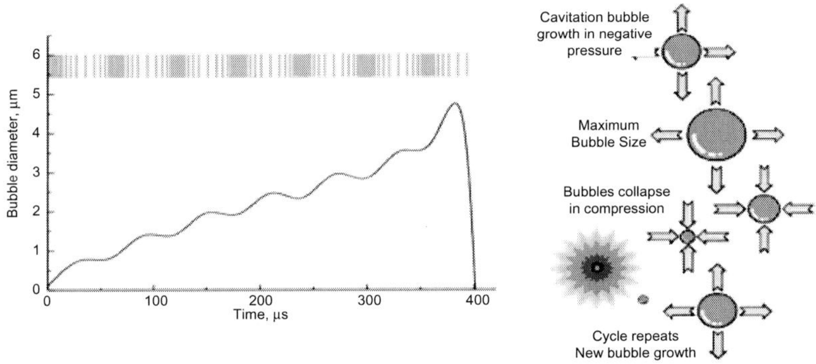

Figure 11.1 Schematics of the development of bubble radius in an ultrasonic field (left) and of the bubble growth and collapse (right). Reprinted from Suslick and Crum (1997). Copyright: the Royal Society of Chemistry and Wiley.

In the study based on the "hot-spot theory" (Margulis, 1985), volatile solutes undergo thermal cleavage with the formation of radicals or excited species within a cavitation bubble. These species may induce subsequent reactions with less-volatile substrates at the bubble shell. Thus, the effect of intense ultrasound is based on the generation of free radicals during cavitation (Margulis, 1994). Moreover, the interfacial region around cavitation bubbles has very high temperature, pressure, and (possibly) electric field gradients (Margulis, 1992, 1994). Liquid motion in this vicinity also generates very large shear and strain gradients. These are caused by the very rapid streaming of solvent molecules around the cavitation bubble, as well as by the intense shock waves emanating upon collapse. Thus, the physical effects include (I) improvement of mass transport from turbulent mixing and acoustic streaming, (II) the generation of surface damage at liquid-solid interfaces by shock waves and microjets, (III) the generation of high-velocity interparticle collisions in slurries, and, as a result, the fragmentation of solids to increase the surface area (Suslick et al., 1990). To date, studies on the influence of ultrasound onto metal surfaces and particles have been mostly focused on surface erosion not analyzing the inner structure of the treated surface, but more focused onto the outer surface processes. However, the metals for analysis were chosen mostly to differ with melting points which are important in the process of, for example, particle collision. Thus, it was shown that metal powders behave differently from inorganic particles when exposed to sonication. Although inorganic particles are degraded in the presence of ultrasound, metallic (tungsten) particles decrease in size and then tend to agglomerate with prolonged sonication as shown by Doktycz and Suslick (1990). This allows one to conclude that the maximum collision temperature reached during sonication is less than 3410°C, as there was no collision of tungstate particles observed. The impact of velocity of movement of small metal particles was estimated to be 100–500 m/s. Suslick's hypothesis that the metallic particles were fusing as a result of localized heating at the collision site has been discredited by Margulis (1992), who contends that the particles would then have to travel at speeds much faster than that of sound in liquids (approximately 1500 m/s).

In our recent publications (Shchukin et al., 2008, 2010a,b), we applied high-intensity ultrasound to the modification of both metal surfaces and metal particles (Fig. 11.2), to fabricate the surface mesoporous structures and metal sponges by the ultrasound treatment. Different intensity and duration of ultrasound were applied to find the best conditions for metal sponge formation in the case of different samples. The plates and particles were varied as follows: active metals (aluminum, magnesium, zinc), base metals (Ni, Fe, Cu, Ti), and noble metals (platinum, gold, silver) as well as their alloys. Moreover, the melting temperature, chemical reactivity of metals, and conditions of sonochemical exposure were taken into account. In addition, certain characteristics can be impacted to the metal surfaces via functionalizing the surface metal sponges with active species. Active species can be chosen in order to improve both the aesthetic properties of the metal surface and its functionality, for example, corrosion resistance (introducing corrosion inhibitors, rare earth metal salts: cerium salts, yttrium salts; transition metals: cobalt, molybdenum, and nickel) and formation of bioactive surfaces (loading drugs, biocides, and so on).

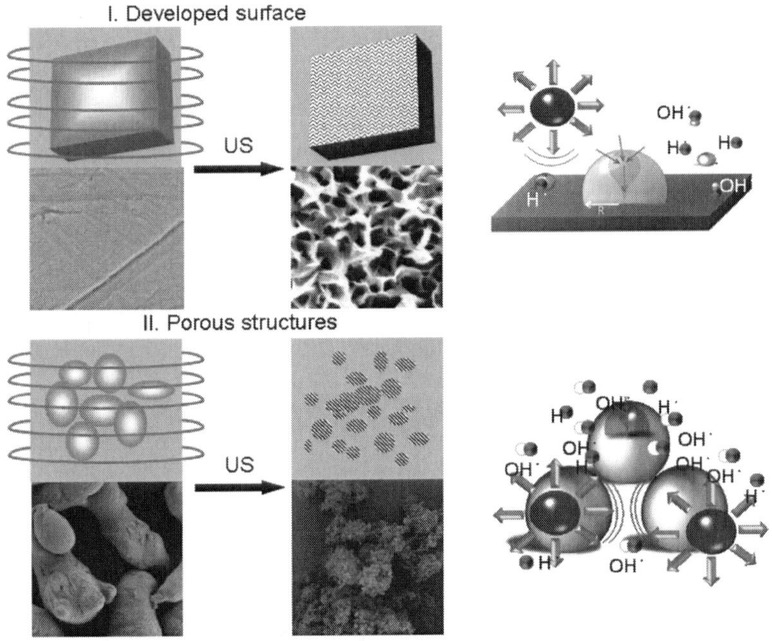

Figure 11.2 Scheme of metal modification by ultrasound.

11.2 Ultrasound-Driven Development of Metal Surfaces

High-intensity ultrasound irradiation has been used for a long time in metal technology, including metal crystallization, forming, and finishing (Mason, 1990). The application of ultrasound to a metal melt generally leads to metals with improved grain refinement and homogeneity. In our first work (Shchukin et al., 2008), we demonstrated that the controlled ultrasonic treatment of aluminum alloy plates leads to the formation of a rough metal surface providing very good adhesion of anticorrosion protective coating. We showed that the collapse of cavitation bubbles near aluminum surface causes (i) surface cleaning (destruction and removal of impurities), (ii) significant increase of roughness of the interfacial layer and, therefore, adhesion of polyelectrolytes to the surface, and (iii) overlapping of developed aluminum structures. Further, we systematically studied the effect of ultrasound irradiation (intensity up to 57 W/cm^2, frequency 20 kHz) on the surface morphology and properties (wettability, adhesion, and corrosion resistance) of different metal alloys. Aluminum, magnesium, and iron alloys were used to demonstrate the versatility of the novel approach to surface treatment. As ultrasonic treatment is not an exotic process and applicable even on large scale in industrial manufacturing, controlling the process may lead to new applications making use of the specially designed surface.

The effect of ultrasound in material science is due to the interplay of conceptually different mechanisms: (i) physical impact due to a pressure wave and a jet impinging on the metal surface and (ii) the extreme temperatures and pressures creating the highly reactive radicals derived from water and oxygen having a chemical impact on the metal surface (Lindlay and Mason, 1987; Suslick, 1997, 1999). Figure 11.3(a) schematically illustrates the ultrasound-driven modification of metal plates. First stage (Fig. 11.3.Ia) is jet imprinting on the metal surface. We suppose that the metals with lower melting point should be more sensitive to the influence of jets; thus at the beginning stage, we expect that the modification of magnesium should be faster than aluminum and iron and the modification of aluminum is faster than iron.

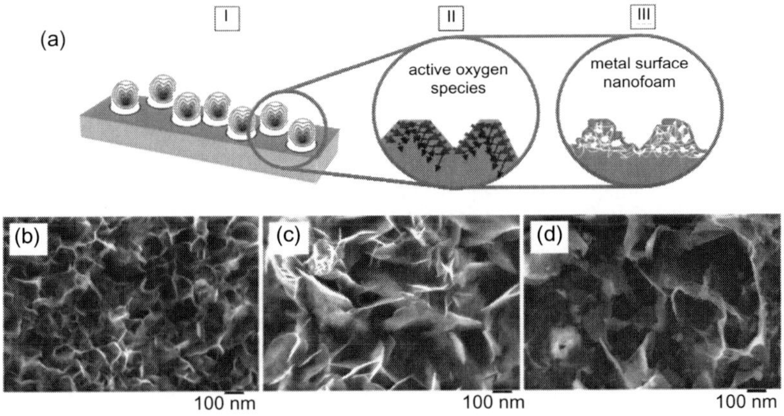

Figure 11.3 (a) Scheme of formation of mesoporous surfaces, and SEM images of (b) aluminum, (c) magnesium alloy, and (d) stainless steel after 40 min of sonication at 57 W/cm^2. Reprinted from Skorb et al. (2010c). Copyright the Royal Society of Chemistry.

The second stage (Fig. 11.3.IIa) is relying on chemical impact and metal oxidation ability. Highly reactive radicals derived from water and oxygen (active oxygen species) begin to modify the unprotected surface that was formed by jets, simultaneously stabilizing the porous structure due to the formation of novel protective oxide layer. The last stage (Fig. 11.3.IIIa) is the stabilization of the formed porous structure. It should be also noted that all processes are extremely fast and difficult to study separately and thus conclusions were made by observing just a variation of a studied metal spectrum. The SEM images (Fig. 11.3b–d) of metal surfaces sonicated at 57 W/cm^2 indicate the formation of highly developed metal surfaces. Surface roughness (R_{ms}) increases from 13 nm to 118 nm in 60 min for aluminum and from 18 nm to 211 nm for steel. The difference in the surface morphology after sonication is caused by different chemical (surface oxidation by free radicals) and physical (melting points, durability) properties and, therefore, different interfacial response to cavitation. This porous sponge nanostructure is also well visible in the TEM image (Fig. 11.4a).

The intensity and duration of ultrasound treatment dramatically influence the development of porous metal structures. Figure 11.5 shows the SEM images of the surfaces of aluminum plates sonicated at different intensity and time. Samples obtained at an intensity of

30 W/cm^2 (Fig. 11.5a,c) and 57 W/cm^2 (Fig. 11.5b,d) exhibit different surface morphology. The morphology of the plates sonicated at 30 W/cm^2 (Fig. 11.4b, 11.5a) does not show any characteristics of nanosponge interfacial layer. This surface has typical morphology of the anodized aluminum oxide layer. On the contrary, the samples prepared at 57 W/cm^2 (Fig. 11.5b) have the morphology of porous surface metal nanosponge. Therefore, increase in the ultrasonic power results in more intense treatment of the metal surface forming a porous metal layer on the top. The thickness of the modified layer estimated by TEM analysis of ultramicrotomed plates is about 200 nm with the foam cell size nearly 7 nm.

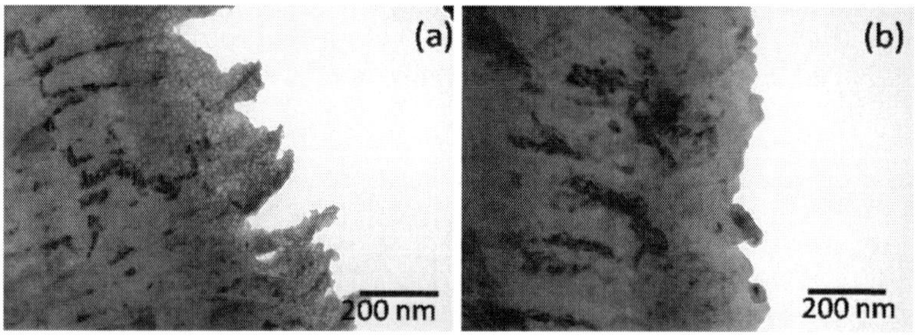

Figure 11.4 TEM images of an aluminum alloy sonochemically treated for 40 min at intensities of (a) 57 W/cm^2 and (b) 30 W/cm^2. Reprinted from Skorb et al. (2010c). Copyright the Royal Society of Chemistry.

The water contact angle measurements evidenced that the ultrasound-modified surfaces exhibit higher hydrophilicity than the initial one. It is very sensitive to the geometrical structure of a solid surface: an intrinsically hydrophobic surface becomes more hydrophobic as its roughness increases and, vice versa, the hydrophilicity shows an increase with roughness for the intrinsically hydrophilic surface.

The monitoring of corrosion activity by scanning vibrating electrode technique of a steel plate with and without ultrasound treatment shows that untreated samples are very sensitive to the corrosion, which develops very fast in contact with NaCl solution. The steel samples after high-intensity ultrasound irradiation exhibit dramatically different behavior. The increase in corrosion activity is very low; after 12 h of immersion in 0.1 M NaCl, the anodic activity is nearly 8 μA/cm^2.

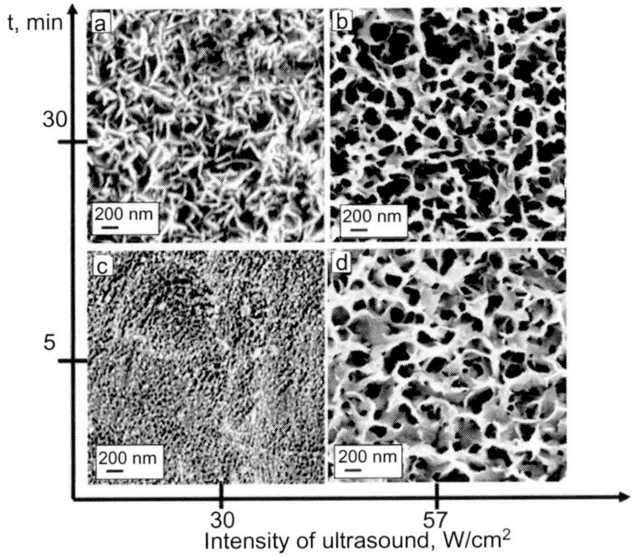

Figure 11.5 SEM images of the aluminum alloy plates after sonication at different intensities and duration: 30 min at intensity 30 W/cm^2 (a); 30 min at intensity 57 W/cm^2 (b); 5 min at intensity 30 W/cm^2 (c); 5 min at intensity 57 W/cm^2 (d). Reprinted from Skorb et al. (2010c), Copyright the Royal Society of Chemistry.

Our latest results showed that the hydrophilicity of the surfaces affects bubble nucleation on the metal surface (Fig. 11.6) (Belova et al., 2010). Thus, it was shown that by patterning a surface, a model of heterogeneous nucleation of cavitation bubbles can be formulated. The ultrasonic impact on surfaces was investigated by varying surface energies, temperature, and ultrasonic power. To verify these models, more quantitative analysis of the growth kinetics will be needed, which is under investigation now. However, chemical patterning for changing the surface energy is also the most suitable one to study further the bubble growth: these bubbles would be expected to grow toward diameters of 100 μm, and the contact area may be limited well below this size, thus allowing the three-phase boundary to be studied. This provides an additional control of surface treatment (or way of preparation) by ultrasound. Further dedicated and quantitative studies should thus have a way to make use of ultrasound to modify the geometry and chemistry of surfaces and films, to leach these preferentially, or to nucleate bubbles on microparticles with a defined surface.

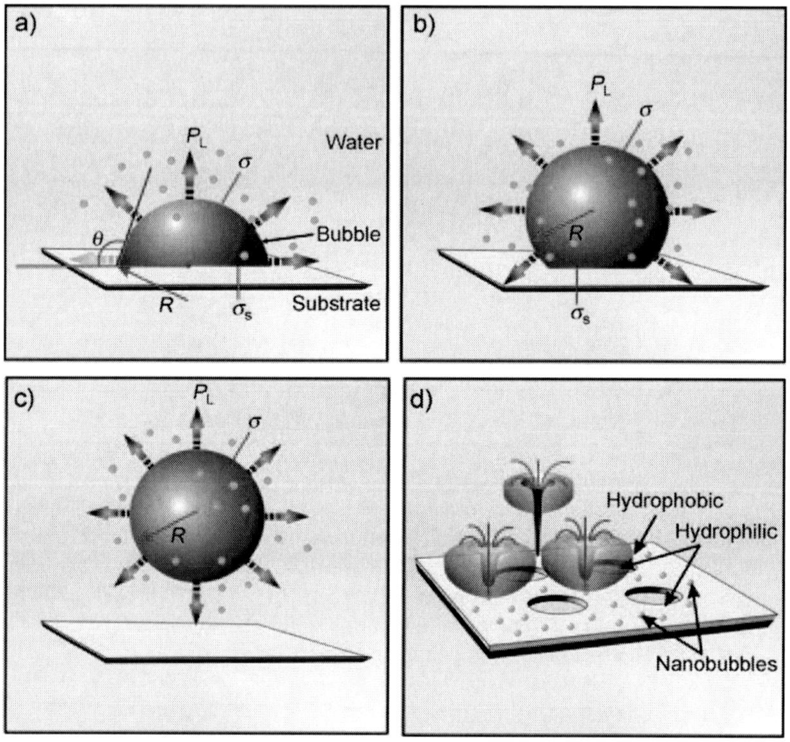

Figure 11.6 Comparison of nucleation and growth of a cavitation bubble at (a) a hydrophobic and (b) a hydrophilic surface, and (c) in bulk. The pressure P performing the volume work is given by the vapor pressure of the liquid, the gas pressure, and the acoustic pressure being in excess of the external pressure P_L. (d) Mechanism of the selective gas-bubble nucleation on the hydrophobic/hydrophilic microstructured surface. Reprinted from Belova et al. (2010). Copyright Wiley.

In conclusion, we have shown that high-intensity ultrasound surface treatment results in the formation of highly developed hydrophilic surfaces. The corrosion stability increases after high-intensity ultrasound metal surface treatment. Systematic investigation of ultrasound effects on various types of alloys shows that the nature of the metal determines the properties of the treated surface.

11.3 Sonochemical Formation of Metal Sponges

The successful formation of sponges under ultrasonic treatment depends on the formation of a stable oxide nanolayer on the metal surface. Aluminum, magnesium, zinc, and titanium are rapidly covered by a layer of oxide on their surface when the metal is exposed to air. Noble metals (platinum, gold, silver) do not react with oxygen in air. Structural and morphological changes of these two groups of metals were studied to reveal the mechanism of formation of mesoporous metal sponges by ultrasound. SEM images of sonicated zinc (Fig. 11.7a,e), magnesium (Fig. 11.7b,f), aluminum (Fig. 11.7c,g), and aluminum–nickel (1:1) alloy particles (Fig. 11.7d,h) show that the size and the surface of the oxide-forming metals and their alloys are dramatically changed by 60 min sonication at an intensity of 57 W/cm^2. Increasing the specific surface area of the particles also indicates the changes inside the deep surface layers. The ultrasound-stimulated modification of metal particles results in the following: particle decomposition (particle size is approximately 20 times reduced); formation of developed surface (due to formation of a new oxide film); and formation of a porous inner structure (increasing surface area).

The mechanism of modification of metal structure under ultrasound irradiation is complex and involves a variety of aspects related to melting, solidification of metals, selective etching, and oxidation of metal surface. It is known that transformation of liquid into solid involves the creation of curved solid/liquid interfaces (leading to capillary effects) and the microscopic flow of heat (Luche, 1994; Thompson and Doraiswamy, 1999). The form of resulting surface microstructure depends on the cooling conditions varying from porous to the dense flat structure. Special conditions provided by ultrasound of high-intensity shell affect crystallization process in the metal particles and, therefore, have influence on the metal structure. Surface oxidation during ultrasound irradiation is another important aspect in the formation of porous structure and having a crucial role in the stabilization of the metal structure casted under ultrasound irradiation.

Ultrasound modification of Zn, Al, and Mg particles starts with the oxidation of surface of metal and formation of a rough metal oxide (hydroxide) layer. The XRD patterns evidence that bayerite $Al(OH)_3$

with a mixture of bohmite AlO(OH) is formed during sonication. The ^{27}Al MAS NMR spectra (Fig. 11.8) of the samples treated by ultrasound exhibit two signals at 1200 ppm and –67 ppm that can be assigned to Al and Al^{3+}, respectively (Weissler, 1959). It is important to note that during ultrasound irradiation, the formation of a porous metallic structure covered by a thin oxide layer takes place, which is seen from well-observed peaks of aluminum in XRD pattern and NMR spectra even after long-term ultrasound treatment. The results of 3D field ion microscopy also confirm the metallic structure of the treated surface covered with dense oxide (hydroxide) layer.

Figure 11.7 SEM images of the surfaces of initial metal particles: aluminum (a), magnesium (b), nickel (c), and titanium (d) and after 60 min of sonication at 57 W/cm^2: aluminum (e), magnesium (f), nickel (g), and titanium (h). Reprinted from Skorb et al. (2010a). Copyright the Royal Society of Chemistry.

Development of Al surface and inner structure (pore size and distribution) differs a lot from the Zn and slightly from the Mg. The surface of the aluminum particles after 1 min of sonication has a cell-like structure of aluminum oxide. Similar structure of the surface can be obtained by hydrothermal treatment of aluminum in boiling water. With 3 min of sonication, decomposition of particles and formation of channel-like structure takes place. The initial 100 μm aluminum particles change into 200 nm thick lamellas in 60 min. The TEM image also shows the formation of 50–100 nm channels between lamellas. It is interesting to note that the metal oxide layer formed for the first minute of sonication is removed by longer (>1 min) treatment. The surface of porous aluminum lamellas is covered by a smooth thin aluminum oxide. Probably, this morphology of aluminum oxide is more stable and can resist the impact of ultrasonic cavitation. After 5 min of sonication, the formation of porous inner structure became noticeable on the electron micrographs. Roughness of the surfaces of the aluminum particles is increased. The BET isotherms show that in the first 5 min, the formation of mesopores is stimulated. After 10 min of ultrasound treatment, we observe both the micropores and mesopores. The average pore diameter estimated by BET is about 4 nm for the 60 min of treated particles. Further increase in the irradiation time leads to the regression of structural development. SEM images and BET analysis demonstrate a decrease in the porosity, surface smoothening, and fusion of the particles after 2 h of sonication.

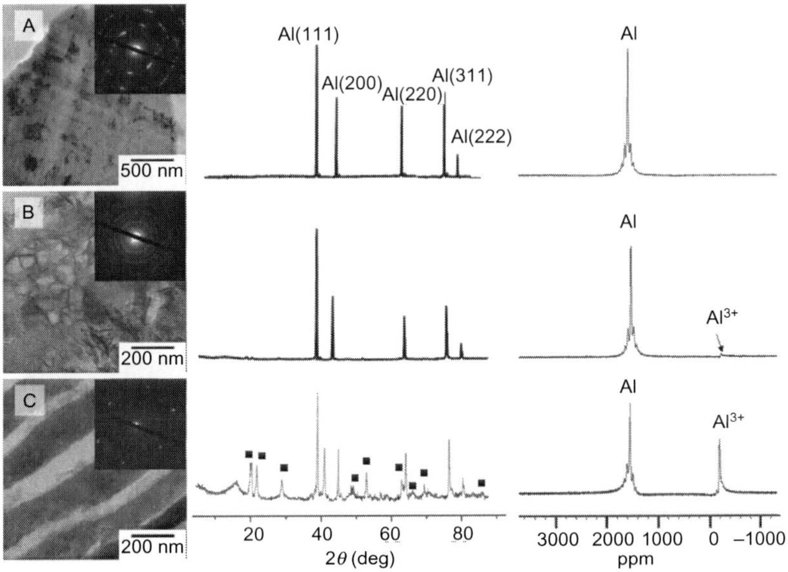

Figure 11.8 TEM images, XRD (black squares show the peaks assigned to bayerite), and ^{27}Al MAS NMR spectra of the initial aluminum particles (a), after 30 min (b) and 60 min sonication (c). Reprinted from Skorb et al. (2010a). Copyright the Royal Society of Chemistry.

The XRD patterns and BET analysis of magnesium after ultrasonic treatment prove the formation of magnesium sponges and development of magnesium hydroxide phase on the metal surface. The surface area for 30 min of sonicated Mg and Al are 68.7 m^2/g and 53.9 m^2/g, respectively. Furthermore, average pore diameter is nearly 14 nm in the case of magnesium and nearly 4 nm in the case of aluminum.

Zinc-based sponges exhibit a completely different surface morphology and inner structure. The SEM images demonstrate ultrasound-stimulated formation of zinc oxide on the surface of zinc particles. It is clear that the formation of oxide layer can be controlled by the intensity and duration of sonication. Around 2 μm zinc oxide crystals formed during short-time sonication exhibit tubular morphology. Long-term treatment (>30 min) stimulates the formation of stable oxide nanorods with length nearly 100 nm and diameter approximately 20 nm. The maximum surface area for 15 min of sonicated Zn particles is 22 m^2/g. Around 30-min and 90-min sonicated samples exhibit nonporous structure. Sonicated Zn particles might have a different structure due to the features of zinc oxide formed by sonication. The porosity of the sonicated Zn is rapidly decreased with sonication time due to the transformation of tubular ZnO into ZnO nanorods.

The metals with higher (comparing to Mg and Al) melting points (Ni, 1723 K and Ti, 1941 K) exhibit changes in the interfacial layer during longer sonication time. The BET and TEM data have no evidence of formation and development of inner structure.

The noble metals (Au, Ag, Pt), due to their low tendency to oxidation, exhibit the highest resistance to ultrasound irradiation (among the metals investigated here) and form neither porous structure nor develop surface after sonication, even in the cases in which their melting point is actually lower than that of one of the metals that can be modified slightly (such as Ni or Ti). Therefore, surface oxidation has a crucial role in the stabilization of the metal structure developed under ultrasound irradiation.

Moreover, in order to test the importance of free radicals in the formation of structure, the sonication experiments were performed in an ionic liquid without any dissolved gas. Suppression of reactive free radical formation in ionic liquid results in the drastic decrease of sponge formation process. In order to demonstrate the importance of oxide layer in the formation of porous metal structure, the pH values of aqueous system (by 0.1 M solutions of HCl and NaOH) were adjusted to suppress the formation of a metal oxide layer. Ultrasonic treatment of aluminum particles at acidic and basic pH prevents sponge formation, as confirmed by SEM images and BET data.

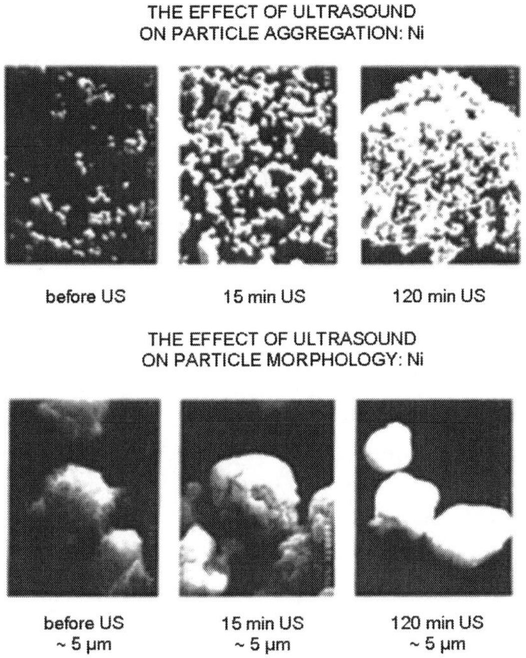

Figure 11.9 The effect of ultrasonic irradiation on particle aggregation and surface morphology of Ni powder. The magnification of the scanning electron micrograph in the upper spectra is 10-fold less than in the lower spectra. Reprinted from Suslick and Casadonte (1987). Copyright the American Chemical Society.

We summarized the characteristic aspects of ultrasound modification of the used metals and detailed the particular features for each of them. Ultrasound modification of metal particles starts with oxidation of the surface of a metal and formation of a rough metal oxide (hydroxide) layer. With several minutes of sonication, the decomposition of particles and formation of a channel-like structure takes place. The initial hundred microns metal particles change into 200 nm thick lamellae (shown in TEM images for Al and Mg).

The XRD patterns evidence that surface oxidation and formation of metal oxide layers have a crucial role in the modification of metal structure and formation of porous inner structure. Furthermore, the intensity of the signals is decreased and slightly shifted due to grain destruction under ultrasound irradiation.

Notwithstanding the fact that the particle collision and following agglomeration (Doktycz and Suslik, 1990) was tested for transition metals but not for noble ones, it is expected that the effect of ultrasonic treatment should be the same for the noble metals, as their melting temperature is lower than tungsten. In particular, a series of transition metal powders was used to probe the maximum temperatures and speed reached during such interparticle collisions. Metal particles that were irradiated in hydrocarbon liquids with ultrasound underwent collisions at roughly half the speed of sound and generated local effective temperatures between 2600°C and 3400°C at the point of impact for particles with an average diameter of nearly 10 μm. SEM images for Ni samples irradiated for 0, 15, and 120 min have been shown in Fig. 11.9. As the malleable particles are irradiated, profound changes in the particle aggregation and morphology are observed. The surface of Ni powder is rough, but upon sonication, it is smoothened rapidly. At the same time, the extent of aggregation increases dramatically. Authors attribute both effects due to the interparticle collisions driven by the turbulent flow created by the ultrasonic field. An increase in the aggregation accounts for the eventual decrease in the observed surface area and probably also causes the small diminution in activity observed after a lengthy sonication. Moreover, this change in surface morphology is associated with a dramatic change in the surface composition. Auger electron spectra depth profiles obtained on Ni samples before and after sonication show that in the case of sonication of such metals in hydrocarbon liquids, a thick oxide layer exists before the ultrasonic irradiation, and although after 1 h of ultrasonic irradiation in octane, it becomes thinner. The total surface area differs dramatically for the sponges formed as shown by Skorb et al. (2010c). For aluminum sponge, maximum porosity reached 80 m^2/g and it was at the level approximately 1.00 m^2/g in the case of nickel sponge. However, in both cases, the porosity had a tendency to increase with time, whereby it reached a maximum and after that it decreased.

It should also be noted that if organic solutions are used, the metals that formed sponges in aqueous solutions (Al, Mg) or have a tendency to total oxidation (Zn) show the same rules as "hard" metals. Irradiation of Zn powder with an average diameter of 5 μm in freshly distilled decane (17 mL, 20% by weight) under Ar at 20 kHz and 50 W/cm^2, for 30 min at 15°C results in small particles connect into one another due to interparticle melting. It was suggested that the effect originates from the rapid cooling of the effectively molten collision zone, as the colliding particles rebound immediately after impact.

11.4 Ultrasound-Assisted Design of Metal Nanocomposites

A one-step method was developed to produce metal nanocomposites from metal alloys under ultrasound irradiation. Ultrasound irradiation was exploited for the formation of nanoscale composites. Thus, if alloy particles consisting of a resistant-to-ultrasound compound and a sensitive one are treated, one would consequently expect that novel nanocomposite structures are developed. We tested Al/Ni, Al/Pd (Fig. 11.10a) and Al/Pd/Cu (Fig. 11.10b) systems. Indeed, the treatment of aqueous alloy dispersions with highly intense ultrasound leads not only to the formation of porous structure but also results in phase segregation yielding ultrasound-resistant nickel, palladium, and copper nanoparticles homogeneously distributed in the porous matrix as evidenced by the TEM micrographs. Simultaneously, BET analysis even shows the more porous structure formation in the case of nanocomposites (the achieved surface area in this case, 280 m^2/g) for the Al/Pd particles and 210 m^2/g for those of Al/Pd/Cu after 60 min of modification. The phase segregation process in multimetal systems stimulates the formation of mesoporous structures by sonication. Such nanocomposites could be used as themselves (see part 1.5 below) as well as precursors for sponges made from "hard" or noble metals.

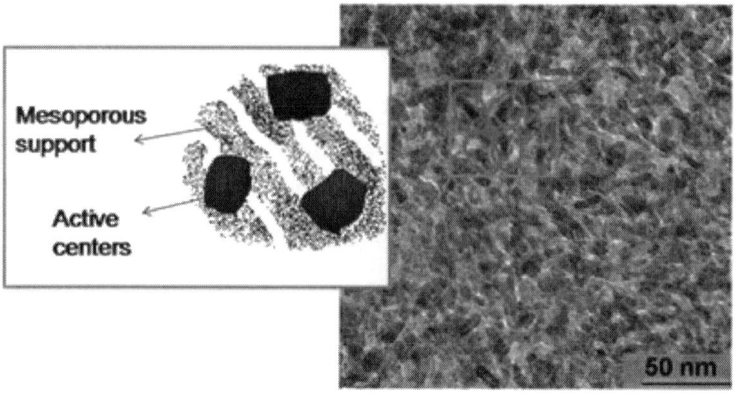

Figure 11.10 One-step formation of mesoporous system from alloy: TEM of palladium particles distributed in the porous aluminum support formed after high-intensity ultrasound (57 W/cm^2) treatment of Pd/Al alloy (the inset shows the sketch image of the Pd/Al nanocomposite). Reprinted from Skorb et al. (2010a). Copyright the Royal Society of Chemistry.

11.5 Advanced Functional Materials

11.5.1 Formation of Surface Superhydrophobic "Lotus" Structures

The aluminum plates sonicated at high intensity exhibit a very pronounced resistance to the corrosion as shown during the monitoring of the aluminum surface by SEM after stopping the corrosion test. Scanning vibrating electrode technique measurements show that anodic activity on the surface of aluminum before and after sonication treatment at 57 W/cm^2 is practically the same, that is, nearly 60 μA/cm^2. The explanation for this is as follows: before sonication, the corrosion process involves the whole surface of aluminum; on the contrary, in the case of sonicated aluminum surfaces, the corrosion peak is well-defined and localized on the intermetallic inclusions in the aluminum alloy. It should also be noted that in the case of sonicated metal surfaces that could be characterized as hydrophilic [water contact angles for aluminum (Fig. 11.11) and steel after 60 min of sonication at 57 W/cm^2 are 7° and 11°, respectively].

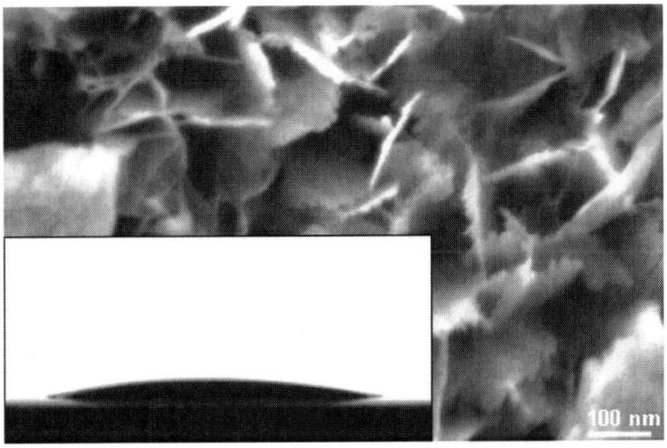

Figure 11.11 SEM image of aluminum mesoporous surfaces and its contact angle: formation of hydrophilic surface.

However, superhydrophilicity can help in the formation of superhydrophobic surfaces involving surface modification of our developed structure (Chen et al., 2006) by the attachment of laurate under low-intensity sonication to the porous surface and the formation of layered double hydroxide laurate. The formed structure is shown in Fig. 11.12. Treatment by sodium laurate induced the hydrophobicity resulting in water contact angle of 163° (Fig. 11.12, inset). The SVET (scanning vibrating electrode technique) data of corrosion resistance (Fig. 11.12b) exhibit very high corrosion resistance with simultaneous self-healing ability (Skorb et al., 2009). Three-dimensional current density map (Fig. 11.12b, inset) shows that there is no corrosion current above the surface detected even after 12 h of immersion in 0.1 M NaCl solution. Thus, the attachment of layered double hydroxide laurate to the ultrasonically treated aluminum surface results in the formation of a very corrosion stable hydrophobic surface.

11.5.2 Self-healing Anticorrosion Coatings Based on pH-Sensitive Polyelectrolyte/Inhibitor Sandwich-Like Nanostructures

Anticorrosion coatings have to be in close molecular contact with the surface to prevent the occlusion of corrosive species at the metal/coating interface (Andreeva et al., 2008). The typical aluminum surface is covered by a 3–7 nm thick natural oxide film. This thin layer is not sufficient to protect against corrosion agents and does not yield good adhesion to subsequent layers of the coating. Therefore, the aluminum surface is always pretreated before use. The most extensively used surface pretreatment procedures are based on aggressive chemicals including chromate solutions (Spadaro et al., 2006). It is generally expected that the surface pretreatment produces a porous oxide layer on the metal surface with roughness sufficient for mechanical interlocking.

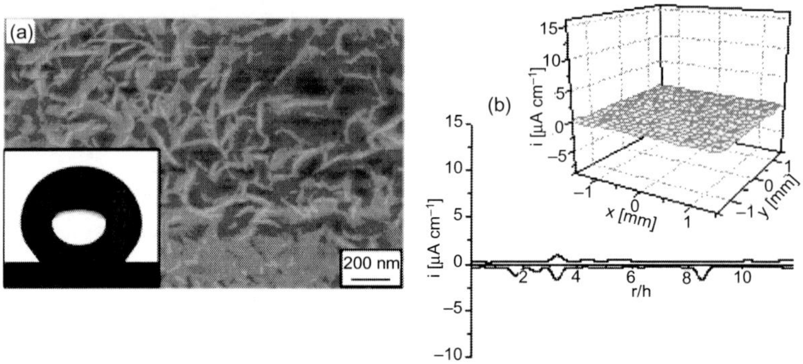

Figure 11.12 SEM image of an aluminum plate covered by layered double hydroxide laurate (a), inset shows water contact angle onto this surface. (b) Time monitoring of the anodic and cathodic activity on the aluminum surface covered by layered double hydroxide laurate, inset shows current density maps of the surface after 12 h immersion in 0.1M NaCl. Reprinted from Skorb et al. (2011). Copyright the Royal Society of Chemistry.

Aluminum alloy samples (AA2024) were degreased in isopropyl alcohol flow and rinsed in purified water. The surface of the aluminum plates was etched by ultrasound (Ultrasonic Processor CV 33 Sonics Inc., 20 kHz, 500 W) within 10 min in pure water. One substantial benefit of ultrasound in the pretreatment processing is the partial removal of the natural oxide layer from aluminum surfaces. Therefore, the new active uniform oxide layer is rapidly formed on the aluminum surface. A highly micro-rough surface of aluminum oxide prepared by sonication (treatment at 500W for 10 min) is shown in Fig. 11.13a. The infrared reflection absorption (IRRA) spectrum shows significant increase in the hydroxyl group concentration in the new oxide film. The highly developed surface area and the increased concentration of the surface hydroxyl groups provide high capacity and adsorption activity to aluminum alloys.

Layers of polyelectrolytes and inhibitor were formed by LbL deposition (Caruso et al., 1998) on freshly sonicated aluminum alloys. Scheme of sandwich-like structure and SEM image of the formed surface are presented in Fig. 11.13b,d. Thus, deposited polyelectrolytes are macromolecules carrying a relatively large number of functional groups that either are charged or, under suitable conditions, can become charged. The films can be enriched in one polymer compared with the other by fabrication in a pH-regime where one of the polymers is weakly charged whereas the other is strongly charged. Active species deposited as a component of the polyelectrolyte nanonetwork or as a dopant can be released on demand. The sensitivity of the polyelectrolyte film to a variety of physical and chemical conditions (such as pH-shift or mechanical impact, and so on) of the surrounding media provides the ability of regulated release of the inhibitor species entrapped into multilayers.

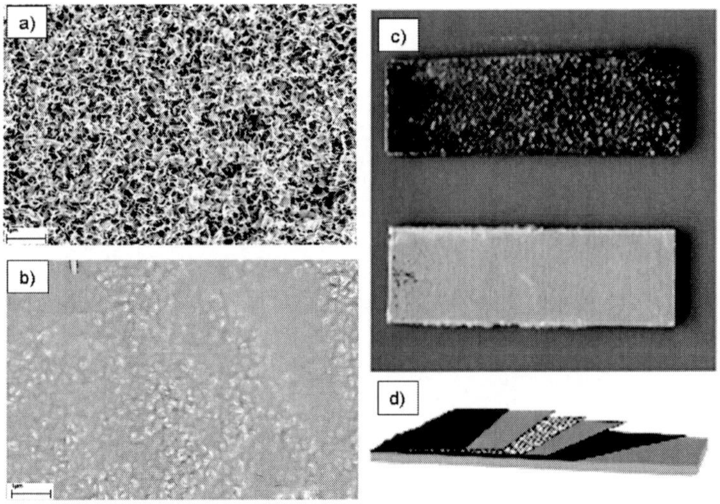

Figure 11.13 SEM images show the surface texture of the treated (a) and treated sample of aluminum alloy AA2024 covered by polymer/inhibitor complex (b,d) and storing for 24 h in 1 M salt solution (c: top—untreated aluminum, bottom—sample b). Reprinted from Andreeva et al. (2008). Copyright Wiley.

The schematic mechanism of a smart self-healing process is shown in Fig. 11.14: (i) the polyelectrolytes have pH-buffering activity and can stabilize the pH at the metal surface in the corrosive media; (ii) the inhibitors are released from polyelectrolyte multilayers only after the start of the corrosion process and directly in the rusted area preventing the corrosion propagation; (iii) polyelectrolytes forming the coating are relatively mobile and have the tendency to seal and to eliminate the mechanical cracks of the coating.

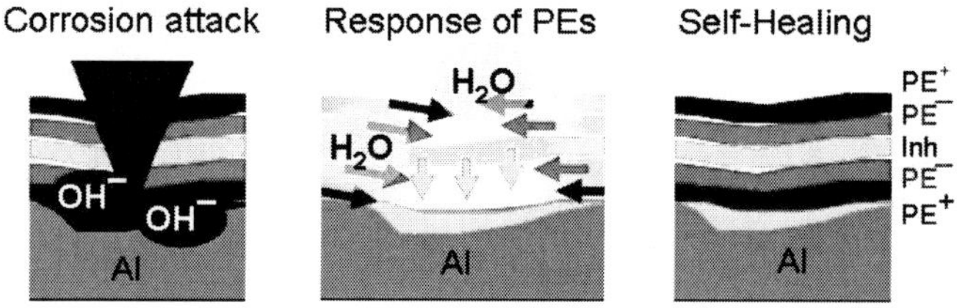

Figure 11.14 Schematic mechanism of self-healing action of a "smart" polyelectrolyte anticorrosion coating. Corrosion attack causes pH change of the system stimulating response of the polyelectrolyte coating: pH buffering, rearrangement of polymer chains, and release of corrosion inhibitor. PE+, positively charged polyelectrolyte; PE^-, negatively charged polyelectrolyte; Inh, corrosion inhibitor. Reprinted from Andreeva et al. (2008). Copyright Wiley.

The long-term coating stability in a corrosive environment was studied by dipping the samples in aqueous NaCl solution at 20°C. The introduced coatings possess active and passive corrosion protection combining effective barrier properties with the possibility of self-healing of defects in the coating. Figure 11.13 shows the pictures of samples coated by the polymer/inhibitor complex (bottom) and without coating (top). Corrosion defects are observed after 12 h of immersion in 0.1 M NaCl on the unmodified aluminum, whereas the sample with the polymer/inhibitor complex does not exhibit any visible signs of corrosion attack even after 21 days of immersion.

11.5.3 Sonochemically Designed Cerium-Rich Anticorrosion Nano-Network

Salts of transition and rare-earth metals (Ce, Co, Mo) have been proposed as alternatives to carcinogenic chromate inhibitors due to their low toxicity, economic efficiency, and sufficient cathodic inhibition of metal corrosion. However, a number of the developed cerium-containing coatings exhibit moderate anticorrosion activity mostly due to their poor adhesion to the substrate and continuous leakage of the inhibitor from the surface. Recently, several concepts were developed to incorporate cerium into aluminum alloys and to provide triggered release on demand (Jakab and Scully, 2005; Hamdy, 2006).

The application of ultrasound to Al substrate in Ce (III) aqueous solution resulted in the formation of a cerium/aluminum (Ce/Al) oxide outermost protective layer. Ultrasound-driven surface modification includes the formation of a developed interfacial metal layer, incorporation and homogeneous distribution of cerium ions on the AA2024 substrate followed by surface oxidation and formation of a Ce/Al oxide nano-network.

Two types of coating can be prepared by ultrasound treatment of aluminum plates in an aqueous solution of cerium (III) nitrate varying the intensity of sonication: a μm-thick Ce/Al oxide layer and thin Ce/Al oxide layer (Fig. 11.15). At low intensity of sonication (30 W/cm^2), an extremely dense and smooth Ce oxide interfacial layer can be formed within 5 min of sonication. However, it can be seen that a longer sonication (40 min) leads to the rupturing of the coating. The cracks are clearly observed after 40 min of sonication. Different behavior can be monitored at 57 W/cm^2 intensity of ultrasound treatment. The higher intensity results in more pronounced changes in the morphology of the metal surface. In 5 min of sonication, a dense Ce oxide layer with huge cracks similar to those described above is formed. Further increase in the sonication time (20 min) results in the transition into a structured interface wherein cerium and aluminum oxides form a nano-network. After 30 min of sonication, the metal surface is homogeneously covered by the protective Ce/Al oxide nano-network layer, which acts better for corrosion protection than the pure Ce oxide layer. Additional sonication does not cause any further changes in surface morphology. The dependence of the atomic concentration of cerium on sonication time and intensity was assessed by EDS. The SEM images and EDS spectra demonstrate that activation and modification of the metal surface is strongly influenced by time and intensity of sonication. At higher intensity (57 W/cm^2), the cerium/aluminum oxide layer is formed very fast in

5 min of sonication. Further irradiation enhances coating integrity by forming a more compact interfacial layer. At lower intensity (30 W/cm^2), the cerium content is exponentially increased and reaches maximum after 50 min of treatment. Thus, ultrasound treatment at low intensity allows more equilibrium condition during coating formation.

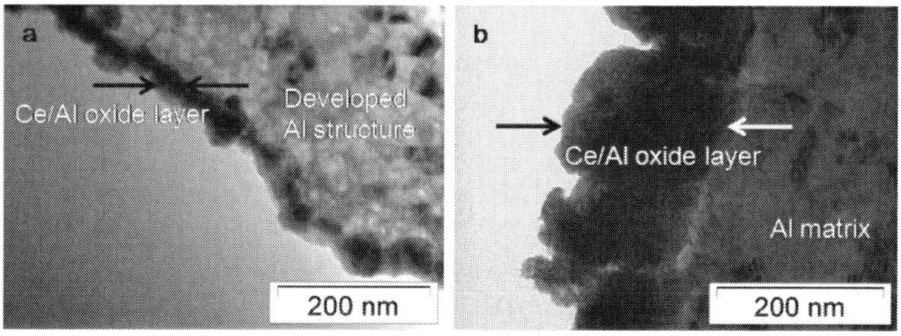

Figure 11.15 Transmission electron microscopy (TEM) image of the cross section of the aluminum alloy aa2024 covered by a cerium/aluminum nanonetwork (a) and a dense cerium oxide layer (b). The dense cerium oxide layer is shown by the arrows. Reprinted from Skorb et al. (2010b), *Langmuir*, in press. Copyright the American Chemical Society.

The thickness of the modified layers was estimated by TEM of the cross-sections of the plates and is approximately 200 nm for a sample prepared at 30 W/cm^2 and 20 nm for one prepared at 57 W/cm^2 (Fig. 11.15, the dark Ce oxide layers are highlighted by arrows). Therefore, SEM, EDS, and TEM measurements are in good agreement and demonstrate the formation of two different types of coating depending on the intensity of treatment: a 200 nm thick dense Ce oxide coating and a developed aluminum surface covered by 20 nm thick Ce oxide.

The structural and morphological differences of two types of coatings result in dramatically different stability and anticorrosion properties. The plates sonicated at higher intensity demonstrate very high resistance to corrosion. Moreover, this type of coating has well pronounced self-healing effect (Shchukin et al., 2006). There is no such self-healing effect detected in the case of unmodified aluminum alloy. Well-defined corrosion activity is observed on the aluminum surface sonicated at an intensity of 30 W/cm^2 and the adhesion of other types of coating in the case of dense cerium rich coating is very low. On the contrary, the samples covered with a Ce/Al oxide nano-network (formed at higher intensity) exhibit a very good stability. Furthermore, the rough surface is a perfect substrate for the deposition of further components (sol–gel coatings, primer and top-coating) of anticorrosion protection. The proposed environmentally friendly method of metal surface modification demonstrated for the aluminum alloy 2024 is universal and can be applied to other metals and alloys in order to supply the surface with anticorrosion, antifouling, and so on.

11.5.4 Applications of Metal Sponges with Defined Porosity

Metal sponges with tunable structures have gained interest for a wide range of applications from filters to reservoirs for energy conversion and storage. Porous metals are designed to exhibit specific characteristics combining the beneficial properties of the base metal and sponge scaffold: resistance to corrosion, mechanical strength, high-temperature performance, thermal and electrical characteristics (Banhart, 2001). Through a process of casting, powder metallurgy, and sputter deposition, porous materials with large pores (200 μm to 2 mm) can be produced. Approaches to fabricate mesoporous metals with controlled porosity are based on electroplating into the self-assembled liquid crystal surfactants (Attard et al., 1997), colloidal crystals (Velev et al., 1999), porous block copolymers (Iyer et al., 2007), or anodic porous alumina (Masuda and Fukuda, 1995). Nanoporous metals have been synthesized by chemical reduction of the corresponding ions using plasma spraying technique (Erlebacher et al., 2001). Some of these methods have the advantages of precise control over the size and structure of the final metal sponge. However, they are multistage, expensive, and time-consuming to implement. Thus, it is very important to develop novel, general, and economically effective approaches to the formation of porous metal structures with the pore size in the nanometer scale, which can be applied to a broad range of metals and alloys. In the previous part, we have shown the great success of ultrasound in the formation of metal sponges with defined structure. This part attributes more to possible applications.

11.5.4.1 Preparation of catalysts

The treatment of an aluminum–nickel alloy dispersion with high-intense ultrasound in aqueous medium leads not only to the formation of sponge-like structure but also results in the phase segregation yielding nickel nanoparticles distributed in the pores as evidenced by the TEM micrograph given in Fig. 11.16 (left) (see also Section 11.4). The sample of AlNi alloy consists mostly from Al_3Ni_2 intermetallide with an admixture of Al_3Ni. Upon ultrasound treatment, bayerite appears as the main oxide product, while the content of Al_3Ni_2 exhibits the pronounced decrease. Nickel metal, however, is not evident in the diffraction pattern (being amorphous). Being in an oxide form, nickel would be rather crystalline. The XPS measurements confirmed the formation of metal nickel phase. Formation of the mesoporous nickel–aluminum structure that was suggested as a catalyst was confirmed by BET analysis, the porosity of which could be effectively controlled by the conditions of the ultrasonic process. Sonication (~50 W/cm ^{2} at 20 kHz in an aqueous solution) of a commercial Al:Ni (1:1) alloy particles results in the formation of aluminum sponge-like structures with larger surface area (to 200 m^2/g) and narrow pore size distribution (~4 nm). Hydrogenation of ketones (Sharma et al., 2007) was chosen as a model catalytic reaction to demonstrate the catalytic effects of the resulting Al/Ni mesoporous system (Fig. 11.16, top right). It is known that the preparation method and the structures of support exert an important influence on the catalytic activity of Ni (Hou and Yashima, 2004). The system prepared for 60 min of sonication (the highest porosity) exhibits more than 99% conversion in an aqueous solution (Fig. 11.16, bottom right) and a very high selectivity in the hydrogenation of ketones due to a very narrow pore size distribution in mesoporous metal system formed by ultrasound. Furthermore, the catalyst prepared by sonication exhibits stability unusual for Raney catalyst and did not require an oxygen-free storage condition.

Another type of nanocomposite material was prepared by the sonication of aluminum or magnesium particles in the presence of Au, Ag, or Pt colloids. Figure 11.17 is the TEM image of mesoporous Al particles with 1 wt% Au particles incorporated by sonication. Combination of different particles (e.g., gold and platinum) is also possible. For particles incorporation, the layer-by-layer assembly incorporation approach suggested by Möhwald et al. (2009) could be used. The precipitated nano-Au/intercalated product could be separated by centrifugation or filtration, washed, and dried.

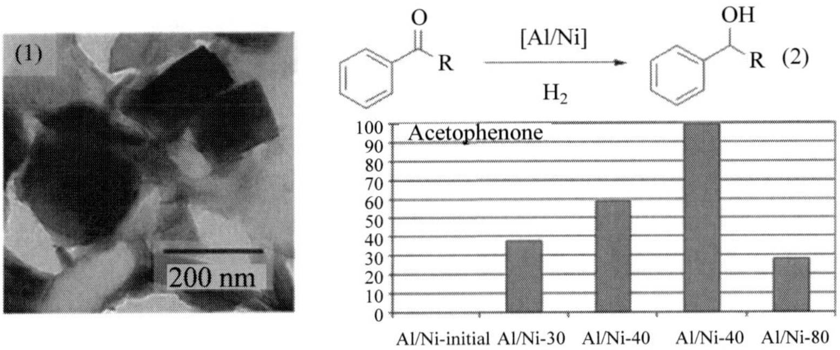

Figure 11.16 One-step formation of Al/Ni mesoporous system: TEM microphotograph of nickel particles distributed in the porous aluminum support (1); effect of sonication time on the efficiency of Al/Ni catalyst in the hydrogenation of acetophenone (2). Reprinted from Skorb et al. (2010a). Copyright the Royal Society of Chemistry.

Sponges formed by the method suggested by Suslick and Casadonte (1987) also exhibit very high catalytic activity. Simple Ni powder is an extremely inactive catalyst for the hydrogenation of alkenes. It was shown that ultrasonic pretreatment of Ni powder for 1 h (~50 W/cm ^{2} at 20 kHz in octane) gives optimal activity, which decreases with longer irradiation. It was shown that the hydrogenation activity is quite general and shows little dependence on the choice of alkene (i.e., similar rates are observed with 1-nonene, 1-decene, *cis*-5-decene, *trans*-5-decene, cyclooctene, and cyclohexene). There was no reduction of ketones or aldehydes observed in the case of described system.

11.5.4.2 Hybrid materials construction

It is possible to carry out radical polymerizations without the addition of initiators under ultrasound irradiation simultaneously with the formation of metal sponges (Fig. 11.17). Such structures are very interesting because of a sponge skeleton formed by metal and unique properties of polymer.

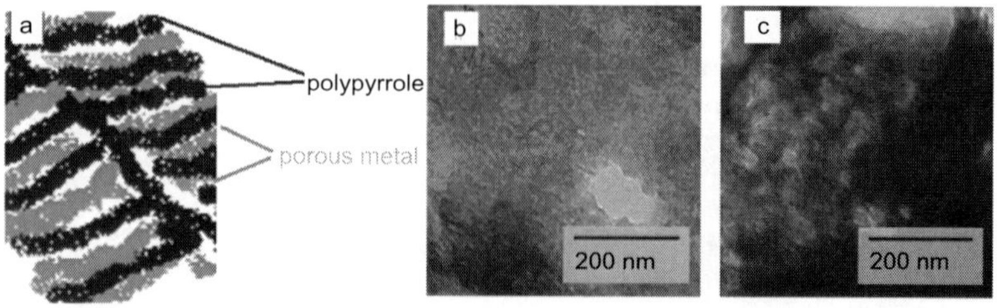

Figure 11.17 Schematic illustration of the structure of metal polymer composite synthesized in one step (a); and TEM (b,c) image providing an evidence of complete filling of pores in metal matrix with polymer (polypyrrole): aluminum particles, 15 min sonication at 57 W/cm^2 in ethanol solution (b); ethanol + pyrrole solution (c). Reprinted from Skorb et al. (2011). Copyright the Royal Society of Chemistry.

11.5.4.3 Implant development

Biomaterials used in dentistry as implantable devices in bone must possess adequate mechanical and physicochemical properties to achieve a high degree of biocompatibility and biofunctionality. Titanium and its alloys are widely used for orthopedic and dental implants because they possess low density, good mechanical properties (elastic modulus, toughness, and fatigue strength), and biological and chemical inertness. Currently, oral implants are manufactured from commercially pure titanium or Ti–6Al–4V (90% titanium, 6% aluminum, 4% vanadium) alloy. Although alloys have superior physical and mechanical properties, commercially pure titanium is universally considered as the material of choice in implant dentistry because of its exceptionally good corrosion resistance and biocompatibility (Sykaras et al., 2000). However, the stiffness of titanium is generally not well adapted to bone and this can lead to stress shielding of the residual bone, which may result in detrimental resorptive bone remodeling (Turner et al., 1986). Until recently, implants were produced by machining titanium rods, followed by most manufacturers, by modification of the titanium surfaces by applying different surface treatments or coatings to improve stability and enhance osteointegration. In the last few years, the development of direct laser metal sintering processes has substantially broadened the field of application of titanium alloys and allowed implants to be produced more economically than by traditional techniques (Traini et al., 2008). The sonochemical alloy powder treatment is also one of the alternatives to improve the properties of powder, which using the rapid prototyping technique could be formed to three-dimensional freeform objects, including dental implants, directly from their CAD models (Fig. 11.18). In conclusion, metal sintering could be an efficient means of construction of dental implants with a functionally graded material, which is better adapted to the elastic properties of the bone. Such implants should minimize stress-shielding effects and improve long-term performance.

Figure 11.18 1 SEM images at 32× magnification of dental implants produced by DLMS. (A) is an implant after sintering procedure. (B) is an implant after sintering and etching process. Reprinted from Traini et al. (2008). Copyright Elsiever.

11.5.4.4 Containers with controllable loading/release properties

Figure 11.19 shows the 3D reconstructions of the transmission and fluorescence images of confocal scanning fluorescence microscopy (CSFM) of the aluminum particles loaded with a water-soluble antibiotic doxorubicin hydrochloride. The CSFM image of microreservoirs loaded with a water-soluble antibiotic doxorubicin hydrochloride is an example of application of metal sponges as "smart" containers. Pore system of the sponges could be effectively loaded even with low molecular weight chemicals. Simultaneously, the pore of the reservoirs should be closed by the procedure suggested by Andreeva et al. (2008). It was shown that the developed approach permitting the formation of metal sponges can be extended to flat metal surfaces that allow the production of highly porous structures at the surface of the macroscopic metal samples. These structures represent a system of microreservoirs (Fig. 11.20), which can be used for storage of different chemical agents (e.g., corrosion inhibitors in their interior). To prevent the leakage of chemicals stored in the pores, polyelectrolyte multilayers were deposited onto the pore surface via layer-by-layer adsorption of oppositely charged polyelectrolytes. The local change in pH accompanying the corrosion process results in conformational transitions in the polyelectrolyte multilayer releasing the corrosion inhibitor just in

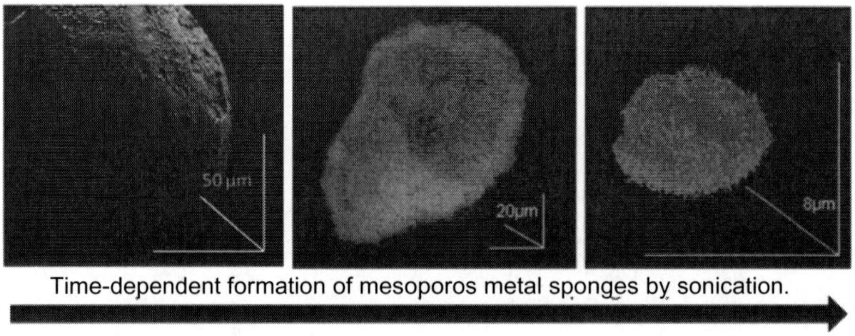

Figure 11.19 3D confocal microscopy reconstruction of the metal sponge initial (left) loaded with doxorubicin hydrochloride: (middle) the metal particle after 3 min (the reconstruction of both transmission and fluorescent mode) and (right) 60 min of sonochemical exposure at 57 W/cm^2. Reprinted from Skorb et al. (2011). Copyright the Royal Society of Chemistry.

the vicinity of the corrosion pit. This "self-healing" effect is clearly demonstrated by the scanning vibration electrode technique that allows measuring the current density over selected areas of the sample, thus monitoring local anodic activity in the corrosion zones. The current density maps shown in Fig. 11.20 evidence that a hybrid protective layer is formed at the surface of aluminum by sonication followed by loading of resultant pores with corrosion inhibitor (mercaptobenzothiazole), and deposition of polyelectrolyte layers ensures complete suppression of corrosion (Fig. 11.20).

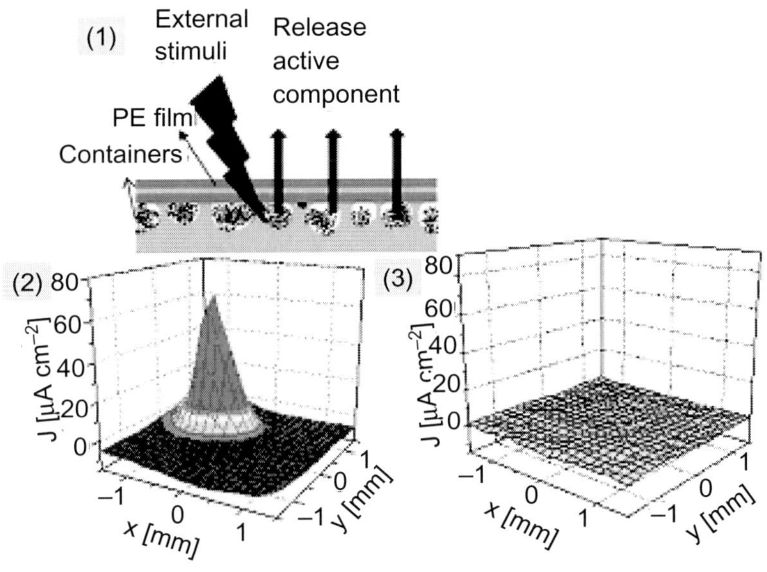

Figure 11.20 Schematic representation of the self-healing corrosion protection layer at the surface of the aluminum plate. The corrosion inhibitor is encapsulated in the pores at the metal surface by applying polyelectrolyte multilayer. The corrosion process initiates the release of corrosion inhibitor (1). The current density maps obtained by SVET for aluminum plates 55 immersed in 0.1 M NaCl for 12 h (2 and 3). Before measurements, the sample (2) was exposed at 57 W/cm^2 ultrasound for 30 min; (3) exposed at 57 W/cm^2 ultrasound for 30 min, soaked in the mercaptobenzothiazole solution, and modified by deposition of polyelectrolyte multilayer; the 60 complete suppression of corrosion is observed in the latter case. Reprinted from Skorb et al. (2011). Copyright the Royal Society of Chemistry.

11.6 Conclusions

High-power ultrasound results from the interplay of conceptually different mechanisms. There is physical impact due to a pressure wave and a jet impinging on the surface of a particle and deep into it, although the extreme temperatures and pressures create highly reactive radicals derived from water and oxygen. Obviously, the formation of mesoporous surfaces and metal sponges demonstrated here rely on both the mechanisms.

Simultaneously, metal surfaces are oxidized and the pore structure is stabilized. Modification of metals by ultrasound depends on the nature of metal. The metals with relatively high reactivity (e.g., Zn, Al, Mg) can be involved in oxidation during sonochemical process. The control over the oxidation process allows the formation of porous metal structures consisting of metal skeleton stabilized with oxide layer. The noble metals do not form sponges mostly due to their low tendency to oxidation. In the case of metal alloys, phase segregation occurs upon sonication yielding metal nanoparticles with higher resistance to ultrasound treatment distributed in the pores of the second, less stable alloy component.

Some possible applications of mesoporous metal surfaces and metal sponges were discussed. In particular, the following materials can be effectively used in different fields of nanotechnology and nanoengineering: (i) formation of metal-polymer interpenetrating nanocomposite ("lotus"-like structure, pH-sensitive polyelectrolyte/inhibitor sandwichlike nanostructures, hybrid materials construction), (ii) formation of heterogenic nanocatalysts by using initial metal alloys and by incorporation of metal nanoparticles into sponge matrix; (iii) implant development; (iv) formation of micrometer thick sponge-like surface structure on metal plates to enhance their properties.

References

Andreeva, D. V., Fix, D., Moehwald, H. and Shchukin D. G. (2008). Self-healing anticorrosion coatings based on pH-sensitive polyelectrolyte/inhibitor sandwich-like nanostructures, *Adv. Mat.*, **20**, pp. 2789–2794.

Arenas, M. A., Conde, A. and de Damborenea, J. J. (2002). Cerium: a suitable 495 green corrosion inhibitor for tinplate, *Corros. Sci.*, **44**, pp. 511–520.

Attard, G. S., Bartlett, P. N., Coleman, N. R. B., Elliott, J. M., Owen, J. R. and Wang, J. H. (1997). Mesoporous platinum films from lyotropic liquid crystalline phases, *Science*, **278**, pp. 838–840.

Bang, J. H. and Suslick, K. S. (2010). Applications of ultrasound to nanostructured materials synthesis, *Adv. Mater.*, **22**, pp. 1039–1059.

Banhart, J. (2001). Manufacture, characterisation and application of cellular metals and metal foams, *Prog. Mater. Sci.*, **46**, pp. 559–632.

Belova, V., Möhwald, H. and Shchukin, D. G. (2009a). Ultrasonic intercalation of gold nanoparticles into clay matrix in the presence of surface-active materials. Part I: Neutral polyethylene glycol, *J. Phys. Chem. C*, **113**, pp. 5381–5389.

Belova, V., Möhwald, H. and Shchukin, D. G. (2009b). Ultrasonic intercalation of gold nanoparticles into clay matrix in the presence of surface-active materials. Part II: Negative sodium dodecylsulfate and positive cetyltrimethylammonium bromide, *J. Phys. Chem. C*, **113**, pp. 6751–6760.

Belova, V., Gorin, D. A., Shchukin, D. G. and Mohwald, H. (2010). Selective ultrasonic cavitation on patterned hydrophobic surfaces, *Angew. Chem. Int. Ed.*, **49**, pp. 7129–7133.

Berlan, J., Trabelsi, F., Delmas, H., Wilhelm, A. M. and Petrignani, J. F. (1994). Oxidative degradation of phenol in aqueous media using ultrasound. *Ultrason. Sonochem.*, **1**, pp. S97–S102.

Caruso, F., Caruso, R. and Mohwald, H. (1998). Nanoengineering of inorganic and hybrid hollow spheres by colloidal templating, *Science*, **282**, pp. 111–114.

Chen, H. Y., Zhang, F. Z., Fuand, S. S. and Duan, X. (2006). Hydroxide monolayer films with curved hexagonal crystals as superhydrophoboic materials. *Adv.Mater.*, **18**, pp. 3089–3093.

Cravotto, G. and Cintas, P. (2007). Reconfiguration of stereoisomers under sonomechanical activation. *Angew. Chem. Int. Ed.*, **46**, pp. 5476–5478.

Delius, M., Brendel, W. and Heine, G. (1988). A mechanism of gallstone destruction by extracorporeal shock waves, *Naturwissenschaften*, **75**, pp. 200–201.

Didenko, Y. and Suslick, K. S. (2002). The energy efficiency of formation of photons, radicals, and ions during single bubble cavitation. *Nature*, **418**, pp. 394–397.

Didenko, Y. T. and Suslick, K. S. (2005). Chemical aerosol flow synthesis of semiconductor nanoparticles, *J. Am. Chem. Soc.*, **127**, pp. 12196–12197.

Doktycz, S. J. and Suslik, K. S. (1990). Interparticle collisions driven by ultrasound. *Science*, **247**, pp. 1067–1069.

Elder, S. A., Kolb, J. and Nyborg, W. L. (1954). Small-scale acoustic streaming effects in liquids. *J. Acoust. Soc. Am.*, **26**, pp. 933–934.

Elder, S. A. (1959). Cavitation microstreaming. *J. Acoust. Soc. Am.*, **31**, pp. 54–64.

Erlebacher, J., Aziz, M. J., Karma, A., Dimitrov, N. and Sieradzki, K. (2001). Evolution of nanoporosity in dealloying, *Nature*, **410**, pp. 450–453.

Jakab, M. A. and Scully, J. R. (2005). On demand release of corrosion inhibiting ions from amorphous Al–Co–Ce alloys, *Nat. Mater.*, **4**, pp. 667–670.

Lide, D. R. (ed.), Handbook of Chemistry and Physics (2004–2005). Boca Raton, FL, CRC Press, 85th edn., pp. 8–47.

Hamdy, A. S. (2006). Advanced nano-particles anti-corrosion ceria based sol-gel coatings for aluminum alloys, *Mater. Lett.*, **60**, pp. 2633–2637.

Harvey, E. N., Barnes, D. K., McElroy, W. D., Whitely, A. H., Pease, D. C. and Cooper, K. W. J. (1944). Bubble formation in animals. *Cell Compar. Physiol.*, **24**, pp. 1–22.

Hou, Z. Y., Yashima, T. (2004). Mesoporous Ni/Mg/Al catalysts for methane reforming with CO_2, *Applied Catalysis A*, **261**, 205–209.

Iyer, A. M., Greish, K., Fang, J., Murakami, R., Maeda, H. (2007). High-loading nanosized micelles of copoly(styrene-maleic acid)-zinc protoporphyrin for targeted delivery of a potent heme oxygenase inhibitor. *Biomaterials*, **28**, 1871–1881.

Lindlay, J. and Mason, T. J. (1987). Sonochemistry; Part I, Part II. *Chem. Soc. Rev.*, **16**, pp. 239–313.

Luche, J.-L. (1994). Effect of ultrasound on heterogeneous systems, *Ultrason. Sonochem.*, **1**, pp. S111-S118.

Margulis, M. A. (1992). Fundamental aspects of sonochemistry. *Ultrasonics*, **30**, pp. 152–155.

Margulis, M. A. (1994). Fundamental problems of sonochemistry and cavitation. *Ultrason. Sonochem.*, **1**, pp. S87–S90.

Margulis, M. A. (1985). Sonoluminescence and sonochemical reactions in cavitation fields. A review. *Ultrasonics*, **23**, pp. 157–169.

Makino, K., Mossoba, M. M. and Riesz, P. (1982). Chemical effects of ultrasound on aqueous solutions. Evidence for •OH and •H by spin trapping. *J. Am. Chem. Soc.*, **104**, pp. 3537–3539.

Makino, K., Mossoba, M. M. and Riesz, P. (1983). Chemical effects of ultrasound on aqueous solutions. Formation of hydroxyl radicals and hydrogen atoms. *J. Phys. Chem.*, **87**, pp. 1369–1377.

Mason, T. J. and Lorimer, J. P. (1988). *Sonochemistry: Theory, Applications, and Uses of Ultrasound in Chemistry*, Ellis Horwood, Chichester and John Wiley and Sons, New York, p. 252.

Mason, T. J. *Chemistry with Ultrasound*, Elsevier Science Publishers Ltd, Amsterdam, 1990, vol. 28, p. 195.

Masuda, H. and Fukuda, K. (1995). Ordered metal nanohole arrays made by a two-step. Replication of hoeycomb structures of anodic alumina, *Science*, **268**, pp. 1466–1468.

Merkel, R., Nassoy, P., Leung, A., Ritchie, K. and Evans, E. (1998). Energy landscapes of receptor-ligand bonds explored with dynamic force spectroscopy, *Nature*, **397**, pp. 50–53.

Naude, C. F. and Ellis, A. T. (1961). On the mechanism of cavitation damage by non-hemispherical cavities in contact with a solid boundary, *J. Basic Eng.*, **83**, pp. 648–656.

Rayleigh, L. (1917). On the pressure developed in a liquid during the collapse of a spherical cavity, *Pil. Mag.* **34**, pp. 94–98.

Sharma, G., Mey, Y., Lu, Y., Ballauff, M., Irrgang, T., Prosh, S., Kempe, R. (2007). Spherical polyelectrolyte brushes as carriers fpr platinum nanoparticles in heterogeneous hydrogenation reactions, *J. Catalysis*, **246**, 10–14.

Shchukin, D. G. and Möhwald, H. (2006). Sonochemical nanosynthesis at engineered interface of cavitation microbubble, *Phys. Chem. Chem. Phys.*, **8**, pp. 3496–3506.

Shchukin, D. G. and Möhwald, H. (2007). Self-repairing coatings containing active nanoreservoirs, *Small*, **3**, pp. 926–943.

Shchukin, D. G., Skorb, E. V. and Sviridov, D. (2008). Hybrid silica-zirconia films loaded with titania nanoparticles and titania-based nanocontiners: Novel materials for thin-film photocatalysts and photocontrollable coatings in *Molecular and Nanoscale Systems for Energy Conversion*, Ed., E. Zaikov, Nova Science Publishers, Inc., pp. 75–87.

Shchukin, D. G., Andreeva, A. V., Skorb, E. V. and Möhwald, H. (2010a). Emerging concepts in interfacial chemistry of hybrid materials: Nanocontainer-based self-healing coatings in *The Supramolecular Chemistry of Organic-Inorganic Hybrid Materials*, Ed., K. Rurack, John Wiley & Sons, Inc., Hoboken, NJ, pp. 639–652.

Shchukin, D. G., Radziuk, D. and Möhwald, H. (2010b). Ultrasonic fabrication of metallic nanomaterials and nanoalloys, *Annu. Rev. Mater. Res.*, **40**, pp. 345–362.

Silberrad, D. (1912). Propeller erosion, *Engineering*, pp. 33–35.

Skorb, E. V., Fix, D., Andreeva, D. V., Shchukin, D. G. and Möhwald, H. (2009). Surface-modified mesoporous SiO_2 containers for corrosion protection, *Adv. Funct. Mater.*, **19**, pp. 2373–2379.

Skorb, E. V. and Shchukin, D. G. (2010a). New corrosion protection concepts in *Anticorrosive Coatings*, Ed., J. Sander, Vincentz Network GmbH, 2010, pp. 159–199.

Skorb, E. V., Shchukin, D. G., Mohwald, H. and Andreeva, D. V. (2010b). Sonochemical design of cerium-rich anticorrosion nanonetwork on metal surface, *Langmuir*, **26**, 16973–16979.

Skorb, E. V., Shchukin, D. G., Mohwald, H. and Andreeva, D. V. (2010c). Ultrasound-driven design of metal surface nanofoams, *Nanoscale*, **2**, pp. 722–727.

Skorb, E. V., Fix, D., Shchukin, D. G., Möhwald, H., Sviridov, D. V., Mousa, R., Wanderka, N., Schäferhans, J., Pérez, N. P., Fery, A. and Andreeva, D. V. (2011). Sonochemical formation of metal sponges, *Nanoscale*, **3**, pp. 985–993.

Spadaro, C., Dispenza, C. and Sunseri, C. (2006). The influence of the nature of the surface oxide on the adhesive fracture energy of aluminium bonded joints as measured by T-peel tests. *J. Phys. Condens. Matter.*, **18**, pp. S2007-S2018.

Suslick, K. S., Gawlenowski, J. J., Schubert, P. F. and Wang, H. H. (1983). Alkane sonochemistry. *J. Phys. Chem.*, **87**, pp. 2299–2301.

Suslick, K. S. and Johnson, R. E. (1984). Sonochemical activation of transition metals. *J. Am. Chem. Soc.*, **106**, pp. 6856–6858.

Suslick, K. S., Hammerton, D. A. and Raymond, E. C. (1986). The sonochemical hot spot. *J. Am. Chem. Soc.*, **108**, pp. 5641–5642.

Suslick, K. S. and Casadonte, D. J. (1987). Heterogeneous sonocatalysis with nickel powder. *J. Am. Chem. Soc.* ,**109**, pp. 3459–3461.

Suslick, K. S., Casadonte, D. J., Green, M. L. H. and Thompson, M. E. (1987). Effects of high intensity ultrasound on inorganic solids. *Ultrasonics*, **25**, pp. 56–59.

Suslick, K. S., Casadonte, D. J. and Doktycz, S. J. (1989). The effect of ultrasound on nickel and copper powders, *Solid State Ionics*, **32/33**, pp. 444–452.

Suslick, K. S. (1990). Sonochemistry, *Science*, **247**, pp. 439–1445.

Suslick, K. S., Doktycz, S. J. and Flint, E. B. (1990). On the origins of sonochemistry and sonoluminescence, *Ultrasonics*, **28**, pp. 280–290.

Suslick, K. S., Choe, S. B., Cichowlas, A. A. and Grinstaff, M. W. (1991). Synthesis of amorphous iron, *Nature*, **353**, pp. 414–416.

Suslick, K. S. and Crum, L. A. (1997). Sonochemistry and sonoluminescence, *Encyclopedia of Acoustics*, Wiley-Interscience, New York, pp. 271–282.

Suslick, K. S. and Price, G. J. (1999). Applications of ultrasound to materials chemistry, *Annu. Rev. Mater. Sci.*, **29**, pp. 295–326.

Sykaras, N., Iacopino, A. M., Marker, V. A., Triplett, R. G. and Woody, R. D. (2000). Implant materials, design and surface topographies: their effect on osseointegation. A literature review. *Int. J. Oral Maxillofac. Implants*, **15**, pp. 675–690.

Thompson, L. H. and Doraiswamy, L. K. (1999). Sonochemistry: Science and Engineering, *Ind. Eng. Chem. Res.*, **38**, pp. 1215–1249.

Thornycroft, J. I., Barnaby, S. W. (1895). Torpedo-boat destroyers. *Min. Proc. Inst. Chem. Eng.*, **122**, pp. 51–69.

Traini, T., Mangano, C., Sammons, R. L., Mangano, F., Macchi, A. and Piattelli, A. (2008). Direct laser metal sintering as a new approach to fabrication of an isoelastic functionally graded material for manufacture of porous titanium dental implants, *Dental materials*, **24**, pp. 1525–1533.

Turner, T. M., Sumner, D. R., Urban, R. M., Rivero, D. P. and Galante, J. O. (1986). A comparative study of porous coatings in a weight-bearing total hip-arthroplasty model. *J Bone Joint Surg Am*, **68**, pp. 1396–1409.

Xingwen, Y., Chunan, C., Zhiming, Y., Derui, Z. and Zhongda, Y. (2001). Corrosion behavior of rare earth metal (REM) conversion coatings on aluminum alloy LY12, *Corros. Sci.*, **43**, pp. 1283–1294.

Velev, O. D., Tessier, P. M., Lenhoff, A. M. and Kaler, E. W. (1999). Class of porous metallic nanostructures, *Nature*, 1999, **401**, pp. 548–549.

Weissler, A. (1959). Formation of hydrogen peroxide by ultrasonic waves: Free radicals. *J. Am. Chem. Soc.*, **81**, pp. 1077–1081.

Zhang, F., Zhao, L., Chen, H., Xu, S., Evans, D. G. and Duan, X. (2008). Corrosion resistance of superhydrophobic layered double hydroxide films on aluminum. *Angewandte Chemie Int. Edition*, **120**, pp. 2466–2469.

Chapter 12

Numerical Simulations of Nucleation and Aggregation of $BaTiO_3$ Nanocrystals Under Ultrasound

Kyuichi Yasui and Kazumi Kato
National Institute of Advanced Industrial Science and Technology (AIST), 2266-98 Anagahora, Shimoshidami, Moriyama-ku, Nagoya 463-8560, Japan
k.yasui@aist.go.jp

Dang et al. (2009, 2010a,b) reported that ultrasound-assisted synthesis of $BaTiO_3$ nanoparticles resulted in oriented aggregation of nanocrystals. Such aggregates are called mesocrystals and the electron diffraction pattern of a mesocrystal is the same as that from a single crystal. In order to understand the mechanism of mesocrystal formation, numerical simulations of nucleation and aggregation of $BaTiO_3$ nanocrystals have been performed under the experimental condition of Dang et al. It is suggested that electric dipole–dipole interaction may result in the mesocrystal formation

Cavitation: A Novel Energy-Efficient Technique for the Generation of Nanomaterials
Edited by Sivakumar Manickam and Muthupandian Ashokkumar

ISBN 978-981-4411-54-7 (Hardcover), 978-981-4411-55-4 (eBook)
www.panstanford.com

12.1 Introduction

12.1.1 Acoustic Cavitation and Sonochemistry

Ultrasound is the inaudible sound above 20 kHz in frequency. It is the propagation of pressure oscillations in liquid, gas, or solid. In liquid in which gas is dissolved, strong pressure oscillations of ultrasound cause the appearance of tiny gas bubbles of a few micrometers in radius (Leighton, 1994; Young, 1999). The bubbles expand during the rarefaction phase of ultrasound and subsequently collapse during the compression phase. The phenomenon is called acoustic cavitation. Cavitation takes place not only by ultrasound but also by the drop in pressure by fluid motion such as in the Venturi tube, which is called hydrodynamic cavitation.

In liquid irradiated by ultrasound, a bubble expands dramatically during the rarefaction phase of ultrasound if the pressure amplitude of ultrasound (acoustic amplitude) is larger than the threshold pressure to overcome the pressure due to surface tension and if the equilibrium bubble radius is smaller than the linear resonance radius (Yasui et al, 2008b). Under the condition, a bubble violently collapses after its expansion. The violent collapse is due to the spherically shrinking geometry and the inertia of the surrounding liquid inwardly flowing (Yasui, 2011). The violent bubble collapse is called the Rayleigh collapse. The speed of the Rayleigh collapse sometimes reaches the sound velocity in the liquid.

At the end of the Rayleigh collapse, a spherical shock wave is emitted from the bubble into the surrounding liquid. In Fig. 12.1, the spherical shock wave was observed by a video camera under the illumination of pulsed laser light of 7 ns duration (Holzfuss et al., 1998). The spherical shock wave is seen as a circle with its center being the position of the collapsed bubble. The speed of the shock wave propagation is more than the sound speed in the liquid. The maximum pressure of the shock wave is more than 1×10^4 bar near the bubble and quickly decays with the propagation distance with a faster decay rate than the usual r^{-1}, where r is the distance from the bubble center.

At the end of the violent bubble collapse, the temperature and pressure inside the bubble dramatically increase to thousands of

Kelvin and thousands of bars or more, respectively. In Fig. 12.2(a), the bubble radius is shown as a function of time for one acoustic cycle, which is calculated by the modified Rayleigh–Plesset equation (Yasui et al., 2005). For details of the modified Rayleigh–Plesset equation see Yasui (2011). In Figs. 12.2(c) and (d), the enlarged view at around the end of the violent collapse is shown. It is seen that the temperature inside a bubble increases to 6500 K at the end of the collapse. As a result, nitrogen is burned and water vapor is dissociated inside a bubble. Many chemical products such as HNO_3, HNO_2, O, H_2O_2, HO_2, and OH are created inside bubble. OH radical is especially important because it is a strong oxidant. Many of the OH radicals created inside a bubble diffuse out of the bubble during the violent bubble collapse, while some others do during the bubble expansion as seen in Fig. 12.2(b). The amount of OH radicals diffusing out of a bubble per acoustic cycle is $10^5 \sim 10^6$ molecules at 52 kHz and $1.3 \sim 1.55$ bar in ultrasonic frequency and pressure amplitude, respectively (Didenko and Suslick, 2002).

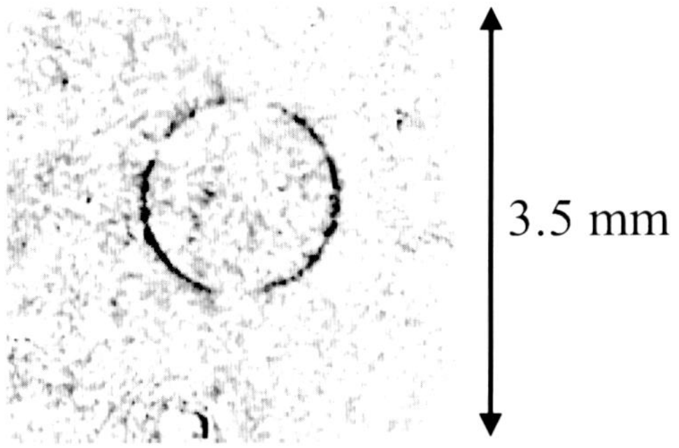

Figure 12.1 A spherical shock wave emitted from a pulsating bubble at the center of the sphere at 480 ns after the bubble collapse. Reprinted figure with permission from Holzfuss, J., Ruggeberg, M., and Billo, A. (1998). Shock wave emissions of a sonoluminescing bubble, *Phys. Rev. Lett.*, **81**, pp. 5434–5437 [http://prl.aps.org/abstract/PRL/v81/i24/p5434_1]. Copyright (1998) by the American Physical Society

Chemical reactions induced or accelerated by ultrasound are called sonochemical reactions. Many of sonochemical reactions are induced by oxidants such as OH radicals and H_2O_2 created inside cavitation bubbles. Some other sonochemical reactions may be caused by the shock waves emitted by cavitation bubbles. There are

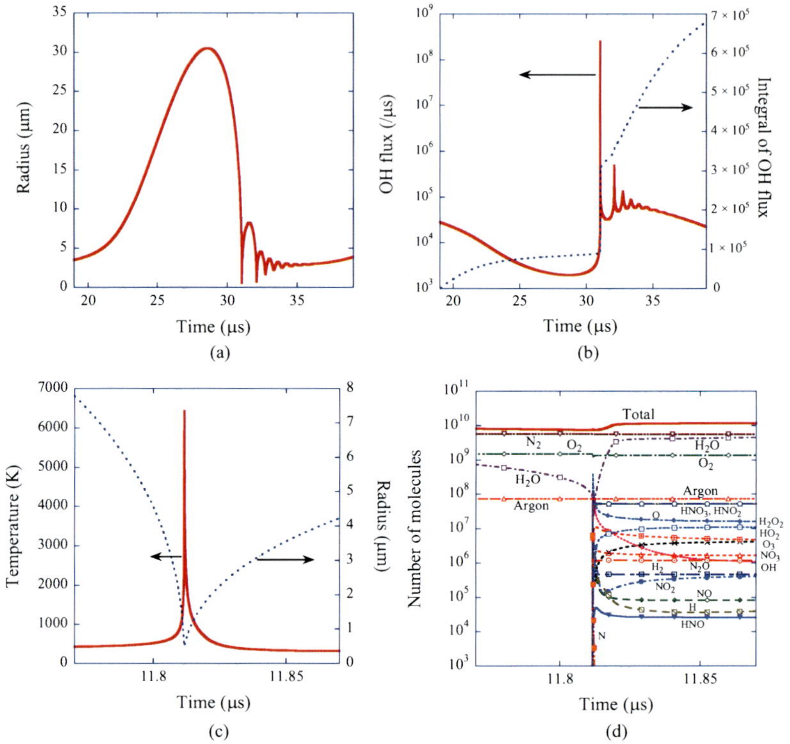

Figure 12.2 The results of numerical simulations when an air bubble in water at 3°C is irradiated by an ultrasonic wave of 52 kHz and 1.52 bar in frequency and pressure amplitude, respectively. The equilibrium bubble radius is 3.6 μm. (a) The bubble radius–time curve for one acoustic cycle. (b) The OH flux from the interior of a bubble into the surrounding liquid (solid line) and its time integral (dotted line) as a function of time for one acoustic cycle. (c) The bubble radius (dotted line) and the temperature inside a bubble (solid line) as a function of time only for 0.1 μs at around the end of the bubble collapse. (d) The number of molecules for various chemical species with the same time axis with that of (c). Reprinted with permission from Yasui, K., Tuziuti, T., Sivakumar, M., and Iida, Y. (2005). Theoretical study of single-bubble sonochemistry, *J. Chem. Phys.*, **122**, 224706. Copyright 2005, American Institute of Physics.

two other important phenomena of cavitation bubbles. One is the liquid jet impact on a solid surface (Fig. 12.3) (Ohl et al., 2006). Due to the broken symmetry of the fluid dynamical environment around a bubble near a solid surface, a liquid jet penetrates the bubble at the violent bubble collapse. The liquid jet finally hits the solid surface and causes the cleaning of the surface as well as its damage. It may also induce some chemical effect on the solid surface. The other is the fluid motion around a pulsating bubble called microstreaming (Fig. 12.4) (Elder, 1959). Microstreaming is stronger especially when a bubble undergoes non-spherical shape oscillations (Leong et al., 2011). The fluid motion may accelerate the diffusion of solutes and may affect some chemical reactions.

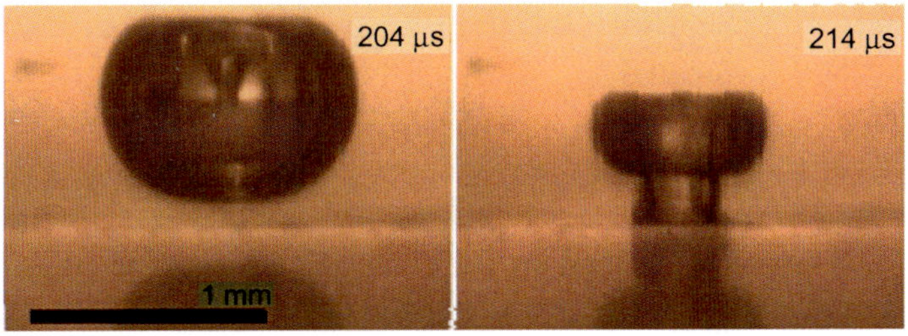

Figure 12.3 A liquid jet within a laser-induced cavitation bubble near a solid boundary (left) and impacting on the boundary (right). Reprinted with permission from Ohl, C. D., Arora, M., Dijkink, R., Janve, V., and Lohse, D. (2006). Surface cleaning from laser-induced cavitation bubbles, *Appl. Phys. Lett.*, **89**, 074102, Copyright (2006), American Institute of Physics.

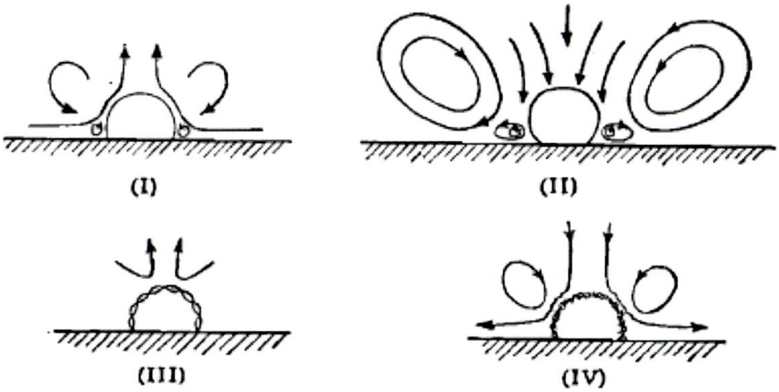

Figure 12.4 Several patterns of microstreaming around a weakly pulsating bubble on a solid surface under ultrasound. Reprinted with permission from Elder, S. A. (1959). Cavitation microstreaming, *J. Acoust. Soc. Am.*, **31**, pp. 54–64. Copyright (1959), Acoustical Society of America.

Under an ultrasonic horn, many cavitation bubbles are produced. Their pulsations strongly interact with each other due to the following two mechanisms. One is the acoustic emissions from surrounding bubbles that act as sources of sound due to their pulsation. The acoustic emissions change the instantaneous pressure at each bubble and influence the pulsation of a bubble. The effect is called the bubble–bubble interaction (Yasui et al., 2008a). The other is the shielding of a driving acoustic wave by the surrounding bubbles. This effect generally reduces the effective acoustic pressure on a bubble. How these effects influence the rate of OH radical production, the intensity of shock waves, and so on has not yet been quantitatively understood.

Many bubbles move downward from the horn tip whereas some others move upward (Moussatov et al., 2003; Yasui et al., 2008a). The fluid flows downward under an ultrasonic horn mostly due to acoustic streaming, which is the fluid flow caused by the radiation pressure of a driving acoustic wave with its intensity decreasing with distance (Mitome et al, 2000). Another reason of the fluid flow is the drag force of bubbles moving downward. Many bubbles move downward mostly due to the radiation force of a driving acoustic wave. Many of them are attracted toward each other due to the radiation force between bubbles called secondary Bjerknes force (Yasui, 2011). The primary Bjerknes force is the name given to the radiation force on a bubble from a standing acoustic wave. The detailed understanding on the trajectories of bubbles under an ultrasonic horn is one of the important subjects in this field to be studied in the future (Mettin, 2007).

According to Gedanken (2004), nanomaterials were obtained in almost all the sonochemical reactions leading to inorganic products. Sonochemically synthesized nanomaterials are amorphous when the precursor is volatile such as $Fe(CO)_5$, which results in the formation of amorphous iron nanoparticles (Suslick et al., 1991). When the precursor is non-volatile, sonochemically synthesized nanomaterials are mostly crystalline nanoparticles, whereas for some other cases, they are amorphous or semi-crystalline nanoparticles (Sivakumar, 2011). The detailed mechanism of nanoparticle formation under ultrasound, however, is unknown at present.

In the present study, numerical simulations of sonochemical production of nanoparticles have been performed in order to study the detailed mechanism of their formation.

12.1.2 Mesocrystal Formation Under Ultrasound

Barium titanate ($BaTiO_3$), a ferroelectric ceramic material, is used extensively as the dielectric in ceramic capacitors, particularly multilayer ceramic capacitors (MLCCs), due to its high dielectric constant and low loss characteristics (Venigalla, 2001). Other applications for $BaTiO_3$-based materials include positive temperature coefficient of resistance (PTCR) thermistors, electro optic and piezoelectric devices, electroluminescent panels, integrated passive filters, and coatings for shielding and electrostatic control (Mori et al., 1995; Uchino, 2000; Qi et al., 2004).

In the MLCC, many dielectric layers and internal electrodes are alternately stacked (Kishi et al., 2003). The capacitance C of the MLCC is given by Eq. (12.1).

$$C = \varepsilon_r \cdot \varepsilon_0 \cdot (L-1) \cdot s/w \qquad (12.1)$$

where ε_r is the relative dielectric constant of the dielectric ceramic, ε_0 is the dielectric constant of vacuum, L is the number of the internal electrode layers, s is the area of each internal electrode, and w is the thickness of the dielectric layer. In order to achieve large capacitance with small size of MLCC, thinning of the dielectric layer is required. An ideal $BaTiO_3$ nanoparticle for the preparation of dielectric layers in MLCCs should have uniform small size and shape, high purity, homogeneous composition and cation distribution, controllability of particle shape, and weak agglomeration (Pithan et al., 2005).

One of the possibilities for the synthesis of such ideal $BaTiO_3$ nanoparticles may be the sonochemical method (Xu et al., 2006; Dang et al., 2009, 2010a, b). In the following, the experiment of Dang et al. (2009, 2010a, b) on sonochemical synthesis of $BaTiO_3$ nanoparticles in aqueous solution is briefly reviewed.
The reaction formula in their experiment is as follows:

$$BaCl_2 + TiCl_4 + 6H_2O \rightarrow Ba(OH)_2 + Ti(OH)_4 + 6HCl \qquad (12.2a)$$

$$6HCl + 6NaOH \rightarrow 6NaCl + 6H_2O \qquad (12.2b)$$

$$Ba(OH)_2 + Ti(OH)_4 \rightarrow BaTiO_3 + 3H_2O \qquad (12.2c)$$

Distilled water was sparged with argon gas for 30 min. $BaCl_2$ was dissolved in it. Then, $TiCl_4$ was added to the solution, resulting in a Ti-based gel suspension. The molar ratio of Ti to Ba was 1. A 5 M NaOH aqueous solution was added at room temperature for peptization to make the pH 14. The suspension was irradiated with ultrasound of 20 kHz using an ultrasonic horn in open air for 20 min at 80°C. Then $BaTiO_3$ nanoparticles were formed. After the synthesis, the precipitate was centrifugally separated and washed with deionized water twice. Then, it was dried at 100°C for 2 h in vacuum.

In Fig. 12.5(a)–(c), the photographs of the sonochemically synthesized $BaTiO_3$ particles by transmission electron microscopy (TEM) are shown with increasing magnification in the order (a) to (c) (Dang et al., 2010a). In Fig. 12.5(a), it is seen that the particle size distribution is narrow around 250 nm in diameter. In Fig. 12.5(b), it is seen that a $BaTiO_3$ particle is an aggregate of primary particles. The primary particle is a nanocrystal of 5–10 nm in diameter (Fig. 12.5c). The particles have been confirmed as $BaTiO_3$ by X-ray diffractometry (XRD) (Dang et al., 2009).

In Fig. 12.5(e), the selected area electron diffraction (SAED) pattern from the circled area in Fig. 12.5(b) is shown. Surprisingly, the SAED pattern from an aggregate is the same as that from a single crystal. It means that the crystal axes of the nanocrystals are aligned in an aggregate. This kind of aggregates is called mesocrystal that has been intensively studied since 2005 as a new class of crystalline materials (Cölfen and Antonietti, 2008; Song and Cölfen, 2010, 2011). Several mechanisms of mesocrystal formation have been proposed such as biomineralization, roles of organic additives, alignment by capillary forces, hydrophobic forces, a mechanical stress field, magnetic fields, dipole and polarization forces, external electric fields, and minimization of the interfacial energy. The mechanism is, however, still unclear.

In the experiment of Dang et al. (2009, 2010a,b), the same reaction was performed with mechanical stirring without ultrasound for comparison. $BaTiO_3$ particles were produced only after 8 h at the same temperature (80°C). Furthermore, the particle size was 1–2 μm, which is much larger than that (0.2–0.4 μm) of the sonochemically synthesized one. The particles synthesized with mechanical stirring are not aggregates but irregular-shaped single crystals (Fig. 12.5d). It is concluded that the effect of ultrasound is the acceleration of the chemical reactions as well as the change in the size and morphology of the $BaTiO_3$ particles. The mechanism of the effect of ultrasound is, however, still unclear.

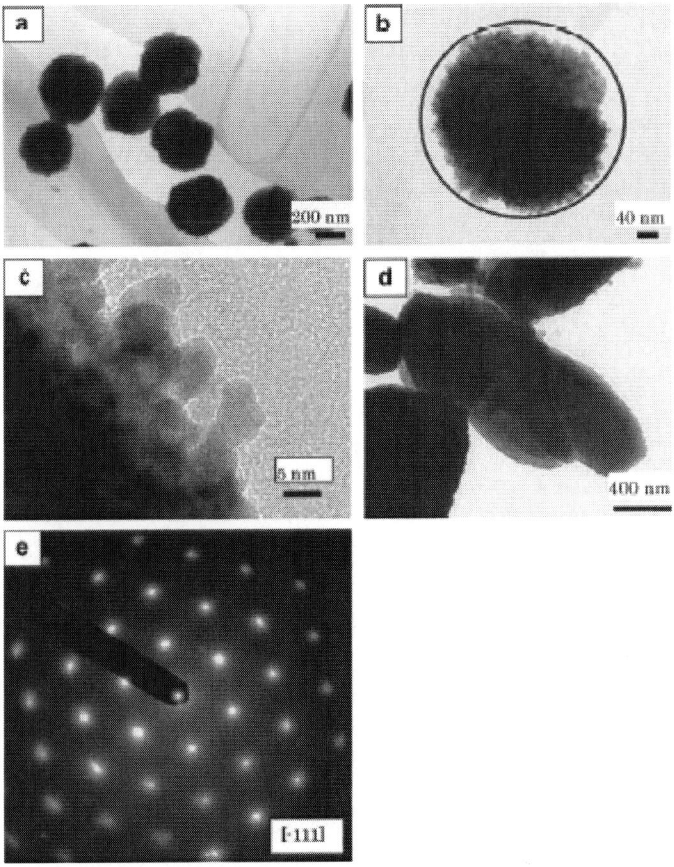

Figure 12.5 TEM photographs of the sonochemically synthesized $BaTiO_3$ particles (a–c) and the $BaTiO_3$ particles synthesized with mechanical stirring under the same condition (d). The electron diffraction pattern of the circled area in (b) is shown in (e). Reprinted from *Ultrasonics Sonochemistry,* vol. 17, Dang, F., Kato, K., Imai, H., Wada, S., Haneda, H. and Kuwabara, M. (2010). A new effect of ultrasonication on the formation of $BaTiO_3$ nanoparticles, pp. 310–314, Copyright (2010), with permission from Elsevier.

Finally, the effect of the initial concentration of $BaCl_2$ (or equivalently $TiCl_4$) is reviewed (Dang et al., 2009). The average particle (aggregate) size produced under ultrasound decreased from 400 nm to 100 nm, as the initial concentration of $BaCl_2$ increased from 0.05 mol/L to 0.4 mol/L. Furthermore, the morphology of particles changed with the initial concentration of $BaCl_2$. At 0.05 mol/L, the shape of an aggregate was confetti-like. For 0.1 L and 0.2 mol/L, the shape of an aggregate was sphere like. For 0.4 mol/L, hollow sphere particles of about 150 nm in diameter that were agglomerates of particles of about 50 nm in diameter were observed, aside from the sphere-like particles of 100 nm in diameter.

12.2 Particle Size

12.2.1 Theoretical Model

As the first step to understand the mechanism of mesocrystal formation under ultrasound, the particle (aggregate) size distribution has been numerically studied (Yasui et al., 2011). In the present section, the theoretical model used in the numerical simulations is described.

There are two mechanisms in the formation of $BaTiO_3$ particles (Eckert et al., 1996). One is the in situ reaction mechanism in which $BaTiO_3$ is formed directly on a Ti-based gel. The other is the dissolution–precipitation reaction mechanism that the starting material is dissolved into the liquid and that $BaTiO_3$ is generated by chemical reaction in the liquid phase. Then, $BaTiO_3$ crystals are nucleated from the solution and $BaTiO_3$ nanoparticles are formed. In the in situ reaction mechanism, the shape of $BaTiO_3$ nanoparticles crucially depends on the morphology of the precursor (a Ti-based gel). In general, $BaTiO_3$ nanoparticles have irregular shape. On the contrary, the dissolution–precipitation reaction mechanism may result in singular shape of $BaTiO_3$ nanoparticles. Thus, we have assumed that sonochemical synthesis is based on the dissolution–precipitation mechanism and that mechanical stirring without ultrasound may result in the in situ reaction mechanism. One of the roles of ultrasound may be the dissolution of a Ti-based gel by physical or chemical effect of acoustic cavitation. It may lead to the dissolution–precipitation reaction, while without ultrasound chemical reaction may slowly take place on a Ti-based gel.

Then, the sonochemical production of $BaTiO_3$ nanoparticles consists of three steps. The first is the chemical reaction (12.2a–c) in the liquid phase. The second is the nucleation of $BaTiO_3$ nanocrystals.

The last is the aggregation of nanocrystals. The chemical reaction rate is described by Eq. (12.3) assuming the second-order reaction for the reaction (12.2a-c).

$$[\mathrm{BaTiO_3}]_\mathrm{t} = c_0\left(1 - \frac{1}{1 + k_2 c_0 t}\right) \tag{12.3}$$

where $[BaTiO_3]_t$ is the concentration of $BaTiO_3$ in the liquid at time t neglecting the effect of nucleation of $BaTiO_3$ crystals, c_0 is the initial concentration of $BaCl_2$ (or equivalently that of $TiCl_4$), k_2 is the rate constant, and t is the time (t = 0 when the reaction starts). The acceleration of the chemical reaction by ultrasound should be taken into account in the determination of the rate constant (k_2). The value is unknown at present, however, and is a fitting parameter in the present study as 5×10^{-4} (m^3/mol s).

The nucleation rate (J) of $BaTiO_3$ nanocrystals is estimated by Eq. (12.4) (Kordylla et al., 2009).

$$J = Z \cdot k^* \cdot C_0 \exp\left(\frac{-\Delta G^*_\mathrm{het}}{k_\mathrm{B} T}\right) \tag{12.4}$$

where Z is the Zeldovich factor, k* is the diffusion controlled frequency factor, C_0 represents the number of the nucleation sites on the bubble surfaces per unit volume of the liquid, ΔG^*_het is the nucleation work decreased by the presence of the liquid/gas interfaces (the bubble surfaces), k_B is the Boltzmann constant, and T is the temperature. In Eq. (12.4), the role of ultrasound is taken into account through the values of C_0 and ΔG^*_het. In other words, cavitation bubbles work as the nucleation sites for $BaTiO_3$ crystals at the liquid/gas interfaces (Kordylla et al., 2009). However, the value of C_0 is unknown and it is an adjustable parameter in the present numerical simulations. On the contrary, ΔG^*_het is estimated by the equations given in the paper of Kordylla et al. (2009).

With regard to aggregation of $BaTiO_3$ nanocrystals, two models have been tested. One is the normal model that any pair of aggregates (or primary particles) agglomerate when they collide (Smoluchowski model) (Elimelech et al., 1995). The other is a new model that only primary particles (nanocrystals) aggregate with other particles (aggregates or primary particles) when they collide. In other words, an aggregate attaches to a primary particle but never attaches to another aggregate.

The rate of aggregation $r(a_1, a_2)$ between two particles of a_1 and a_2 in radius is estimated by Eq. (12.5) if they can aggregate [Elimelech et al., 1995].

$$r(a_1,a_2)=B\sqrt{\frac{E_0}{\eta}}(a_1+a_2)^3 n(a_1)n(a_2) \tag{12.5}$$

where B is a constant, E_0 is the rate of turbulent energy dissipation per unit mass of the liquid, η is the kinematic viscosity, and $n(a_1)$ is the number density of particles (aggregates) of a_1 in radius. In Eq. (12.5), the role of ultrasound is taken into account in E_0. In the experiment of Dang et al. (2009), it is estimated to be 1500 W/kg, as 100 mL solution was irradiated with about 150 W ultrasound. Lower values of E_0 do not reproduce the experimental results with mechanical stirring without ultrasound because the mechanism of $BaTiO_3$ formation would be completely different from that with ultrasound as explained above. The intensity of ultrasound should also affect the number of bubbles [C_0 in Eq. (12.4)]. The kinematic viscosity (η) is estimated by the viscosity (μ) and the density (ρ) of the solution as $\eta = \mu/\rho$. In the present numerical simulations, ρ = 1000 kg/m^3 is assumed. The viscosity μ is assumed to be as 100, 200, and 400 mPa·s for the initial $BaCl_2$ concentration of 0.1, 0.2, and 0.4 mol/L, although they were not experimentally measured.

In the following, the number density of aggregates consisting of i primary particles is denoted as $n(i)$. Then, the rate of the change of $n(i)$ is given by Eq. (12.6).

For $i = 1$,

$$\frac{dn(1)}{dt}=J-B\sqrt{\frac{E_0}{\eta}}\times n(1)\sum_{i=1}^{N_{max}} n(i)[a(1)+a(i)]^3 \tag{12.6a}$$

where J is the nucleation rate of primary particles [Eq. (12.4)] and N_{max} is the number of constituent primary particles in the largest aggregate present in the numerical simulation.

For $i \geq 2$,

$$\frac{dn(i)}{dt}=B\sqrt{\frac{E_0}{\eta}}[S_i-n(i)L_i] \tag{12.6b}$$

where S_i and L_i are given as follows.

For the normal model of aggregation,

$$S_i = \sum_{N<i} n(i-N)n(N)[a(i-N)+a(N)]^3 \tag{12.7a}$$

$$L_i = \sum_{N'=1}^{N_{\max}} n(N')[a(i)+a(N')]^3 \tag{12.7b}$$

For the new model of aggregation,

$$S_i = n(1)n(i-1)[a(1)+a(i-1)]^3 \tag{12.8a}$$

$$L_i = n(1)[a(1)+a(i)]^3 \tag{12.8b}$$

where $a(i)$ is the radius of an aggregate consisting of i primary particles and estimated by Eq. (12.9).
For $i \geq 2$,

$$a(i) = d \times \sqrt[3]{\frac{3i}{4\pi}} \tag{12.9a}$$

For $i = 1$,

$$a(1) = \frac{d}{2} \tag{12.9b}$$

where d is the diameter of a primary particle ($d \approx 5$ nm).

The population density $[P(a)]$ as a function of radius (a) of an aggregate is estimated by Eq. (12.10) derived by the relationship $P(a)da = n(i)di$ and Eq. (12.9).

$$P(a) = \frac{\pi}{2} \times \frac{[a(i)]^2}{[a(1)]^3} \times n(i) \tag{12.10}$$

The unit for the population density $[P(a)]$ is m^{-4} because $P(a)$ is the number density of aggregates per unit radius.

In the present numerical simulations, all the processes of chemical reaction, nucleation, and aggregation have been simultaneously calculated. For more details of the model, see Yasui et al. (2011).

12.2.2 Results and Discussions

In Fig. 12.6, the particle (aggregate) size distribution calculated by the normal model of aggregation is shown for the initial $BaCl_2$ concentration of 0.1 mol/L (Yasui et al., 2011). In the calculated result, the peak in the size distribution is at about 5 nm, which is the size of the primary particle (nanocrystal). It disagrees with the experimental result of the average particle (aggregate) size of about 200 nm (Dang et al., 2009, 2010a,b).

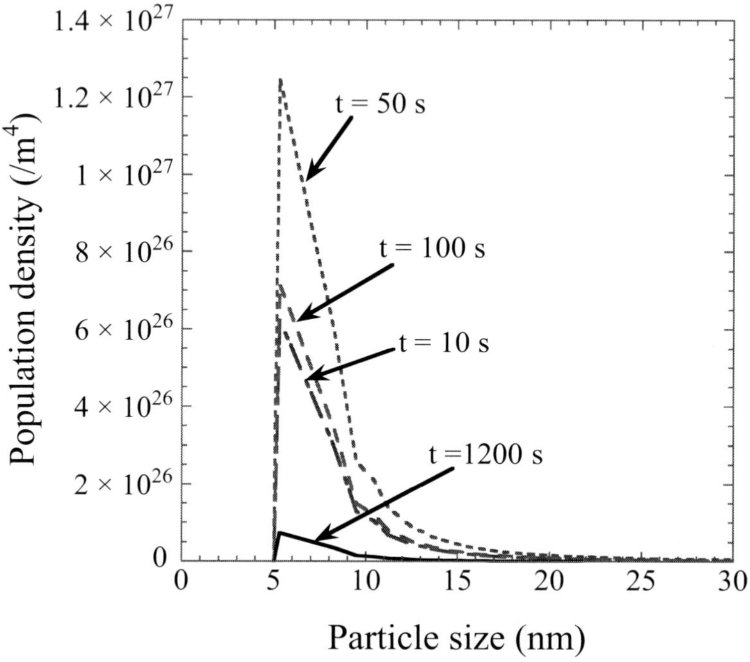

Figure 12.6 The particle (aggregate) size distribution numerically simulated by the normal model (Smoluchowski model) of aggregation. The initial $BaCl_2$ concentration is 0.1 mol/L.

In Fig. 12.7(a), the particle (aggregate) size distribution calculated by the new model is shown under the same condition as that of Fig. 12.6. In the calculated result, some aggregates are larger than 200 nm while the peak is at around 100 nm. It nearly agrees with the experimental result. It suggests that only primary particles (or possibly only small aggregates) aggregate with other particles.

In Fig. 12.7(b), the calculated particle (aggregate) size distribution for the initial $BaCl_2$ concentration of 0.4 mol/L is shown with the other conditions being the same as Fig. 12.7(a). It is seen that the average particle (aggregate) size is smaller than that for 0.1 mol/L. It agrees with the experimental results. The reason for the smaller particle size for higher initial concentration is both higher

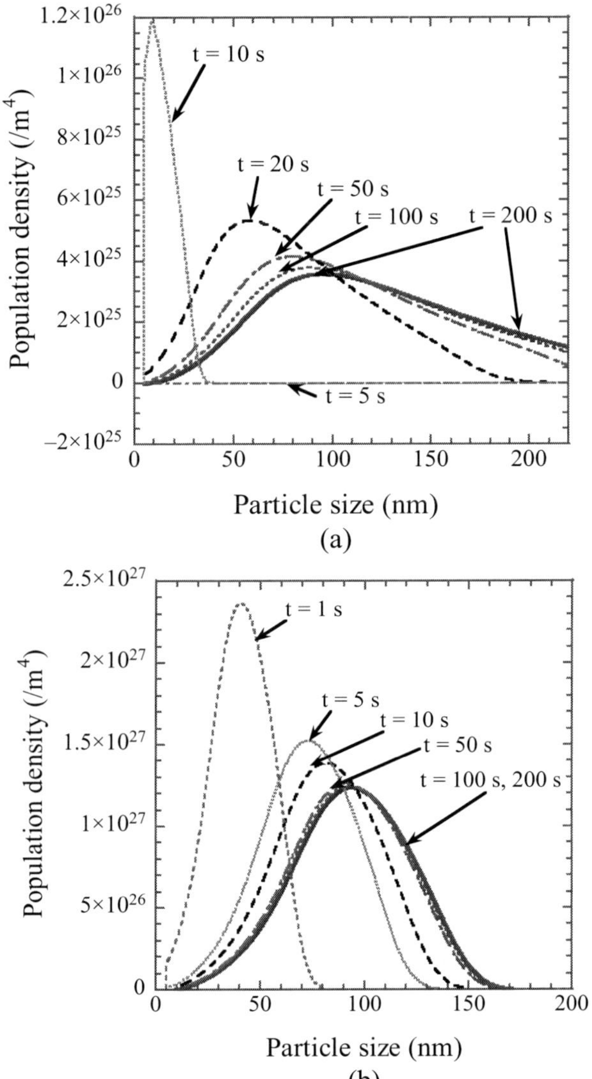

Figure 12.7 The particle (aggregate) size distribution numerically simulated by the new model of aggregation that only primary particles aggregate with other particles. For the initial $BaCl_2$ concentration of 0.1 mol/L (a) and 0.4 mol/L (b). Reprinted from *Ultrasonics Sonochemistry*, vol. 18, Yasui, K., Tuziuti, T., and Kato, K. (2011). Numerical simulations of sonochemical production of $BaTiO_3$ nanoparticles, pp. 1211–1217, Copyright (2011), with permission from Elsevier.

viscosity of the solution resulting in lower rate of aggregation [Eq. (12.5)] and shorter reaction time due to Eq. (12.3) resulting in shorter time for aggregation. The effect of viscosity is seen in Fig. 12.8. Lower viscosity results in the formation of larger aggregates under the same initial $BaCl_2$ concentration, as the aggregation rate is higher according to Eq. (12.5). The dependence of the morphology of particles (aggregates) on the initial $BaCl_2$ concentration is, however, still unclear.

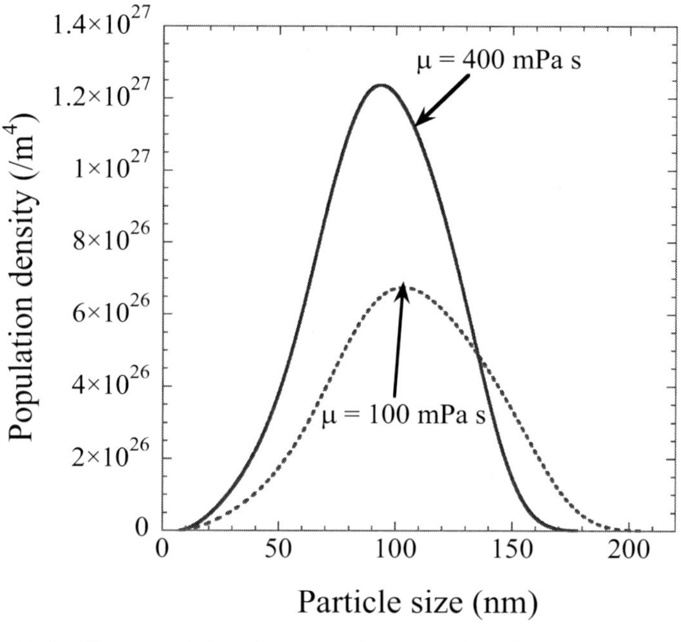

Figure 12.8 The particle (aggregate) size distribution numerically simulated by the new model of aggregation for the initial $BaCl_2$ concentration of 0.4 mol/L with the liquid viscosities of 100 mPa s and 400 mPa s. Reprinted from *Ultrasonics Sonochemistry*, vol. 18, Yasui, K., Tuziuti, T., and Kato, K. (2011). Numerical simulations of sonochemical production of $BaTiO_3$ nanoparticles, pp. 1211–1217, Copyright (2011), with permission from Elsevier.

There still remain many unsolved problems. Why do only primary particles (or possibly only small aggregates) attach with other aggregates (or primary particles)? Why do primary particles (or aggregates) attach each other with their crystal axes aligned?

Why is the chemical reaction accelerated by ultrasound? What is the role of ultrasound (or acoustic cavitation) on the morphology of particles?

In the next section, electric dipole–dipole interaction model is discussed in order to study the mechanism of oriented attachment of $BaTiO_3$ nanocrystals as well as the mechanism of aggregation restricted to primary particles or small aggregates.

12.3 Mesocrystal Formation

12.3.1 Theoretical Model

In the present section, collisions between two particles (aggregates) are discussed in order to study the mechanism of oriented attachment of nanocrystals as well as that of repulsion between two large aggregates (Yasui and Kato, 2012).

In Fig. 12.9, a conceptual figure of a collision between a primary particle (nanocrystal) and an aggregate is shown. The sphere at the left side represents an aggregate. On the surface of the aggregate, one of the constituent primary particles (nanocrystals) is explicitly depicted as a small cube. The cube at the right side represents a primary particle. It is assumed that a primary particle has an electric dipole moment as well as any constituent primary particles of an aggregate. With regard to an aggregate, most of the electric dipole moments of constituent primary particles cancel each other because they are in anti-parallel. As a result, only the dipole moment of a primary particle on the surface of an aggregate is considered. This assumption is also based on the very short Debye length in the solution at pH 14 such as 0.3 nm. It means that electrostatic force between a pair of dipoles work only when they nearly attach each other (at the distance smaller than 0.3 nm). An aggregate and a primary particle collide and finally attach with each other with their electric dipoles aligned as shown in Fig. 12.9.

In the present model, three kinds of interaction forces between two particles are assumed. One is the electric dipole–dipole interaction. Another is the double-layer interaction, which is the repulsive force between charged surfaces of two particles. In an electrolyte solution, the repulsive force is a little bit complicated by the existence of the counter ions near the charged surfaces and is called the double-layer interaction. The other is the hydrogen bond between hydroxyl groups on each surface of particles. There are several hydroxyl groups (–OH) on the surface of metal oxides due to the following equilibrium (Elimelech et al., 1995).

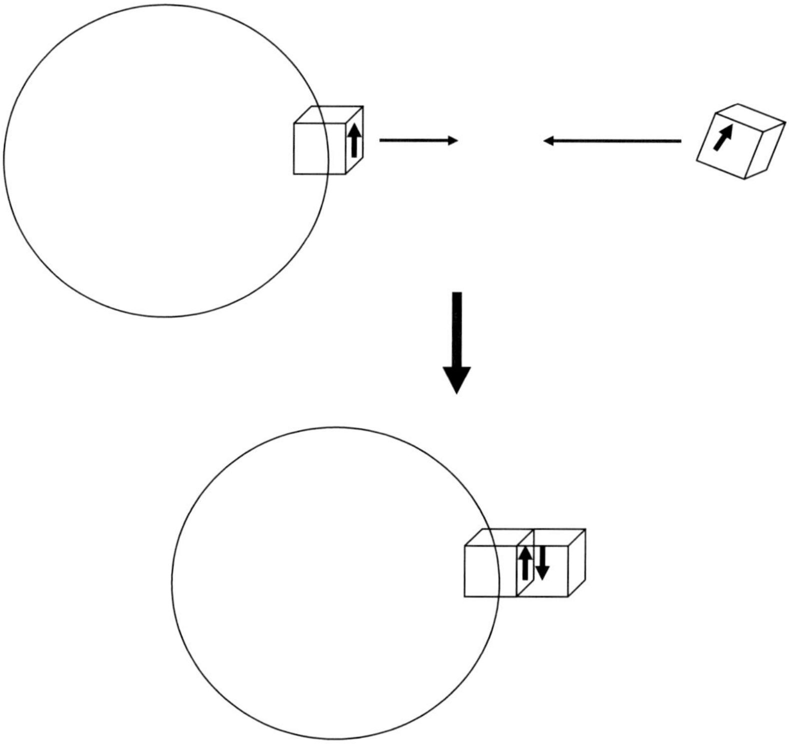

Figure 12.9 A conceptual figure of a collision between a primary particle (nanocrystal) and an aggregate. The cube on the right side is a primary particle. The sphere on the left side represents an aggregate. One of the constituent primary particles is explicitly shown as a cube on the right side on the surface of a spherical aggregate. The small arrows indicate the electric dipoles. Reprinted with permission from Yasui, K. and Kato, K. (2012). Dipole–dipole interaction model for oriented attachment of $BaTiO_3$ nanocrystals: a route to mesocrystal formation, *J. Phys. Chem. C*, **116**, pp. 319–324. Copyright (2012), American Chemical Society.

$$S\text{-}OH_2^+ \rightarrow S\text{-}OH \rightarrow S\text{-}O^- \quad (12.11)$$

where S denotes the solid surface. In addition to the above three interactions, the random thermal force acts on each particle that causes the Brownian motion.

Firstly, the dipole–dipole interaction is discussed. Here, the z-axis is taken on the line between two particles with its origin at the center of the mass. The xy-plane is taken perpendicular to the z-axis. In the present model, both dipole moments of the two particles are assumed on the planes parallel to the xy-plane. The angle between the dipole and x-axis is designated by θ_1 and θ_2 for particle 1 and 2, respectively. Then, the potential energy (V_{dipole}) of the dipole–dipole interaction is given by Eq. (12.12) (Jackson, 1975).

$$V_{\text{dipole}} = \frac{p^2 \cos(\theta_1 - \theta_2) e^{-\kappa h}}{4\pi\varepsilon h^3} \quad (12.12)$$

where p is the dipole moment, κ is the Debye–Hückel parameter and is the inverse of the Debye length, h, is the distance between the surfaces of the two particles, and ε is the permittivity of the aqueous solution. The Debye–Hückel parameter is given by Eq. (12.13) (Elimelech et al., 1995).

$$\kappa = \sqrt{\frac{2e_1^2 n_i}{\varepsilon k_B T}} \quad (12.13)$$

where e_1 is the electron charge [$e_1 = 1.6 \times 10^{-19}$](C)] and n_i is the number concentration of ion in the solution.

Next, the double-layer interaction is discussed. The potential energy (V_{double}) of the double-layer interaction between two particles is given by Eq. (12.14) (Elimelech et al., 1995).

$$V_{\text{double}} = \frac{128\pi a_1 a_2 n_i k_B T \gamma_1 \gamma_2 e^{-\kappa h}}{(a_1 + a_2)\kappa^2} \quad (12.14)$$

where a_1 and a_2 are the radius of particles 1 and 2, respectively, and γ_1 and γ_2 are a function of zeta potential of particles 1 and 2, respectively.

Finally, hydrogen bond between the surfaces of the two particles is discussed. It is very crudely assumed that the hydrogen bond is expressed by the Coulomb interaction between a positive charge of $+0.3e_1$ at H atom and a negative charge of $-0.3e_1$ at O atom (Barker and Watts, 1969). Then, the potential energy ($V_{\text{hydrogen-bond}}$) of the hydrogen bond is given by Eq. (12.15).

$$V_{\text{hydrogen-bond}} = -\frac{(0.3e_l N_{\text{-OH}})^2 e^{-\kappa h}}{4\pi\varepsilon h} \tag{12.15}$$

where $N_{\text{-OH}}$ is the number of hydroxyl groups on the surface of a $BaTiO_3$ nanocrystal (primary particle).

The equation of motion for particle 1 is given by Eq. (12.16).

$$m_1 \frac{dz_1}{dt} = -\frac{\partial(V_{\text{dipole}} + V_{\text{double}} + V_{\text{hydrogen-bond}})}{\partial z_1} + \sqrt{\frac{12\pi a_1 \mu k_B T}{\Delta t}} r_n \cos\phi_r \tag{12.16}$$

where m_1 is the mass of particle 1, z_1 is the z-component of the position of particle 1, t is time, a_1 is the radius of particle 1, μ is the viscosity of the solution, Δt is the time step in the numerical integration of the differential equation (the equation of motion), r_n is the random number between −1 and 1, and φ_r is the random number between 0 and 2π. The last term in the right side of Eq. (12.16) is the random thermal force causing the Brownian motion (Li and Ahmadi, 1992; You and Choi, 2007). In the present numerical simulations, the term was included after the collision of the two particles in order to ensure the occurrence of the collision.

The initial velocities of two particles are assumed as follows considering the velocity of the Brownian motion (Einstein, 1956).

$$\left.\frac{dz_1}{dt}\right|_{t=0} = -\sqrt{\frac{3m_2 k_B T}{m_1(m_1 + m_2)}} \tag{12.17}$$

$$\left.\frac{dz_2}{dt}\right|_{t=0} = \sqrt{\frac{3m_1 k_B T}{m_2(m_1 + m_2)}} \tag{12.18}$$

where m_2 is the mass of particle 2 and z_2 is the z-component of the position of particle 2.

The equation of motion for the direction of the electric dipole moment of particle 1 is given by Eq. (12.19).

$$I_1 \frac{d^2\theta_1}{dt^2} = \frac{p^2 \sin(\theta_1 - \theta_2) e^{-\kappa h}}{4\pi\varepsilon h^3} - 8\pi\mu a_1^3 \dot{\theta}_1 - \frac{\mu_d a_1 N_f \dot{\theta}_1}{|\dot{\theta}_1|} + \sqrt{\frac{16\pi a_1^3 \mu k_B T}{\Delta t}} r_n' \cos\phi_r' \tag{12. 19}$$

where I_1 is the moment of inertia for particle 1, θ_1 is the angle between the dipole of particle 1 and x-axis, p is the dipole moment, μ_d is the dynamic friction coefficient, N_f is the total load (the total force perpendicular to the surface of particle 2 when particle 1 pushes particle 2), r'_n is the random number between −1 and 1, and φ'_r is the random number between 0 and π. The first term in the right side of Eq. (12.19) is the dipole–dipole interaction. The second term is the friction due to viscosity of the liquid (Lamb, 1993). The third term is the dynamic friction when the two particles attach with each other (Maeno and Bogy, 1992). The last term is the random thermal torque causing the rotational Brownian motion (Evans et al., 1982). The last term was included only after a collision took place in order to fix the initial condition of the collision.

12.3.2 Results and Discussions

In Fig. 12.10, the result of numerical simulation for a collision between a primary particle (n = 1) and an aggregate consisting of 200 primary particles (n = 200) is shown where n is the number of constituent primary particles in an aggregate (or primary particle). In diameter, a primary particle is 5 nm, while an aggregate is about 36 nm in this case. It is seen that after a collision, they attach with each other due to the hydrogen bond. Although initially the dipole–dipole interaction is repulsive, the dipole of the primary particle rotates due to the dipole–dipole interaction after the attachment and the two dipoles are finally aligned in an anti-parallel manner ($\theta_1 - \theta_2 = \pi$). In other words, the crystal axes of the two particles are aligned. The small fluctuation around π is due to the rotational Brownian motion.

In Fig. 12.11, the result for a collision between two aggregates (n = 200) is shown. Although the initial dipole–dipole interaction is attractive, the two particles are reflected and never attach with each other due to the repulsive double-layer interaction. The repulsive double-layer interaction is stronger for larger aggregates according to Eq. (12.14). As a result, any pair of aggregates both larger than about n = 180 never attach with each other.

In Fig. 12.12, the total potential energy (V_{dipole} + V_{double} + $V_{\text{hydrogen-bond}}$) is shown as a function of the distance (h) between

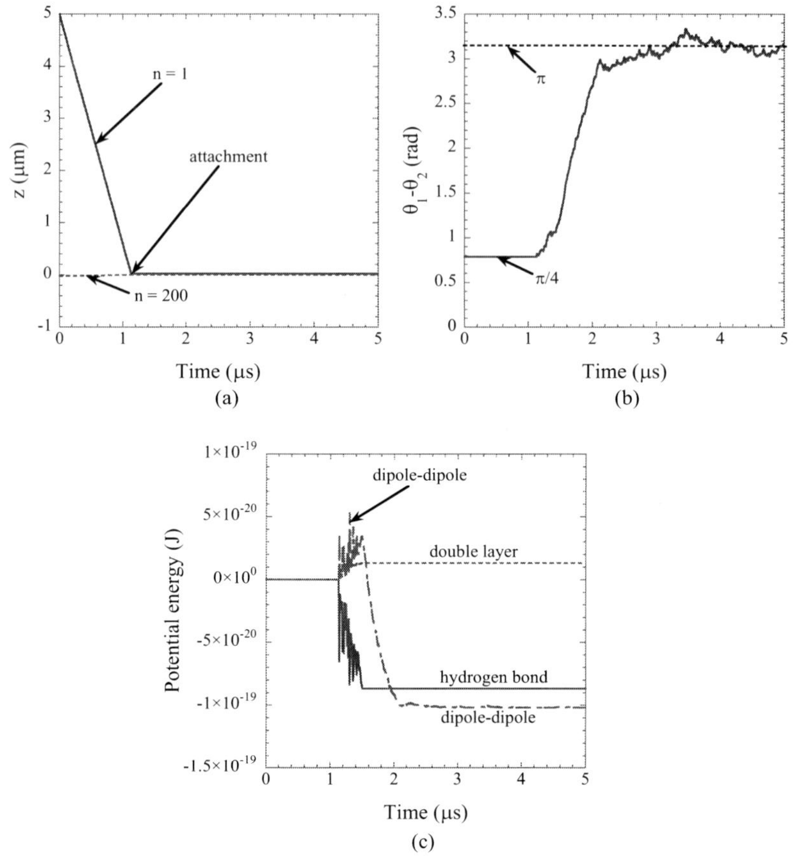

Figure 12.10 The result of numerical simulation for a collision between a primary particle (n = 1) and an aggregate (n = 200). (a) The position (z-component) of each particle as a function of time. (b) The relative angle of the two dipoles as a function of time. The initial value is set to be $\pi/4$. The relative angle of π means that the two dipoles are in anti-parallel. (c) The potential energy as a function of time. Negative value means that the interaction is attractive.

the surfaces of the two particles of equal size when the pair of dipoles is antiparallel. For a pair of primary particles (n = 1), the total energy is always higher than the potential energy resulting in the attachment at the collision [Fig. 12.12(a)]. On the contrary, for a pair of aggregates (n = 200), the maximum potential energy is higher than the total energy resulting in the reflection of the aggregates [Fig. 12.12(b)]. An aggregate moves initially leftward, but at the reflection point shown in Fig. 12.12(b), it stops as the kinetic energy (the total energy minus the potential energy) becomes zero. At the point, repulsive (positive) force acts to the aggregate as $F = -\partial V/\partial h > 0$, where F is the force and V is the total potential. Then, the aggregate starts moving rightward. In Fig. 12.11, the total energy is $1.5k_BT$ according to Eqs. (12.17) and (12.18).

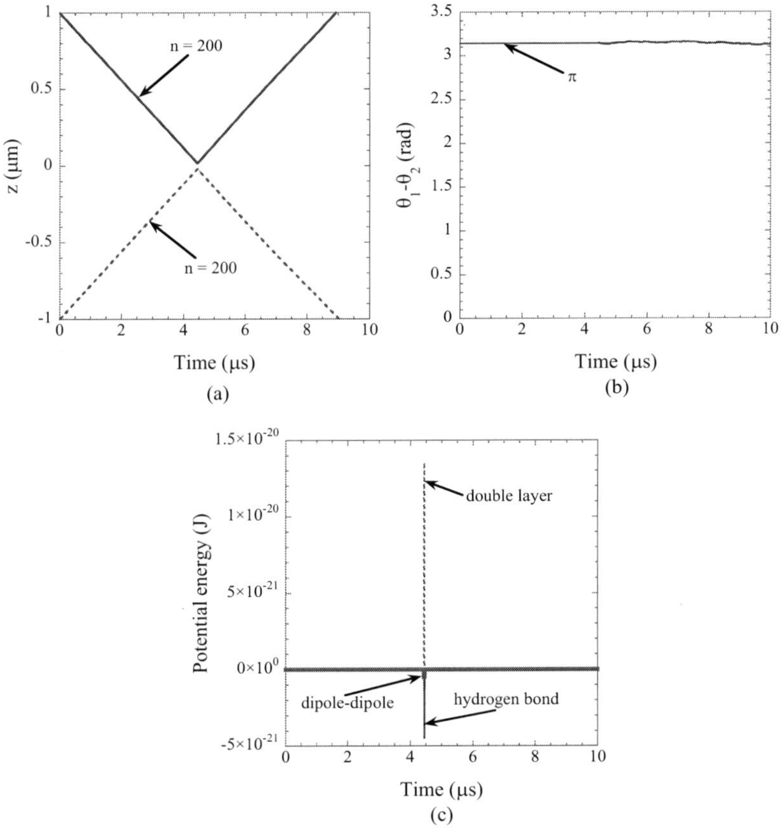

Figure 12.11 The result of numerical simulation for a collision between two aggregates (n = 200). (a) The position (z-component) of each particle as a function of time. (b) The relative angle of the two dipoles as a function of time. The initial value is set to be π, which is the condition for the strongest attraction between dipoles. (c) The potential energy as a function of time.

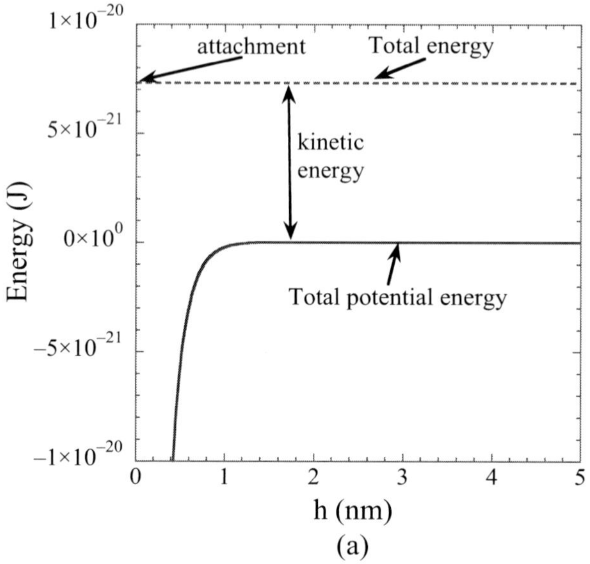

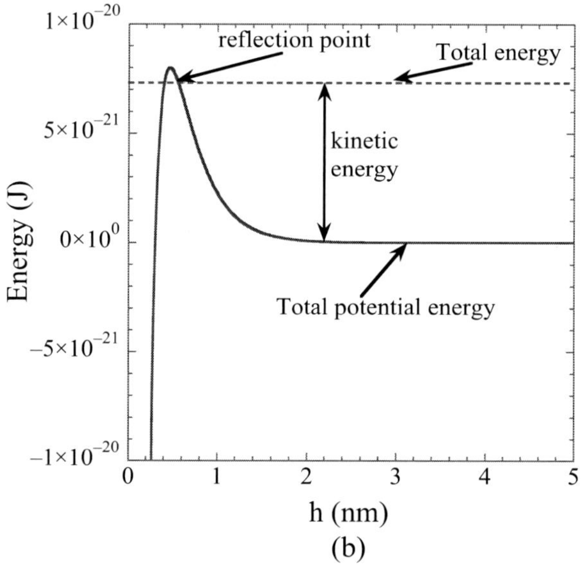

Figure 12.12 The energy as a function of the distance (h) between the surfaces of the two particles of equal size when the pair of dipoles is anti-parallel. (a) For a pair of primary particles (n = 1). They attach on collision. (b) For a pair of aggregates, each consisting of 200 primary particles (n = 200). They are reflected at the collision.

In Fig. 12.13, the maximum of the total potential energy is plotted as a function of the size (n) of an aggregate when a pair of dipoles is antiparallel. Furthermore, each potential energy at the maximum total potential energy is also plotted. The total potential energy increases as n increases because the potential energy of the double-layer interaction dramatically increases according to Eq. (12.14). Both the potential energies of the dipole–dipole interaction and hydrogen bond decrease because the distance (h) for the maximum total potential energy decreases as n increases, although both the energies do not explicitly depend on n.

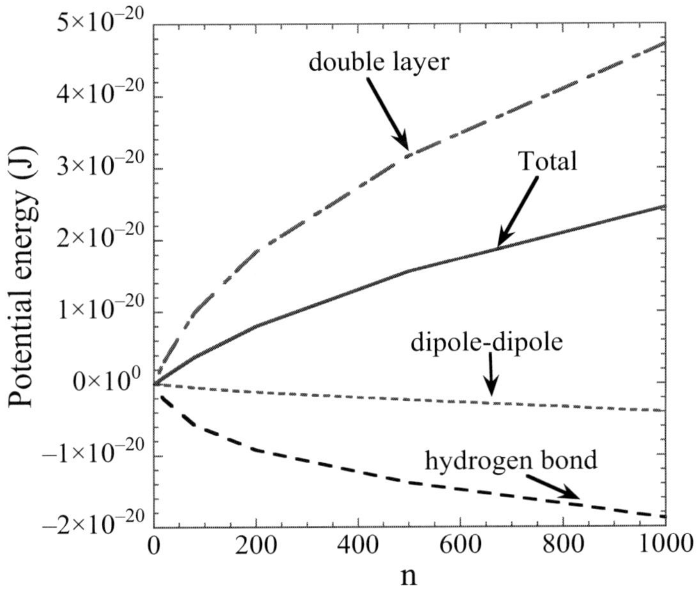

Figure 12.13 The potential energy (at the maximum of the total potential energy) as a function of the size (n) of an aggregate for a pair of aggregates of the equal size when the pair of dipoles is in anti-parallel.

In the present analysis, however, the acoustic radiation force on either of the particles has been completely neglected. The acoustic radiation force between a pair of particles is given by Eq. (12.20) (Nyborg, 1978).

$$F_z = F_0\,(3\cos^2\phi - 1) \tag{12.20a}$$

$$F_\phi = F_0 \sin 2\phi - 1) \tag{12.20b}$$

$$F_0 = \frac{6\pi\rho(\rho_{\text{particle}} - \rho)^2 U_0^2 a_1^3 a_2^3}{(2\rho_{\text{particle}} + \rho)^2 (a_1 + a_2 + h)^4} \tag{12.20c}$$

where ϕ is the angle between the local sound velocity and the particle–particle axis (z-axis), F_z (F_ϕ) is the z-component (ϕ -component) of the radiation force, $\rho_{\text{particle}}(\rho)$ is the density of a particle (liquid), and U_0 is the velocity amplitude of the acoustic wave. Equation (12.20) is applicable even if $|\rho_{\text{particle}} - \rho|/\rho > 1$. When $|\rho_{\text{particle}} - \rho|/\rho << 1$, Eq. (12.20c) is approximated as Eq. (12.21) as widely known (Nyborg, 1978; Weiser et al., 1984; Nyborg, 1989; Coakley and Bazou, 2005).

$$F_0 = \frac{2\pi(\rho_{\text{particle}} - \rho)^2 U_0^2 a_1^3 a_2^3}{3\rho(a_1 + a_2 + h)^4} \tag{12.21}$$

When a pair of particles is on the line parallel to direction of the acoustic wave propagation ($\phi = 0$), the radiation force between them is repulsive (positive). When they are on the line perpendicular to the direction of the acoustic wave propagation ($\phi = \pi/2$), it is attractive (negative). Under the condition of the experiment by Dang et al. (2009, 2010a, b), the magnitude of the radiation force (Eq. 12.20c) is one order of magnitude smaller than those of the dipole–dipole interaction, the double-layer interaction, and the hydrogen bond. For more accurate study, however, the effect of the radiation force should be taken into account.

In summary, a pair of large aggregates does not aggregate due to the repulsive double-layer interaction. In other words, aggregation may take place only when one (or two) of aggregates is smaller than a critical size ($n \approx 180$). It is consistent with the conclusion derived by the numerical simulation of the particle (aggregate) size distribution in the previous section. It suggests that the dipole–dipole interaction model may explain the aggregation process of $BaTiO_3$ nanocrystals in the experiment of Dang et al. (2009, 2010a,b) as well as the oriented attachment and mesocrystal formation.

In the numerical simulations of Figs. 12.10 and 12.11, the dipole moment (p) was assumed as p = 19 (D) = 6×10^{-29} (C m) = 5×10^{-4} (C/m^2) × V_1 (m^3), where V_1 is the volume of a primary particle. For macroscopic $BaTiO_3$ crystal, the spontaneous polarization is 0.16 (C/m^2), which is much larger than the value [5×10^{-4} (C/m^2)] assumed for 5 nm $BaTiO_3$ nanocrystal in the present simulations (Jona and Shirane, 1993). If the spontaneous polarization is the double [1×10^{-3} (C/m^2)] of the assumed value, two particles are reflected and do not attach with each other under the same condition of Fig. 12.10. On the contrary, if the spontaneous polarization is 1/5 [1×10^{-4} (C/m^2)] of the assumed value, the rotational Brownian motion causes the deviation of the direction of the dipole as much as 20%. For quantitative discussions on the spontaneous polarization, however, the exact strength of the hydrogen bond should be known.

It has been reported that the crystal structure of $BaTiO_3$ is cubic below about 20 nm in crystal size (Hoshina et al., 2006). A cubic $BaTiO_3$ crystal has no spontaneous polarization due to the highly symmetric structure (Jona and Shirane, 1993). Recent study has suggested that adsorbate-induced charge screening stabilizes the tetragonal crystal structure of a $BaTiO_3$ nanocrystal and its spontaneous polarization (Yasui and Kato, 2013).

12.4 Conclusion

Two kinds of numerical simulations have been performed in order to study the mechanism of mesocrystal formation in the sonochemical synthesis of $BaTiO_3$ nanoparticles in aqueous solution as reported by Dang et al. (2009, 2010a, b). The first one is on the particle (aggregate) size distribution (Yasui et al., 2011). It has been shown that the experimentally observed particle size is reproduced by numerical simulations when only primary particles (or possibly only small aggregates) attach with other particles. The other one is on collisions between two particles assuming that each particle has electric dipole moment (Yasui and Kato, 2012). It has been shown that by the dipole–dipole interaction, the crystal axes of the two particles are aligned after the attachment. Furthermore, a pair of large aggregates never attaches with each other due to stronger repulsive double-layer interaction. The result is consistent with the conclusion derived in the numerical simulations of the particle size distribution. Thus, it has been suggested that 5 nm $BaTiO_3$ nanocrystal has spontaneous polarization. Adsorbate-induced charge screening may stabilize the spontaneous polarization of a BaTiO3 nanocrystal (Yasui and Kato, 2013). There are, however, many unsolved problems. What is the role of ultrasound and the radiation force? What is the mechanism for the dependence of the morphology of the sonochemically synthesized $BaTiO_3$ particles on the initial $BaCl_2$ concentration? Why was the mesocrystal formation suppressed in ethanol–water mixture solvent with the other conditions being the same as those for aqueous solutions without ethanol except the liquid temperature (Dang et al., 2011)? Numerical studies on the mechanism of formation of other nanoparticles under ultrasound are also required, as it is not necessarily the same as that of $BaTiO_3$ nanoparticles.

Acknowledgment

The authors would like to thank Toru Tuziuti and Feng Dang for useful discussions.

Symbols

a_1: radius of a particle numbered 1
a_2: radius of a particle numbered 2
$a(i)$: radius of an aggregate consisting of i primary particles
$a(1)$: radius of a primary particle (nucleus or nanocrystal) ($\approx 2.5 \times 10^{-9}$ m)
B : a constant in the rate of particle aggregation (assumed as 10^5)
$[BaTiO_3]$: concentration of $BaTiO_3$ in the liquid at time t neglecting the effect of nucleation of $BaTiO_3$ crystals (mol/m^3)
C : capacitance of a multilayer ceramic capacitor (MLCC) (C/V)
C_0 : initial concentration of $BaCl_2$ (or equivalently that of $TiCl_4$) (mol/m^3)
C_0 : number of nucleation sites on the bubble surfaces per unit volume of the liquid (/m^3)
d : diameter of a primary particle (nucleus or nanocrystal) ($\approx 5 \times 10^{-9}$ m)
e_l : electron charge ($=1.6 \times 10^{-19}$ C)
E_0 : rate of turbulent energy dissipation per unit mass of the liquid (W/kg)
F_z : z-component of the acoustic radiation force between two particles (The z-axis is taken on the line between two colliding particles with its origin at the center of the mass.) (N)
F_0: a quantity defined by Eq. (12.20c) that determines the strength of the acoustic radiation force between two particles (N)
h: distance between the surfaces of the two particles (m)

i: number of constituent primary particles in an aggregate
I_1: moment of inertia for particle 1 (kg m^2)
J: nucleation rate of $BaTiO_3$ nanocrystals (/m^3 s)
k^* : diffusion controlled frequency factor in the nucleation rate (/s)
k_B : Boltzmann constant (=1.38 × 10^{-23} J/K)
k_2: rate constant of the chemical reaction (12.2a)–(12.2c) (assumed as 5 × $10^{-4}$$m^3$/mol s)
L: number of the internal electrode layers in a multilayer ceramic capacitor (MLCC)
L_i: a function of i given by Eq. (12.7b) for the normal model of aggregation and by Eq. (12.8b) for the new model of aggregation
m_1: mass of particle 1 (kg)
m_2: mass of particle 2 (kg)
N_f: total load (the total force perpendicular to the surface of particle 2 when particle 1 is attached to and pushes particle 2) (N)
n_i: number concentration of ion in the aqueous solution (/m^3)
N_{max} : number of constituent primary particles in the largest aggregate
N_{-OH}: number of hydroxyl groups on the surface of a $BaTiO_3$ nanocrystal (primary particle) (assumed as 8)
$n(a_1)$: number density of particles (aggregates) of a_1 in radius (/m^3)
$n(a_2)$: number density of particles (aggregates) of a_2 in radius (/m^3)
$n(i)$: number density of particles (aggregates) consisting of $i$primary particles (/m^3)
$n(1)$: number density of primary particles (nuclei or nanocrystals) (/m^3)
p: electric dipole moment of a primary particle ($BaTiO_3$ nanocrystal) (C m)
$P(a)$: population density (number density of aggregates per unit radius) (/m^4)
r_n: random number between –1 and 1
r'_n: random number between –1 and 1
$r(a_1,a_2)$: rate of aggregation between two particles of a_1 and a_2 in radius (/m^3 s)
S: area of each internal electrode in a multilayer ceramic capacitor (MLCC) (m^2)
S_i: a function of i given by Eq. (12.7a) for the normal model of aggregation and by Eq. (12.8a) for the new model of aggregation (/m^3)

t: time (s)
T: temperature (K)
U_0: velocity amplitude of the acoustic wave (m/s)
V_{dipole}: potential energy of the dipole–dipole interaction (J)
V_{double}: potential energy of the double-layer interaction between two particles (J)
$V_{hydrogen\text{-}bond}$: potential energy of the hydrogen bond (J)
V_1: volume of a primary particle (nucleus or nanocrystal) (m^3)
w : thickness of a dielectric layer in a multilayer ceramic capacitor (MLCC) (m)
Z: Zeldovich factor
z_1 : z-component of the position of particle 1 (The z-axis is taken on the line between two colliding particles with its origin at the center of the mass.) (m)
z_2 : z-component of the position of particle 2 (The z-axis is taken on the line between two colliding particles with its origin at the center of the mass.) (m)
γ_1: a function of zeta potential of particle 1 (= $(\exp(y_0/2)-1)/(\exp(y_0/2)+1)$, where $y_0 = e_1\zeta_1/k_BT$ and ζ_1 is the zeta potential of particle 1)
γ_2: a function of zeta potential of particle 2 (= $(\exp(y_0/2)-1)/(\exp(y_0/2)+1)$, where $y_0 = e_1\zeta_2/k_BT$ and ζ_2 is the zeta potential of particle 2)
ΔG^*_{het}: nucleation work decreased by the presence of the liquid/gas interfaces (the bubble surfaces) (J)
Δt : time step in the numerical integration of the differential equation (the equation of motion) (s)
ε: dielectric constant of the aqueous solution (=$61.03\varepsilon_0$ for pure water) ($C^2/N\ m^2$)
ε_r : relative dielectric constant of the dielectric ceramic
ε_0: dielectric constant of vacuum (=8.854×10^{-12} $C^2/N\ m^2$)
η: kinematic viscosity of the liquid (m^2/s)
θ_1: angle between the dipole of particle 1 and x-axis (rad)
θ_2 : angle between the dipole of particle 2 and x-axis (rad)
κ: Debye–Hückel parameter (the inverse of the Debye length) (/m)
μ: viscosity of the liquid (Pa s)
μ_d : dynamic friction coefficient between two particles (assumed as 0.1)
ρ: liquid density (kg/m^3)

$\rho_{particle}$: density of a particle (=6 × 10^3 kg/m^3 for $BaTiO_3$)
ϕ: angle between the local sound velocity and z-axis (The z-axis is taken on the line between two colliding particles with its origin at the center of the mass.) (rad)
ϕ_r : random number between 0 and 2π (rad)
ϕ'_r : random number between 0 and π (rad)

References

Barker, J. A. and Watts, R. O. (1969). Structure of water; A Monte Carlo calculation, *Chem. Phys. Lett.*, **3**, pp. 144–145.

Coakley, W. T. and Bazou, D. (2005). Particle and cell manipulation by radiation force in ultrasound standing waves, in *Bubble and Particle Dynamics in Acoustic Fields: Modern Trends and Applications*, edited by A. A. Doinikov (Research Signpost: Kerala, India), chapter 8, pp. 313–338.

Cölfen, H., and Antonietti, M. (2008). *Mesocrystal and Nonclassical Crystallization* (John Wiley & Sons: Chichester, U.K.).

Dang, F., Kato, K., Imai, H., Wada, S., Haneda, H., and Kuwabara, M. (2009). Characteristics of $BaTiO_3$ particles sonochemically synthesized in aqueous solution, *Jpn. J. Appl. Phys.*, **48**, 09KC02.

Dang, F., Kato, K., Imai, H., Wada, S., Haneda, H., and Kuwabara, M. (2010a). A new effect of ultrasonication on the formation of $BaTiO_3$ nanoparticles, *Ultrason. Sonochem.* **17**, pp. 310–314.

Dang, F., Kato, K., Imai, H., Wada, S., Haneda, H., and Kuwabara, M. (2010b). Oriented aggregation of BaTiO3 nanocrystals and large particles in the ultrasonic-assistant synthesis, *Cryst. Eng. Commun.*, **12,** pp. 3441–3444.

Dang, F., Kato, K., Imai, H., Wada, S., Haneda, H. and Kuwabara, M. (2011). Growth of $BaTiO_3$ nanoparticles in ethanol–water mixture solvent under an ultrasound-assisted synthesis, *Chem. Eng. J.*, **170**, pp. 333–337.

Didenko, Y. T., and Suslick, K. S. (2002). The energy efficiency and formation of photons, radicals and ions during single-bubble cavitation, *Nature (London),* **418**, pp. 394–397.

Eckert Jr., J. O., Hung-Houston, C. C., Gersten, B. L., Lencka, M. M., and Riman, R. E. (1996). Kinetics and mechanisms of hydrothermal synthesis of barium titanate, *J. Am. Ceram. Soc.*, **79**, pp. 2929–2939.

Einstein, A. (1956). *Investigations on the Theory of the Brownian Movement* (Dover, New York).

Elder, S. A. (1959). Cavitation microstreaming, *J. Acoust. Soc. Am.*, **31**, pp. 54–64.

Elimelech, M., Gregory, J., Jia, X. and Williams, R. A. (1995). *Particle Deposition & Aggregation* (Butterworth Heinemann: Woburn, MA, USA).

Evans, M. W., Evans, G. J., Coffey, W. T. and Grigolini, P. (1982). *Molecular Dynamics* (John Wiley & Sons: New York).

Gedanken, A. (2004). Using sonochemistry for the fabrication of nanomaterials, *Ultrason. Sonochem.*, **11**, pp. 47–55.

Holzfuss, J., Ruggeberg, M., and Billo, A. (1998). Shock wave emission of a sonoluminescing bubble, *Phys. Rev. Lett.*, **81**, pp. 5434–5437.

Hoshina, T., Kakemoto, H., Tsurumi, T. and Wada, S. (2006). Size and temperature induced phase transition behaviors of barium titanate nanoparticles, *J. Appl. Phys.*, **99**, 054311.

Jackson, J. D. (1975). *Classical Electrodynamics,* 2nd ed. (John Wiley & Sons: New York).

Jona, F. and Shirane, G. (1993). *Ferroelectric Crystals* (Dover: New York).

Kishi, H., Mizuno, Y. and Chazono, H. (2003). Base-metal electrode-multilayer ceramic capacitors: past, present and future perspectives, *Jpn. J. Appl. Phys.*, **42**, pp. 1–15.

Kordylla, A., Krawczyk, T., Tumakaka, F., and Schembecker, G. (2009). Modeling ultrasound-induced nucleation during cooling crystallization, *Chem. Eng. Sci.*, **64**, pp. 1635–1642.

Lamb, H. (1993). *Hydrodynamics* (Cambridge University Press: Cambridge).

Leighton, T. G. (1994). *The Acoustic Bubble* (Academic Press: London).

Leong, T., Collis, J., Manasseh, R., Ooi, A., Novell, A., Bouakaz, A., Ashokkumar, M., and Kentish, S. (2011). The role of surfactant headgroup, chain length, cavitation microstreaming on the growth of bubbles by rectified diffusion, *J. Phys. Chem. C,* **115**, pp. 24310–24316.

Li, A. and Ahmadi, G. (1992). Dispersion and deposition of spherical particles from point sources in a turbulent channel flow, *Aerosol Sci. Technol.*, **16**, pp. 209–226.

Maeno, T. and Bogy, D. B. (1992). Effect of the hydrodynamic bearing on rotor/stator contact in a ring-type ultrasonic motor, *IEEE Trans. Ultrason. Ferroele. Freq. Control*, **39**, pp. 675–682.

Mettin, R. (2007). From a single bubble to bubble structures in acoustic cavitation, in *Oscillations, Waves and Interactions*, edited by T. Kurz, U. Parlitz, and U. Kaatze (Universitaetsverlag Goettingen, Goettingen), pp. 171–198.

Mitome, H., Hatanaka, S., and Tuziuti, T. (2000). Observation of spatial nonuniformity in a sonochemical reaction field, in *Nonlinear Acoustics at the Turn of the Millennium: ISNA 15*, edited by W. Lauterborn and T. Kurz, *AIP Conference Proceedings*, **524**, pp. 473–476.

Mori, M, Kineri, T., Kadono, K., Sakaguchi, T., Miya, M., Wakabayashi, H., and Tsuchiya, T. (1995). Effect of the atomic ratio of Ba to Ti on optical properties of gold-dispersed $BaTiO_3$ thin films, *J. Am. Ceram. Soc*, **78**, pp. 2391–2394.

Moussatov, A., Granger, C., and Dubus, B. (2003). Cone-like bubble formation in ultrasonic cavitation field, *Ultrason. Sonochem.*, **10**, pp. 191–195.

Nyborg, W. L. (1978). Physical principles of ultrasound, in *Ultrasound: Its Applications in Medicine and Biology*, edited by F. J. Fry (Elsevier: Amsterdam), chapter 1, pp. 1–75.

Nyborg, W. L. (1989). Theoretical criterion for acoustic aggregation, *Ultrasound Med. Biol.*, **15**, pp. 93–99.

Ohl, C. D., Arora, M., Dijkink, R., Janve, V., and Lohse, D. (2006). Surface cleaning from laser-induced cavitation bubbles, *Appl. Phys. Lett.*, **89**, 074102.

Pithan, C., Hennings, D., and Waser, R. (2005). Progress in the synthesis of nanocrystalline $BaTiO_3$ powders for MLCC, *Int. J. Appl. Ceram. Technol.*, **2** (1), pp. 1–14.

Qi, X., Zhou, J., Yue, Z., Gui, Z., Li, L., and Buddhudu, S. (2004). A ferroelectric ferromagnetic composite material with significant permeability and permittivity, *Adv. Funct. Mater.*, **14**, pp. 920–926.

Sivakumar, M. (2011). Sonochemical synthesis of oxides and sulfides, in *Theoretical and Experimental Sonochemistry Involving Inorganic Systems*, edited by Pankaj and M. Ashokkumar (Springer: Dordrecht), chapter 8, pp. 191–211.

Song, R. Q. and Cölfen, H. (2010). Mesocrystals-ordered nanoparticle superstructures, *Adv. Mater.*, **22**, pp. 1301–1330.

Song, R. Q. and Cölfen, H. (2011). Additive controlled crystallization, *Cryst. Eng. Comm.*, **13**, pp. 1249–1276.

Suslick, K. S., Choe, S. B., Cichowlas, A. A., and Grinstaff, M. W. (1991). Sonochemical synthesis of amorphous iron, *Nature (London)*, **353**, pp. 414–416.

Uchino, K. (2000). *Ferroelectric Devices* (Marcel Dekker, New York), pp. 1--304

Venigalla S. (2001). Advanced materials & powders—barium titanate, *Am. Ceram. Soc. Bull.*, **80** (6), pp. 63–64.

Weiser, M. A., Apfel, R. E., and Neppiras, E. A. (1984). Interparticle forces on red cells in a standing wave field, *Acta Acustica*, **56**, pp. 114–119.

Xu, M., Lu, Y., Liu, Y., Shi, S., Qian, T., Lu, D. (2006). Sonochemical synthesis of monosized spherical $BaTiO_3$ particles, *Powder Technol.*, **161**, pp. 185–189.

Yasui, K., Tuziuti, T., Sivakumar, M., and Iida, Y. (2005). Theoretical study of single-bubble sonochemistry, *J. Chem. Phys.*, **122**, 224706.

Yasui, K., Iida, Y., Tuziuti, T., Kozuka, T., and Towata, A. (2008a). Strongly interacting bubbles under an ultrasonic horn, *Phys. Rev. E*, **77**, 016609.

Yasui, K., Tuziuti, T., Lee, J., Kozuka, T., Towata, A., and Iida, Y. (2008b). The range of ambient radius for an active bubble in sonoluminescence and sonochemical reactions, *J. Chem. Phys.*, **128**, 184705.

Yasui, K. (2011). Fundamentals of acoustic cavitation and sonochemistry, in *Theoretical and Experimental Sonochemistry Involving Inorganic Systems*, edited by Pankaj and M. Ashokkumar (Springer: Dordrecht), chapter 1, pp. 1–29.

Yasui, K., Tuziuti, T., and Kato, K. (2011). Numerical simulations of sonochemical production of $BaTiO_3$ nanoparticles, *Ultrason. Sonochem.*, **18**, pp. 1211–1217.

Yasui, K. and Kato, K. (2012). Dipole–dipole interaction model for oriented attachment of $BaTiO_3$ nanocrystals: A route to mesocrystal formation, *J. Phys. Chem. C*, **116**, pp. 319–324.

Yasui, K. and Kato, K. (2013). Influence of adsorbate-induced charge screening, depolarization factor, mobile carrier concentration, and defect-induced microstrain on the size effect of a $BaTiO_3$ nanoparticle, *J. Phys. Chem. C*, **117**, pp. 19632–19644.

You, S. and Choi, M. (2007). Numerical simulation of microscopic motion and deposition of nanoparticles via electrodynamic focusing, *Aerosol Sci.*, **38**, pp. 1140–1149.

Young, F. R. (1999). *Cavitation* (Imperial College Press: London).

Chapter 13

Ultrasonics and Sonochemistry: Some Issues and Future Perspectives

Muthupandian Ashokkumar[a,b] **and Sivakumar Manickam**[c]
[a]*School of Chemistry, University of Melbourne, 1-100 Grattan Street, Parkville VIC 3010, Australia*
[b]*Chemistry Department, King Abdulaziz University, Abdullah Sulayman, Jeddah 22254, Saudi Arabia*
[c]*Department of Chemical and Environmental Engineering, Faculty of Engineering, The University of Nottingham, Malaysia Campus, Jalan Broga, 43500 Semenyih, Selangor Darul Ehsan, Malaysia*
masho@unimelb.edu.au

Although numerous lab-scale trials on the sonochemical synthesis of nanomaterials and other sonochemical reactions have been reported, the development of large-scale reactors is still in progress. The main reason for the lack of large-scale applications involving ultrasonics and sonochemistry is the dependence of reaction efficiencies on a number of experimental parameters in a complex way. This chapter identifies some issues associated with optimizing sonochemical reactions in view of large-scale applications. In particular, the effects of ultrasonic frequency on the efficiency of sonochemical reactions are discussed along with other experimental parameters.

Cavitation: A Novel Energy-Efficient Technique for the Generation of Nanomaterials
Edited by Sivakumar Manickam and Muthupandian Ashokkumar

ISBN 978-981-4411-54-7 (Hardcover), 978-981-4411-55-4 (eBook)
www.panstanford.com

The mechanical effects of ultrasound have been found to be useful in a variety of potential applications, communication and imaging are some that can be mentioned. When ultrasound is passed through a liquid medium, physical and chemical effects are generated due to the interaction of the sound energy with dissolved microbubbles in the liquid (Ashokkumar and Mason, 2007). The readers, after reading various Chapters presented in this book, should be in a position to relate these physical and chemical effects to acoustic cavitation. The physical forces such as turbulence, shock wave, and microstreaming are helpful in enhancing mass transfer leading to increased reaction rates and emulsification processes. The chemical effects relate to the generation of a variety of reducing and oxidizing radicals that can be used for specific chemical reactions.

In this book, a number of leading researchers have summarized their research activities and reviewed specific research work in the area of sonochemical synthesis of nanomaterials. The key issues focused by these researchers are developing a simple and energy efficient synthetic methodology, synthesizing nanomaterials with high catalytic activity, developing nanomaterials with controlled physical and chemical properties, and the use of sonochemically synthesized nanomaterials for specific applications. For sustained development of the sonochemical technique for synthesizing nanomaterials, other aspects that control the efficiency of sonochemical reactions in general need to be addressed in addition to the above-mentioned key issues. Hence, the focus of this chapter is to provide some key criteria that need to be kept in mind when ultrasonics and sonochemistry are used for synthesizing nanomaterials.

For some reactions, the physical forces are as important as the chemical effects. For example, in emulsion polymerization reactions, the shear forces are crucial to generate the emulsion droplets (Price, 2003). It is well known that the shear forces generated by a 20 kHz horn system are much stronger due to the high energy density at the tip of the horn. Despite a very low radical yield at this frequency, emulsion polymerization reactions have been successfully achieved at 20 kHz only. Of course, dual-frequency systems involving 20 kHz and higher frequencies have been used to generate nanoemulsions (Nakabayashi et al., 2011). On the contrary, substantial amounts of primary radicals are generated at very high frequencies (>200 kHz to <1000 kHz). For chemical reactions in which the shear forces are not that important, higher frequencies are suitable (Okitsu et al., 2005). In fact, the yield of radicals as a function of frequency is something that has not been clearly dealt with in the literature. For chemical reactions that primarily depend on the radical yield, a clear understanding of the factors that affect the amount of radicals produced is important.

For a given volume of sonicated solution at a certain acoustic power level (above the cavitation threshold for all frequencies), the number of cavitation bubbles may increase with an increase in frequency. A decrease in the resonance size of the cavitation bubble with an increase in frequency has been predicted theoretically (Leighton, 1994) and proven to be the case experimentally (Brotchie et al., 2009). This leads to two opposing effects. With an increase in frequency, the maximum bubble temperature reached upon bubble collapse decreases; this would lead to a decrease in radical yield per bubble. However, as the number of bubbles increases with an increase in frequency (provided other experimental parameters remain constant), the total radical yield increases with an increase in frequency. A schematic representation of an increase in the number of standing waves as a function of frequency that results in an increase in the number of bubbles is shown in Fig. 13.1.

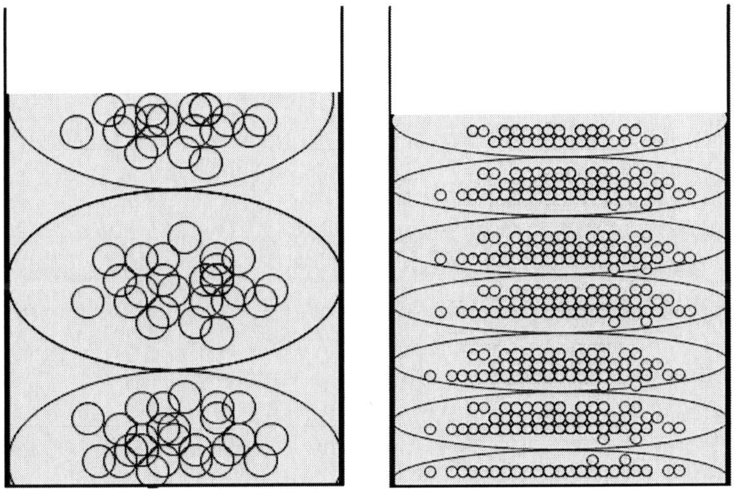

Figure 13.1 A schematic representation showing an increase in the number of active bubbles with an increase in frequency [left: low frequency (e.g., 20 kHz); right: high frequency (e.g., 200 kHz)].

Hence, it can be theoretically predicted that the total radical yield should increase with an increase in frequency. However, experimental observations (Fig. 13.2) do not show such trend introducing another complexity. Several studies (Okitsu et al., 2005) have shown that the sonochemical activity is found to reach an optimal level around 300–600 kHz range followed by a drop-off in the yield at frequencies >600 kHz.

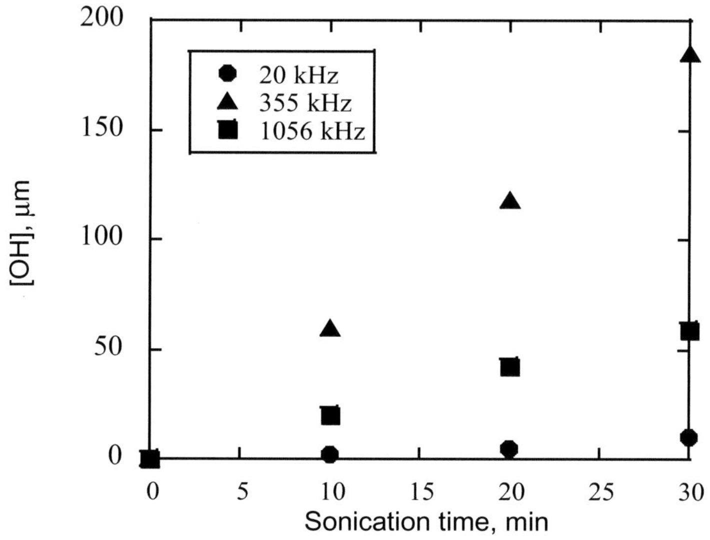

Figure 13.2 OH yield as a function of sonication time at 20 kHz, 355 kHz, and 1056 kHz. OH yield is measured using the iodide oxidation method (Ashokkumar and Mason, 2007).

In order to understand this behavior, another parameter should be considered. As mentioned earlier, the average bubble temperature can be expected to decrease with an increase in frequency due to the reduction in the size of the cavitation bubble. A decrease in temperature may generate a lower amount of radicals per bubble. However, experimentally measured bubble temperatures using MRR method do not show significant difference in temperatures in the range 350 kHz to 1 MHz (Ciawi et al., 2006). The second parameter that has not been given much attention in the literature is the amount of solvent vapor that can evaporate into the bubble during the expansion phase of the acoustic cycle. For example, in aqueous solutions, primary radicals, viz., H and OH, are generated by the homolysis of water vapor present within the bubble. The amount of water that can evaporate into the bubble depends on the time available during the expansion phase of the acoustic cycle. Obviously, the amount of water vapor that can evaporate into the bubble decreases with an increase in frequency due to a decrease in the time available. Consider 20 kHz frequency: the expansion of bubbles occurs during the negative pressure cycle, which can be approximated to 25 μs. This time reduces to 0.5 μs at 1 MHz. It has been shown in the literature that the evaporation of water requires certain time. The evaporation rate of water at 303 K has been reported to be 178 g/s/m^2 (Tronson, 2004). Using these values, the time available for a monolayer of water molecules to evaporate into the bubble during the expansion phase of the bubble can be computed and the results are shown in Fig. 13.3.

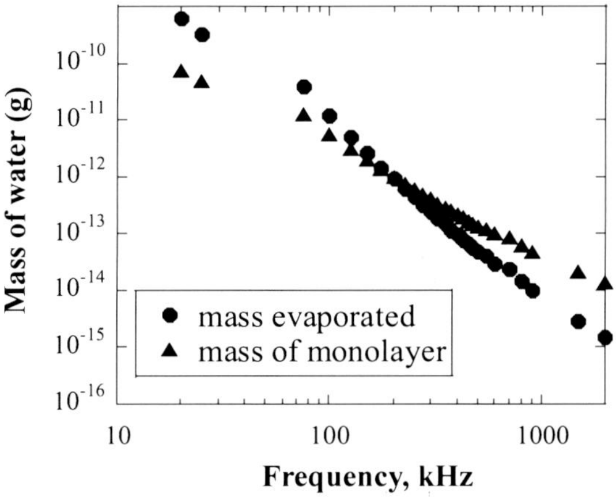

Figure 13.3 Calculated amounts of mass of water that can evaporate during one expansion cycle of ultrasound frequency and the mass available for evaporation in a monolayer surrounding the bubble (Ashokkumar, 2011).

From the data shown in Fig. 13.2 and calculations shown in Fig. 13.3, it is clear that the time available for evaporation (expansion phase of the bubble) plays an important role in determining the yield of radicals. At very high frequencies, the total number of bubbles and bubble temperatures generated may be high. However, due to a low amount of water vapor present inside the bubbles, the amount of radicals generated is lower. Hence, for the frequency effect, three key parameters should be considered, namely, bubble temperature, total number of bubbles, and the amount of solvent vapor present inside the bubbles.

There are a number of other experimental parameters that need to be considered when optimizing sonochemical reactions. Sometimes, a slight change in the arrangement of the experimental setup can affect the reaction efficiency. Consider the sonochemiluminescence (SCL) images shown in Fig. 13.4 recorded at 647 kHz. Both images were recorded under the same experimental conditions except for the placement of a glass lid on the surface of water, which is referred to as the closed system in Fig. 13.4. It can be seen that the placement of a lid at solution/air interface increases the homogeneity of the cavitation field. In the open system, the region closer to the transducer does not show cavitation activity, whereas the closed system shows a homogeneous cavitation activity throughout the liquid volume. This difference may be due to the changes in traveling and standing wave components (Lee et al., 2008), which seems to be an important factor at higher frequencies.

Figure 13.4 Photographs showing SCL: Left—open system, Right—closed system; frequency = 647 kHz; Power 60 W; 2 min exposure.

There are several other basic questions that do not have straightforward answers. Examples include: Does a higher bubble temperature mean higher sonochemical reaction efficiency? Does a higher acoustic power mean higher sonochemical reaction efficiency? The optimization of the efficiency of a simple sonochemical reaction can be a complex process. In particular, when it comes to up-scaling the reaction volume, it cannot simply be a multiplication process. For example, the power and frequency that are optimized for a given volume cannot simply be increased linearly with an increase in the reaction volume. Hence, a number of experimental parameters need to be optimized for a particular application involving large-scale processes. Despite the complexities involved in optimizing the reaction efficiency, ultrasonics and sonochemistry can be found useful in various applications, in particular energy (Neppolian et al., 2010), environment (Neppolian et al., 2009), and health (Cavalieri et al., 2011).

References

Ashokkumar, M., and Mason, T. J. (2007). *Sonochemistry, Kirk-Othmer Encylcopedia of Chemical Technology* (John Wiley & Sons).

Ashokkumar, M. (2011). The characterization of acoustic cavitation bubbles–An overview, *Ultrason. Sonochem.*, **18,** pp. 864–972.

Brotchie, A., Grieser, F. and Ashokkumar, M. (2009). Effect of frequency and power on bubble-size distribution in acoustic cavitation, *Phys. Rev. Lett.*, **102**, pp. 0843021–0843024.

Cavalieri, F., Zhou, M., Caruso, F. and Ashokkumar, M. (2011). One-pot ultrasonic synthesis of multifunctional microbubbbles and microcapsules using synthetic thiolated macromolecules, *Chem. Comm.*, **47**, pp. 4096–4098.

Ciawi, E., Rae, J., Ashokkumar, M. and Grieser, F. (2006). Determination of temperatures within acoustically generated bubbles in aqueous solutions at different ultrasound frequencies, *J. Phys. Chem. B*, **110**, pp. 13656–13660.

Lee, J., Yasui, K., Tuziuti, T., Kozuka, T., Towata, A. and Iida, Y. (2008). Spatial distribution enhancement of sonoluminescence activity by altering sonication and solution conditions, *J. Phys. Chem. B*, **112**, pp. 15333–15341.

Leighton, T. G. (1994). *The Acoustic Bubble* (Academic Press: London).

Nakabayashi, K., Amemiya, F., Fuchigami, T., Machida, K., Takeda, S., Tamamitsu, K. and Atobe, M. (2011). Highly clear and transparent nanoemulsion preparation under surfactant-free conditions using tandem acoustic emulsification, *Chem. Comm.*, **47**, pp. 5765–5767.

Neppolian, B., Doronila, A., Grieser, F. and Ashokkumar, M. (2009). Simple and efficient sonochemical method for the oxidation of arsenic(III) to arsenic(V), *Env. Sci. Technol.*, **43**, pp. 6793–6798.

Neppolian, B., Kim, Y., Ashokkumar, M., Yamashita, H. and Choi, H. (2010). Preparation and properties of visible light responsive $Bi_2O_3/TiZrO_4$

photocatalysts for 4-chlorophenol decomposition, *J. Hazardous Mat.*, **182**, pp. 557–562.

Okitsu, K., Ashokkumar, M. and Grieser, F. (2005). Sonochemical synthesis of gold nanoparticles: effects of ultrasound frequency, *J. Phys. Chem. B*, **109**, pp. 20673–20675.

Price, G. J. (2003). Recent developments in sonochemical polymerization, *Ultrason. Sonochem.*, **10,** pp. 277–283.

Rohan, T. (2004). The effect of surface active solutes and ultrasound frequency on sonoluminescence in aqueous solutions (PhD thesis, University of Melbourne).

Index